ADVANCED MICROSCOPY

A Strong Analytical Tool in Materials Science

ADVANCED MICROSCOPY

A Strong Analytical Tool in Materials Science

Edited by

Merin Sara Thomas, PhD

Józef T. Haponiuk, PhD

Sabu Thomas, PhD, DSc

Anne George, MD

AAP | APPLE ACADEMIC PRESS

First edition published 2023

Apple Academic Press Inc.
1265 Goldenrod Circle, NE,
Palm Bay, FL 32905 USA

760 Laurentian Drive, Unit 19,
Burlington, ON L7N 0A4, CANADA

CRC Press
6000 Broken Sound Parkway NW,
Suite 300, Boca Raton, FL 33487-2742 USA

4 Park Square, Milton Park,
Abingdon, Oxon, OX14 4RN UK

Apple Academic Press exclusively co-publishes with CRC Press, an imprint of Taylor & Francis Group, LLC

Library and Archives Canada Cataloguing in Publication

Title: Advanced microscopy : a strong analytical tool in materials science / edited by Merin Sara Thomas, PhD, Józef T. Haponiuk, PhD, Sabu Thomas, PhD, DSc, Anne George, MD.
Names: Thomas, Merin Sara, editor. | Haponiuk, Józef T., editor. | Thomas, Sabu, editor. | George, Anne, 1961- editor.
Description: First edition. | Includes bibliographical references and index.
Identifiers: Canadiana (print) 20220244480 | Canadiana (ebook) 20220244510 | ISBN 9781774910429 (hardcover) | ISBN 9781774910436 (softcover) | ISBN 9781003282044 (ebook)
Subjects: LCSH: Microscopy. | LCSH: Materials science.
Classification: LCC QH205.2 .A38 2023 | DDC 502.8/2—dc23

Library of Congress Cataloging-in-Publication Data

CIP data on file with US Library of Congress

ISBN: 978-1-77491-042-9 (hbk)
ISBN: 978-1-77491-043-6 (pbk)
ISBN: 978-1-00328-204-4 (ebk)

About the Editors

Merin Sara Thomas, PhD

Merin Sara Thomas, PhD, is Assistant Professor in the Department of Chemistry, Mar Thoma College, Tiruvalla, Kerala, India. She has been working in the multidisciplinary field of nanotechnology with a focus on biomedical applications of electrospun nanocomposites and filtration applications. Her research interests include biopolymers and synthesis of nanoparticles. She has published six internationally peer-reviewed papers and five book chapters. She earned her PhD at Mahatma Gandhi University, Kottayam, Kerala, India.

Józef Tadeusz Haponiuk, PhD

Józef T. Haponiuk, PhD, has been in the Faculty of Chemistry of Gdansk University of Technology (GUT), Poland, since 1974. He was formerly Head of Polymers Technology Department, which achieved outstanding results in scientific and educational activities under his leadership. It is now one of the leading departments of the university. Dr. Haponiuk holds 10 patents and has 25 patent applications. He is also author or co-author of over 300 original scientific contributions in the area of chemical technology, polymer engineering, polymer characterization, chemistry and technology of polymers and rubber, thermal analysis of polymers, materials science, and biopolymers. He is especially interested in polyurethane technology and the recycling of elastomers and rubber. He actively collaborates with Mahatma Gandhi University in Kottayam, India, and the Lviv Polytechnic, Ukraine. He has received multiple awards and distinctions for his excellent scientific activity. Dr. Haponiuk obtained his PhD from the Technische Hochschule Leuna Merseburg, Germany.

Sabu Thomas PhD, DSc, FRSC

Sabu Thomas, PhD, DSc, FRSC, is the Vice Chancellor, senior most professor and outstanding scientist, and leading academic of Mahatma Gandhi University, India, where he has been serving for the last 34 years. He has held many positions, which include Pro Vice Chancellor; Regional Director of the School of Technology and Applied Sciences; Dean of Technology and Applied Sciences and Syndicate Member; Erudite Professor in the Department of Regenerative Medicine and Translational Sciences at the Calcutta School of Tropical Medicine; Adjunct Professor and AcSIR Director at the School of Chemical Sciences; and Director at the International and Inter University Center for Nanoscience and Nanotechnology. Currently he also holds the position of the Honorary Director of the School of Energy Materials of Mahatma Gandhi University. Dr. Thomas's groundbreaking inventions in polymer nanocomposites, polymer blends, and green bionanotechnological and nano-biomedical sciences have made transformative differences in the development of new materials in the automotive, space, housing, and biomedical fields. In collaboration with India's premier tire company, Apollo Tyres, Professor Thomas's group invented new high performance barrier rubber nanocomposite membranes for inner tubes and inner liners for tires. A prolific author and researcher, Dr. Thomas has won many awards for his work and is acknowledged as one of the most productive researchers in India.

Anne George, PhD

Anne George, MD, is currently Director of the Institute for Holistic Medical Sciences, Kerala, India. Dr. George has worked as a senior house surgeon, resident doctor, gynecologist, and associate professor at various hospitals in India. She has received two international research fellowships, from the University Laval, Canada, and KU Leuven Gasthuisberg, Belgium. She is an expert in the field of anatomy, cancer, Ayurveda, and nanomedicine. Dr. George is a fellow of the American Medical Society and a member of many professional bodies. She has published peer-reviewed research papers and organized several international conferences. She received an MBBS from Trivandrum Medical College, University of Kerala, India; a Diploma in Acupuncture from the University of Vienna, Austria; and MD from Kottayam Medical College, Mahatma Gandhi University, Kerala, India.

Contents

Contributors

Divya Adiga
Department of Cell and Molecular Biology, Manipal School of Life Sciences, Manipal Academy of Higher Education, Manipal, Karnataka, India

Jyoti Bamne
Department of Physics, Barkatullah University, Bhopal, Madhya Pradesh, India
Optical Nanomaterials Lab, Department of Physics, Maulana Azad National Institute of Technology, Bhopal, Madhya Pradesh, India

Shamee Bhattacharjee
Department of Zoology, West Bengal State University, Berunanpukuria, Malikapur, Kolkata 700126, India; E-mail: shamee.zoology@wbsu.ac.in

Sanjiban Chakrabarty
Department of Cell and Molecular Biology, Manipal School of Life Sciences, Manipal Academy of Higher Education, Manipal, Karnataka, India

Arun Kumar Chakraborty
Indian Institute of Engineering Science & Technology, Shibpur, India

P. S. Baby Chakrapani
Centre for Neuroscience, Cochin University of Science and Technology (CUSAT), Kochi 682022, Kerala, India
Department of Biotechnology, Cochin University of Science and Technology (CUSAT), Kochi 682022, Kerala, India

Vivek Chandel
Optical Nanomaterials Lab, Department of Physics, Maulana Azad National Institute of Technology, Bhopal, Madhya Pradesh, India

Pritam Dey
Department of Chemical Engineering, National Institute of Technology Agartala, Jirania 799046, Tripura, India; E-mail: rays.nita@gmail.com

Mainak Ghosal
Indian Institute of Engineering Science & Technology, Shibpur, India

Peter J. Halley
Australian Institute of Bioengineering & Nanotechnology, School of Chemical Engineering, University of Queensland, St. Lucia, Brisbane 4072, Queensland, Australia

Fozia Z. Haque
Optical Nanomaterials Lab, Department of Physics, Maulana Azad National Institute of Technology, Bhopal 462003, India; E-mail: foziazia@rediffmail.com

Aswini Harindran
Department of Physics School of Basic and Applied Sciences, Central University of Tamil Nadu, Tamil Nadu, India

Dania Alphonse Jose
Centre for Neuroscience, Cochin University of Science and Technology (CUSAT), Kochi 682022, Kerala, India
Department of Biotechnology, Cochin University of Science and Technology (CUSAT), Kochi 682022, Kerala, India

R. Vasantha Jothi
Department of Chemistry, Avinashilingam Institute for Home Science and Higher Education for Women, Coimbatore, Tamil Nadu, India

Shama Prasada Kabekkodu
Department of Cell and Molecular Biology, Manipal School of Life Sciences, Manipal Academy of Higher Education, Manipal, Karnataka, India
Centre for DNA Repair and Genome Stability (CDRGS), Manipal School of Life Sciences, Manipal Academy of Higher Education, Manipal, Karnataka, India;
E-mail: spbhat81@gmail.com, shama.prasada@manipal.edu

Senthil Kumar Kandasamy
Department of Chemical Engineering, Kongu College of Engineering, Tamil Nadu, India

Nagamalleswara Rao Kanidarapu
Centre for Disaster Mitigation and Management, Vellore Institute of Technology, Vellore, Tamil Nadu, India

Radhagovinda Koijam
Department of Civil Engineering, National Institute of Technology Agartala, Jirania 799046, Tripura, India

Ponnaganti Sai Krishna
Department of Civil Engineering, National Institute of Technology Agartala, Jirania 799046, Tripura, India

Parveen Kumar
Nanotechnology Lab, H-1 Division, CSIR-Central Scientific Instruments Organisation, Chandigarh 160030, India

V. Madhurima
Department of Physics School of Basic and Applied Sciences, Central University of Tamil Nadu, Tamil Nadu, India

Raju B. Maliger
Australian Institute of Bioengineering & Nanotechnology, School of Chemical Engineering, University of Queensland, St. Lucia, Brisbane 4072, Queensland, Australia;
E-mail: r.maliger@gmail.com

Deba Prasad Mandal
Department of Zoology, West Bengal State University, Berunanpukuria, Malikapur, Kolkata 700126, India

K. M. Mishra
Department of Physics, D.D.U. Gorakhpur University, Gorakhpur, Uttar Pradesh, India

Arvind Mittal
Energy Centre, Maulana Azad National Institute of Technology, Bhopal, Madhya Pradesh, India

Arabinda Mukherjee
Department of Civil Engineering, National Institute of Technology Agartala, Jirania 799046, Tripura, India

Venkata Ratnam Myneni
Department of Chemical Engineering, Mettu University, Mettu, Ethiopia; E-mail: mvratnam81@gmail.com
Department of Chemical Engineering, A.U. College of Engineering, Vizag, Andhra Pradesh, India

Padmini Pandey
Optical Nanomaterials Lab, Department of Physics, Maulana Azad National Institute of Technology, Bhopal 462003, India
Department of Physics, Savitribai Phule Pune University, Pune 411007, Maharashtra, India

Mohammad Ramzan Parra
Optical Nanomaterials Lab, Department of Physics, Maulana Azad National Institute of Technology, Bhopal 462003, India
Department of Physics, Government Boys Degree College, Sopore 193201, Jammu and Kashmir, India

T. C. Prathna
Department of Irrigation and Flood Control, Government of NCT of Delhi, New Delhi 110002, India (formerly with Department of Environmental Engineering and Water Technology, UNESCO-IHE, The Netherlands); E-mail prathna.tc@gmail.com

Srimanta Ray
Department of Chemical Engineering, National Institute of Technology Agartala, Jirania 799046, Tripura, India; E-mail: rays.nita@gmail.com

Rupak Roy
Department of Chemical Engineering, National Institute of Technology Agartala, Jirania 799046, Tripura, India; E-mail: rays.nita@gmail.com

Gaurav Sapra
UIET, Punjab University, Chandigarh 160014, India; E-mail: gaurav.sapra@pu.ac.in

R. Saratha
Department of Chemistry, Avinashilingam Institute for Home Science and Higher Education for Women, Coimbatore, Tamil Nadu, India; E-mail: saratha63@gmail.com

Manu Sharma
UIET, Punjab University, Chandigarh 160014, India

P. K. Sharma
Department of Physics, Barkatullah University, Bhopal, Madhya Pradesh, India

Hafsa Siddiqui
Optical Nanomaterials Lab, Department of Physics, Maulana Azad National Institute of Technology, Bhopal 462003, India

Jaskaran Singh
UIET, Punjab University, Chandigarh 160014, India

Neha Singh
Optical Nanomaterials Lab, Department of Physics, Maulana Azad National Institute of Technology, Bhopal, Madhya Pradesh, India

Nitu Singh
Optical Nanomaterials Laboratory, Department of Physics, Maulana Azad National Institute of Technology, Bhopal, Madhya Pradesh, India

Vaibhav Shukla
Department of Cell and Molecular Biology, Manipal School of Life Sciences, Manipal Academy of Higher Education, Manipal, Karnataka, India

Kajol Taiwade
Department of Physics, Barkatullah University, Bhopal, Madhya Pradesh, India

Vinay Koshy Varghese
Department of Cell and Molecular Biology, Manipal School of Life Sciences, Manipal Academy of Higher Education, Manipal, Karnataka, India

Meena Vangalapati
Department of Chemical Engineering, A.U. College of Engineering, Vizag, Andhra Pradesh, India

Jishnu Padacherri Vethil
Department of Cell and Molecular Biology, Manipal School of Life Sciences, Manipal Academy of Higher Education, Manipal, Karnataka, India

Renu Vig
UIET, Punjab University, Chandigarh 160014, India

Justin J. Cooper-White
Australian Institute of Bioengineering & Nanotechnology, School of Chemical Engineering, University of Queensland, St. Lucia, Brisbane 4072, Queensland, Australia

Abbreviations

AC	activated carbon
ACT-PRESTO	active clarity technique-pressure-related efficient and stable transfer of macromolecules into organs
AD	Alzheimer's disease
AEL	*Ailanthus excelsa Roxb*
AFM	atomic force microscopy
ALS	amyotrophic lateral sclerosis
ASTM	American Society for Testing and Materials
ATOF	acousto-optic tuneable filter
BDST	bed depth service time
BET	Brunauer, Emmett, and Teller
BJH	Barrett, Joyner, and Halenda
BL	buffer layer
BRL	*Bauhinia racemosa Lam*
BSA	bovine serum albumin
C	carbon
CAM assay	choriallantoic membrane assay
CB	conduction band
CCAR	constant contact angle regime
CFDA	carboxyfluorescein diacetate
CFP	cyan fluorescent protein
CKD	cement kiln dust
CLARITY	clear lipid-exchanged acrylamide-hybridized rigid imaging-compatible tissue-hydrogel
CLSM	confocal laser scanning microscope
CNS	colloidal nano-silica
CO_2	carbon dioxide
COD	chemical oxygen demand
CRR	constant radius regime
CRT	cathode ray tube
CSV	chemical solvent
CUBIC	clear, unobstructed brain imaging cocktail
CVD	chemical vapor decomposition

C-A-S-H	calcium aluminosilicate hydrate
C-S-H	calcium silicate hydrate
$Ca(OH)_2$	calcium hydroxide
$Cr_2(SO_4)_3$	chromium sulphate
Cy2	cyanine 2
d	discrete interplanar distance
D	crystalline size
DABCO	1,4-diazabicyclo [2.2.2] octane
DAPI	4′,6-diamidino-2-phenylindole
DEA	diethanolamine
DiA	dihexadecylaminostyryl-N-methylpyridinium iodide
DIC	differential interphase contrast
Dil	1,1'-dioctadecyl-3,3,3',3'-tetramethylindocarbocyanine perchlorate
DiO	3,3'-dioctadecyloxacarbocyanine perchlorate
DiOC6	3,3′-dihexyloxacarbocyanine iodide
DISCO	3D imaging of solvent-cleared organs
DNA	deoxyribonucleic acid
DP	degree of polymerization
DPX	dibutylphthalate polystyrene xylene
DSC	differential scanning calorimetry
DSSC	dye-sensitized solar cell
EBCT	empty contact time of the bed
ECM	extracellular matrix
EDS or EDX	energy dispersive x-ray spectroscopy
EDX	energy-dispersive x-ray analysis
EGFP	enhanced green fluorescent protein
EMT	epithelial to mesenchymal transition
ER	endoplasmic reticulum
ETC	electrophoretic tissue clearing
Et-OH	ethanol
FA	fine aggregate
FAD	flavin adenine dinucleotide
Fe	iron
FE-SEM	field emission scanning electron microscopy
FITC	fluorescein isothiocyanate
FLIM	fluorescence lifetime imaging microscopy

FM4-64	N-(3-triethylammoniumpropyl)-4-(6-(4-(diethylamino) phenyl) hexatrienyl) pyridiniumdibromide
FMN	flavin mononucleotide
FNAC	fine needle aspiration cytology
FRAP	fluorescence recovery after photobleaching
FRET	forster resonance energy transfer
FTD	frontotemporal dementia
FTO	fluorine doped tin oxide
FTIR	Fourier transforms infrared spectroscopy
FWHM	full-width at half maximum peak intensity
Fe_2O_3	iron (III) oxide or ferric oxide
G	glycerol
GFN	Global Footprint Network
GFP	green fluorescent protein
GO3	graphene oxide
H	hydrogen
HBL	Hardwickia binata Roxb
HD	Huntington disease
HHV	higher heating value (MJ/kg)
HRTEM	high range transmission electron microscopy
HNO_3	nitric acid
H/C	hydrogen to carbon ratio
Ia	peak intensity of material in dry air at RT
IF	immunofluorescence
Ig	peak intensity of material in the presence of gas
IgG	immunoglobulin G
IVM	intravital microscopy
JCPDS	Joint Committee on Powder Diffraction Standards
LDOS	local density of states
LFM	lateral force microscopy
L-MBF	laser molecular beam epitaxy
Li_2O	lithium oxide
LSCM	laser scanning confocal microscopy
LSFM	light sheet fluorescence microscopy
LSM	two-photon and light-sheet microscopy
MOS	metal oxide semiconductor
MB	methylene blue
MONPs	metal oxide nanoparticles

MgO	magnesium oxide
MTZ	mass transfer zone
MVD	microvessel density
MW	megawatt
MWCNT	multiwall carbon nanotube
NADH	nicotinamide adenine dinucleotide
NADPH	nicotinamide adenine dinucleotide phosphate
NDDs	neurodegenerative disorders
NFA	National Footprint Account
NPG	N-propyl gallate
NPs	nanoparticles
n-MgO/AC	nano MgO-loaded activated carbon
NH_3	ammonia
NOx	nitric oxide
NH_4OH	ammonium hydroxide
NH_4VO_3	ammonium metavanadate
nS	nano silica
O	oxygen
O/C	oxygen to carbon ratio
OPC	ordinary portland cement
OPD	optical path difference
PACT/PARS	passive clarity technique/perfusion-assisted agent release in situ
PBS	phosphate buffer saline
PEEK	polyether ether ketone
PGS	poly (glycerol-sebacate)
PL	photoluminescence spectroscopy
PLD	pulse laser deposition
PMMA	polymethyl methacrylate
PMT	photomultiplier tube
PPD	P-phenylenediamine
PVDF	polyvinylidene fluoride
PVP	polyvinylpyrrolidone
PXRD	powder x-ray diffraction
QE	quantum efficiency
Raman	Raman spectroscopy
RE	Rare Earth
RFP	red fluorescent protein

RMS	root mean square
RT	room temperature
S	sebacic acid
S	sensitivity
SAEDP	selected area of electron diffraction patterns
SC	stable carbon
SECM	scanning electro-chemical microscopy
SEM	scanning electron microscopy
SDG	sustainable development goals
SDS	sodium dodecyl sulphate
SiO_2	silicon di-oxide
SMO	semiconductor metal oxide
SNAFL	seminaphthofluorescein
SNARF	seminaphtharhodafluor
$SnCl_4$	stannic chloride
SPM	scanning probe microscopy
SPR	surface plasmon resonance
SSA	specific surface area (m^2/g)
STEM	scanning transmission electron microscopy
STM	scanning tunneling microscopy
STS	scanning tunnelling spectroscopy
T	temperature
TDS	total dissolved solids
TEM	transmission electron microscopy
TGA	thermogravimetric analysis
TiO_2	titanium dioxide
TIRF	total internal reflection fluorescence
TRITC	tetramethylrhodamine
TTIP	titanium tetraisopropoxide
TUNEL	terminal deoxynucleotidyl transferase-UTP nick-end labeling
UV	ultra violet
UV-IR	ultraviolet-infrared
UV-Vis	ultraviolet-visible spectroscopy
VB	valence band
Vis	visible
WAXS	wide angle x-ray scattering
WGA	wheat germ agglutinin

WHO	World Health Organization
XPS	x-ray photoelectron spectroscopy
XRD	x-ray diffraction
Y	yttrium nitrate hexahydrate
YFP	yellow fluorescent protein
ZnO	zinc oxide

Preface

The application of a material can be explored only by understanding and analyzing the material. Here comes the role of advanced microscopy, which plays an important role in the analysis of the microstructure, morphology, particle size, etc., of materials. Light microscopy is the basic microscopic technique used for characterization. Advanced microscopic techniques include tools such as scanning electron microscopy (SEM), transmission electron microscopy (TEM), atomic force microscopy (AFM), confocal microscopy, etc. All microscopic techniques are harmonizing other particle size determining tools. Nowadays, advanced microscopic techniques are used for detailed understanding of materials such as semiconductors, metals, polymers, biopolymers, etc.

This book has 14 chapters that cover the methodology and applications of different advanced microscopic techniques in various research fields such as chemistry, nanotechnology, polymers, chemical engineering, and biomedical engineering. This book gives a brief idea regarding advanced microscopic techniques to researchers for better understanding as well as implementation of novel techniques in their research, so that they can contribute to the scientific and serve the world's scientific community.

CHAPTER 1

Confocal Microscopy: Fundamentals and Applications in Life Sciences

VAIBHAV SHUKLA[1], JISHNU PADACHERRI VETHIL[1], VINAY KOSHY VARGHESE[1], DIVYA ADIGA[1], SANJIBAN CHAKRABARTY[1,2], and SHAMA PRASADA KABEKKODU[1,2*]

[1]*Department of Cell and Molecular Biology, Manipal School of Life Sciences, Manipal Academy of Higher Education, Manipal, Karnataka, India*

[2]*Centre for DNA repair and Genome Stability (CDRGS), Manipal School of Life Sciences, Manipal Academy of Higher Education, Manipal, Karnataka, India*

[*]*Corresponding author. E-mail: spbhat81@gmail.com; shama.prasada@manipal.edu*

ABSTRACT

The introduction of the confocal microscope has aided in obtaining the cell's 3D structure by increasing resolution, higher contrast, and greater depth of field of microscopes. Sample processing and handling play a crucial role in obtaining the right quality image or microscopy data. The living sample used for imaging must be made sure that it could endure intense illumination; in the case of fixed samples, the distortion of the specimen's architecture could be avoided by proper preservation. The motive of this chapter is to illustrate the appropriate technique to process samples for a superior quality image. This chapter helps develop expertise in handling microscope and sample preparation techniques.

Advanced Microscopy: A Strong Analytical Tool in Materials Science. Merin Sara Thomas, Jozef T. Haponiuk, Sabu Thomas, & Anne George (Eds.)

1.1 INTRODUCTION

A microscope (from the Ancient Greek: μικρός, mikrós, "small" and σκοπεῖν, skopeîn, "to look" or "see") is an instrument used to observe the structures and objects that cannot be seen by the naked eye. Ever since its development, the microscope has evolved to a greater extent. Magnification and resolution of the microscope were improved over the year, facilitating the development of a microscope with diverse applications. By comparison, the relatively basic microscopes of Antoni van Leeuwenhoek and his contemporaries were far less powerful than even the most basic microscopes in use today. The evolution of biology and biomedical sciences was substantially supported by the invention of confocal microscopy. The new technological improvements made in confocal microscopy have enabled its usage to be simplified and in outreaching to a larger population.[1] In contrary to a conventional microscope, a confocal microscope extends numerous benefits like shallow depth of field, removal of out-of-focus flare, and the potential to bring together sequential optical sections from a thick sample. The introduction of fluorescent probes in confocal microscopes made a major impact on biomedical research, which enabled the imaging of fixed or living cells and tissues. The detection of microscopic cellular entities was made easier with a wide range of fluorochromes found naturally and synthesized artificially.[2] The outline of this chapter is to provide the fundamentals of confocal microscopy with special emphasis on its application in life sciences.

1.1.1 HISTORY OF MICROSCOPY

Historically, microscopy credit goes to Zacharias and Hans Janssen for inventing the microscope in the year 1590. The father–son duo experimented with several lenses in a tube and found that small objects looked significantly enlarged. This paved its way to the genesis of microscopes and telescopes. The father of modern physics and astronomy, Galileo, in 1609 drove his interest in the lens. He came up with the lens principles, which enabled him to come up with a much better focusing device, which was later called a compound microscope by Giovanni Faber in 1625.[3]

Antoni van Leeuwenhoek using his masterpiece single-lens microscope made excellent biological observations and is considered as the father of microbiology. "Animalcules" was a label given by Leeuwenhoek to describe bacterial cells and other microorganisms. Leeuwenhoek observed blood cells in the tadpole capillaries and recorded synchronized action of heartbeat and blood flow in the tadpole.[4] The microscope's potential and its astonishing future were realized once Robert Hooke published his book Micrographia in 1665.[5] It's because Antoni van Leeuwenhoek and Robert Hook's discovery helped the microscope to be evolved as a primary scientific instrument leading to many revolutionary findings.

The modern-day light microscope did not develop much until the 19th century. Further, an in-depth understanding of the light microscope's fundamental limits and technological reinforcements helped in improving the resolution of light microscopes. The theoretical characterization of diffraction, which causes blurring of the sample's images, was studied by George Airy in mid-1800 and named it a point spread function (PSF).[6] Ernst Abbe, Otto Schott, and August Köhler's contributions helped set the basis of the modern, bright light microscope. Overcoming chromatic aberration was the major concern in those days. The invention of the apochromatic lens by Ernst Abbe in 1865 aided to overcome the problem of chromatic aberration and structural anomaly. The collaboration of Ernst Abbe and Otto Schott in 1883 introduced the high-quality lens that produced limited diffraction, which was used for over a century, till the arrival of the 4π microscope.[7,8]

Low contrast and low resolution are the two major challenges encountered while imaging the biological samples. Various methods were used to overcome and or reduce the task. To improve the contrast, researchers in microscopy worked on phase change, polarization, and started using staining and fluorescence. Fluorescence microscope with various advancements in illumination [confocal, fluorescence recovery after photobleaching (FRAP), total internal reflection fluorescence (TIRF), and two-photon and light-sheet microscopy (LSM)], which improved both visualization and quantification of an image.[3]

1.1.2 EVOLUTION OF CONFOCAL MICROSCOPE

The microscope and the field of microscopy played an important role in understanding the biological system's complexity. A confocal microscope influenced researchers to use it to explore cell structure and function to obtain stern images. In 1957, Marvin Minsky set the benchmark in confocal microscopy by patenting his idea about the confocal microscope.[9] After a decade, Egger and Petran technically advanced the confocal microscopy to produce a multiple-beam confocal microscope, which helped to uplift the curiosity to improve the technology further.[10] In the 1970s, Egger, along with Davidovits, continued their research on the confocal microscope at Yale University to build a scanning confocal laser microscope. In 1973, they were able to generate the first confocal laser photomicrographs.[11] In the 1970s, cutting-edge development in computer hardware, software for imaging, and lasers helped advance microscopy in biological research.[12] In the late eighties, confocal laser microscopy was made commercially accessible to realize its importance in biological research.[13,14] Development in optics, electronics, powerful lasers, and photodetectors led to the up-gradation to the better-quality confocal microscope image during the 1990s. Further, fast computing and improved software helped in image processing, which enabled to expand the abilities of the confocal microscope and its comprehensive application in research (Fig. 1.1).

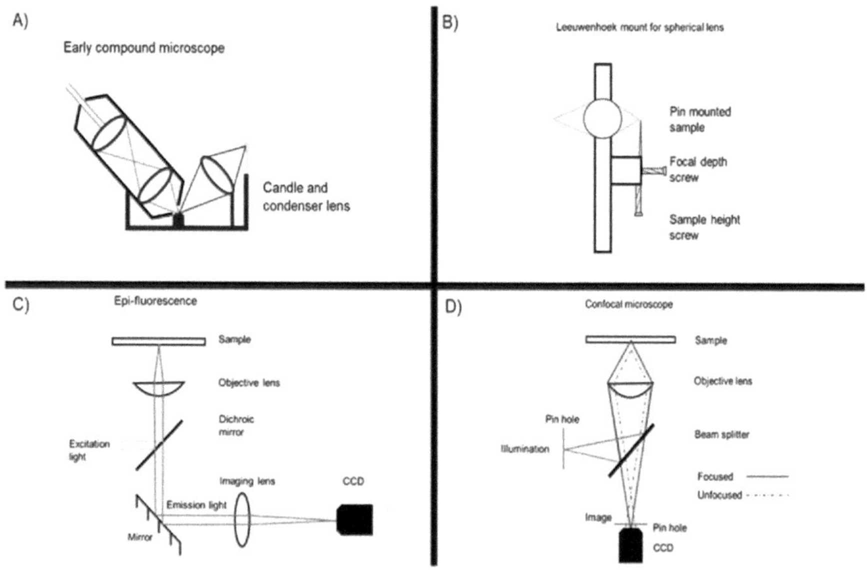

FIGURE 1.1 Optical microscope and its designs. (a) Simple compound microscope used by Hooke. (b) Antonia van Leeuwenhoek's single spherical lens mount system. (c) A simple epi-fluorescence system. (d) Simple modern-day confocal microscope.

1.2 PRINCIPLE OF CONFOCAL MICROSCOPE

A specimen in conventional microscopy is positioned on a microscope stage with the specimen's complete field illuminated simultaneously.

Apart from the brightest and highest intensity at the microscope's objective lens's focal point, it includes illumination from other parts of the sample, not from the microscope's focal plane resulting in background noise. This makes the image more haze, blur, and compromises quality.[15] However, in confocal microscopy, only a desired focal spot is allowed to pass through by using a small pinhole aperture. The screen blocks any other scattered light outside of the focal plane. Fluorescence confocal microscopy uses pinhole to eliminate scattered light resulting in the light collection from a highly focused point. It focuses only on a small portion of the specimen, and incident light scans the entire plane of focus in a raster pattern generating a complete image. The pinhole is placed in a CONjugate FOCAL plane, and hence the name confocal microscope.[16,17] Confocal microscopes have become one of the dominant tools for researchers for active cell structures and functional imaging. Fluorescence confocal microscopy uses dyes that fluoresce when stimulated by light known as "fluorophores," which are added to the specimen depending on the specific applications such as dyes that target nucleus, mitochondria, lysosomes, and many more. These fluorophores increase the signal-to-noise ratio and improve the sensitivity and specificity, allowing sharper target detection. The excitation light for fluorescence confocal microscopy is majorly provided by a laser (Table 1.1), which is a coherent light at a wavelength that will excite the specific fluorophores (Table 1.2). This wavelength excites an electron in an atom to a higher energy state than the ground state. Following this, electrons relax to the ground state, which results in the emission of a photon of different energy and different wavelength to the excitation photon. Both the excitation beam and emission fluorescence pass through the same objective, which has a dichroic mirror with different transmission properties at a different wavelength and reflects incoming high energy and short-wavelength laser light and allows low energy and high wavelength light to pass through the light detector. Confocal microscopy creates optical sections with exceptional contrast and with minute details that were previously hidden by out of focus blur.[18]

The confocal microscopy helps to overcome the restrictions offered by wide-field fluorescence microscopy. In widefield fluorescence microscopy, the entire specimen is stimulated with the light from its source; the emission from the sample is detected by the photodetector, which even includes the unfocused background. Contrary to widefield fluorescence microscopy, confocal microscopy uses laser as a light source and pinhole

TABLE 1.1 Difference between Widefield Fluorescence Microscope and Confocal Microscope.

Characteristics	Widefield fluorescence microscope	Confocal microscope
Fundamental setup	Comprises or fluorescence light Source, solid-state charge-coupled devices (CCDs), and fluorescence; filter cube.	Comprises or laser, a light source pinhole aperture, photomultiplier detector, detector pinhole aperture, and dichrome mirror
Principle	The entire specimen is bathed with light.	Only a single point is illuminated at a time to avoid unwanted scattering of light.
Resolution	Low	High
Regions of out-of-focus information	Blurred and large	Reduced
Optical resolution in z	2–3 μm	0.5 μm
Signal from optical sections	Many signals cannot be seen separately.	Improved *Z*-resolution allows for more accurate signal discrimination.
Acquisition of 3D data	Not possible	Possible
Cost	Relatively inexpensive	Expensive
Hazards	Not hazardous	High-intensity laser irradiation is hazardous.

for eliminating out-of-focus signals. Acquisition of 3D data is possible in confocal microscopy, which is not possible by widefield fluorescence microscopy, and resolution is high for the confocal system compared to the fluorescence microscopy (Fig. 1.2).

TABLE 1.2 Lasers Used in Confocal Scanning Microscopes.

Types of laser	Emission wavelength	
Gas laser	Krypton-argon	488,568,647
	Argon	458,488,514
	UV-argon	488,514
	Helium-neon	546,561,594,633
Semiconductor and crystal	Diode	405
	Cr:LiSAF	780, 1000

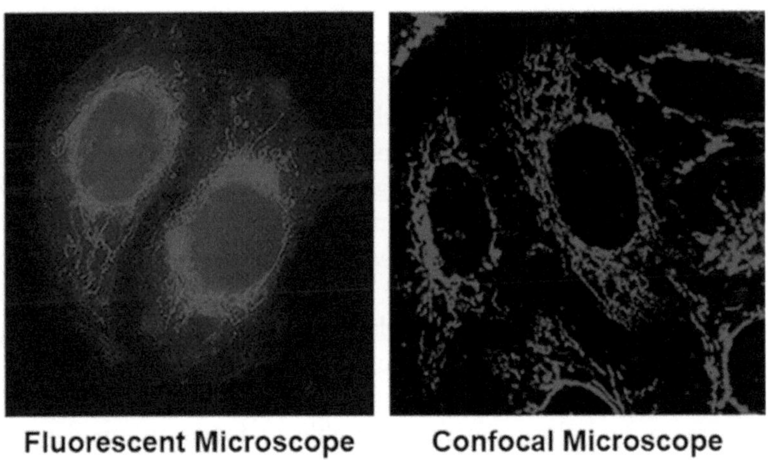

FIGURE 1.2 Comparison between fluorescent microscope and confocal microscope. Image captured by fluorescence microscope and confocal microscope specimen having Mitotracker red. The confocal image showed the spectral profiles of fluorophores are very well separated compared to fluorescence microscopy.

1.3 IMPORTANT COMPONENTS OF A CONFOCAL MICROSCOPE

The basic elements of the confocal microscope are

A. Microscope: Based on the type of specimen used for imaging, the microscope varies. An upright microscope could be used if the specimen to be imaged is fixed or the living specimen is used with immersion objective. Living samples that need to be germ-free throughout the experiment and need to be accessed by the microinjectors an inverted model microscope are useful. Different attachments improved the functioning of confocal microscopy. The Stepper motor enables the sample's movement, which allows the exact focusing of samples and helps to take a stack of optical sections popularly known as Z-series optical sections. The XY stage controller controls the horizontal movement of samples. Vibrations can hinder image acquisition; hence, microscope must be placed on the vibration-free table.

B. Objective lens: The quality of the confocal image greatly depends upon the objective lens. The objective lens must proficiently diffuse light without any aberration as it could worsen the image's quality. Optical aberration is one factor among others that need to be taken care of.[19] A thick specimen leads to the spherical aberration as the light illuminating fails to converge to a single focal point. The image developed could have a loss of intensity from samples due to chromatic aberration, as shorter and longer wavelength does not focus at the same focal point. The working distance of the objective greatly affects the image quality. Magnification also plays a significant role in the procurement of exceptional images, since the image's brightness is inversely proportional to the objective's magnification.

C. Light source: Light source plays an essential role in microscopy. To develop the right image, the perfect light source is the one which would give out high brightness, significantly less noise, lesser beam deviation, and must be able to focus on small spots. For these reasons, lasers are being used in the confocal laser scanning microscope (CLSM). Spinning disk microscopes are not restricted just to the laser. It can also be used with mercury, xenon, or metal halide as the light source.[20] In recent confocal microscopes, lasers (argon ion laser) are used in combinations with acousto-optic tuneable filter (ATOF) for laser attenuation and wavelength selection.[21] ATOF reduces the ambient noise coming from autofluorescence and avoids interference from other channels if samples are labeled with multiple dyes. In the new generation CLSM, the lasers are parted from the scan head, which reduces the vibration arising from the cooling system. Optic fiber is used to transport the laser beam to the scan head.

D. Scan head: Scan head houses the numerous significant components of confocal microscopy. The scanning system does raster scanning and frame scanning for more excellent temporal resolution. A scanning mirror is another element found in CLSM. To acquire images, various mechanisms are made use of such as a rotating polygon mirror, a single galvanometer mirror, two closely coupled galvanometer mirror, four mirror arrangements built by using two concave mirrors, and two galvanometer mirrors.[14] Apertures that are present in the scan head can be of pinhole or slit type. The scanning system does raster scanning and frame scanning for greater temporal resolution.
E. Photodetector: The photodetector's location depends upon the confocal system; it can be either inside or outside the scan head. A photomultiplier tube (PMT) helps in the sensitive detection of the light energy, converts it to electrical energy, and commonly used as a detector.[22] When light falls on phosphor, the electrons emitted are amplified, and the signals are displayed on the monitor. Based on requirements, confocal microscopy can accommodate multiple PMT, which would help in imaging multiple labeled specimens (Figs. 1.3 and 1.4).

1.4 SAMPLE PREPARATION AND PROCESSING

In optical microscopy, CLSM is an advanced microscope mainly because this method allowed visualization of sharp and high-resolution images in both living and fixed samples and tissues with a capacity to acquire sharply characterized optical areas from which 3D images can be created. Today, confocal microscopy, when coupled with the application of newly advanced immunofluorescence reagents, synthetic fluorophores, and fluorescent proteins, is one of the most sophisticated methods available to investigate subcellular structures. The confocal microscopy has been improved by powerful computers, storage technology, interference filters, detectors, laser systems, and fluorophores for high specificity.[23,24]

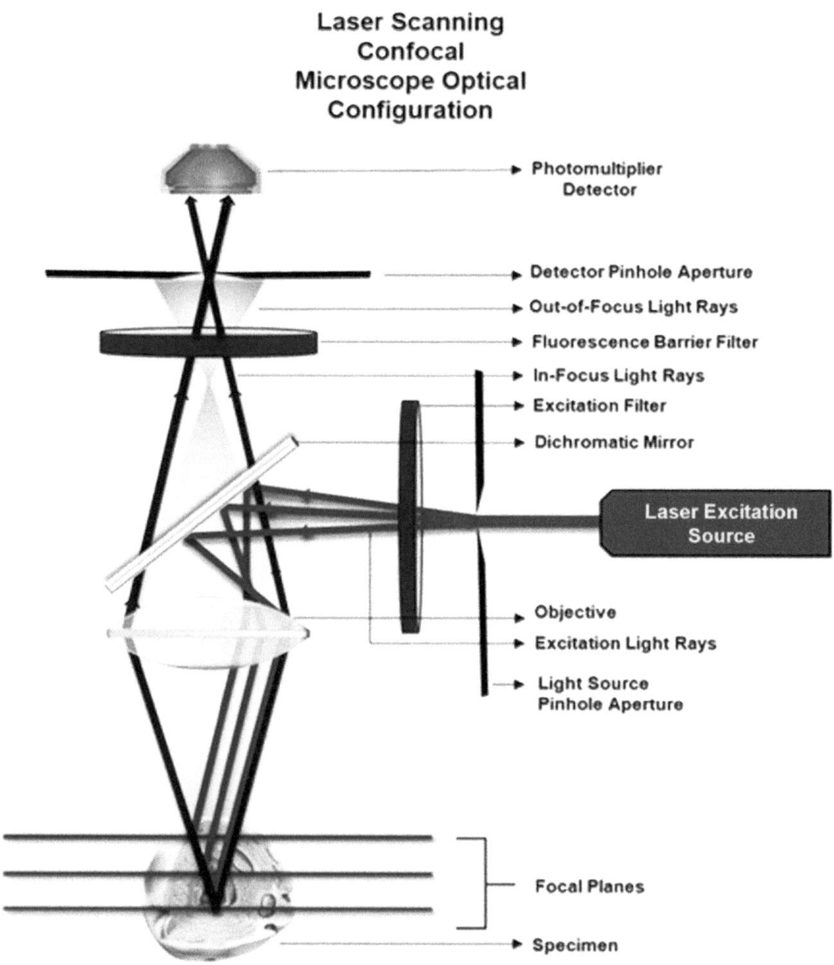

FIGURE 1.3 Schematic diagram of the optical pathway in a laser scanning confocal microscope.

Source: Reprinted from https://www.researchgate.net/figure/Optical-configuration-of-a-laser-scanning-confocal-microscope-Reprinted-from-32_fig3_260280581.

1.4.1 SAMPLE PREPARATION

Like any other types of microscopy, confocal microscopy also requires adequate sample preparation.[25] Transparent candidate materials are the best for confocal microscopy. A transparent solution like commercial mounting medium, glycerin, or water has excellent reflective qualities and ought to be used as a mounting material. In ideal conditions, confocal microscopes allow maximum penetration depth of about 100–200 μm thicker material may need to be physically separated or optically cleared before imaging. Opaque materials cannot be penetrated with the lasers; in any case, strategies are accessible to visualize the surface of opaque objects facing the laser.[26]

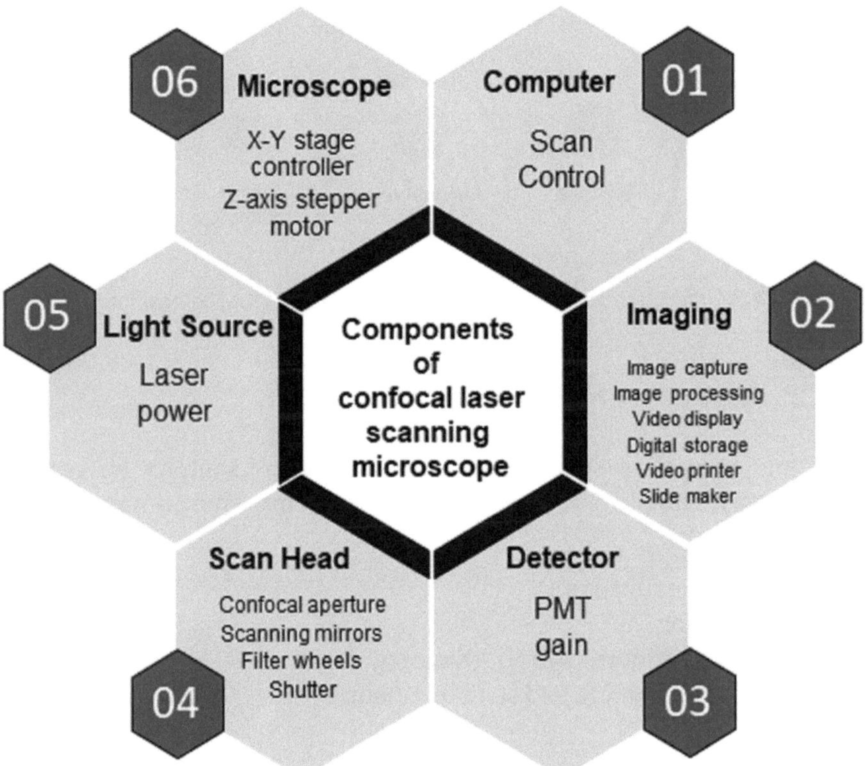

FIGURE 1.4 Confocal laser scanning microscope and its component.

Coverslips can be used to culture cells directly, and it should be cleaned well before seeding cells. A coating of the coverslip with suitable extracellular matrix molecules may be useful to improve adhesion and normal cell dynamics. The magnifying objective lens used in microscopes are mostly made for No. 1.5 coverslips (0.17 mm thick), and any variations in coverslip thickness could lead to spherical aberration. However, experimental conditions can sometimes demand the use of thinner coverslips. Suspension cells can adhere to slides via Cytospin (a low-speed centrifuge that can deposit a uniform monolayer of cells onto a glass slide while maintaining cell morphology).[27,28]

In the case of taking images of live samples then it is crucial to provide all the requirements for its survival and also consider that there is sufficient access by the objective lens to focus on the particular region of interest (ROI) (Fig. 1.5).[26]

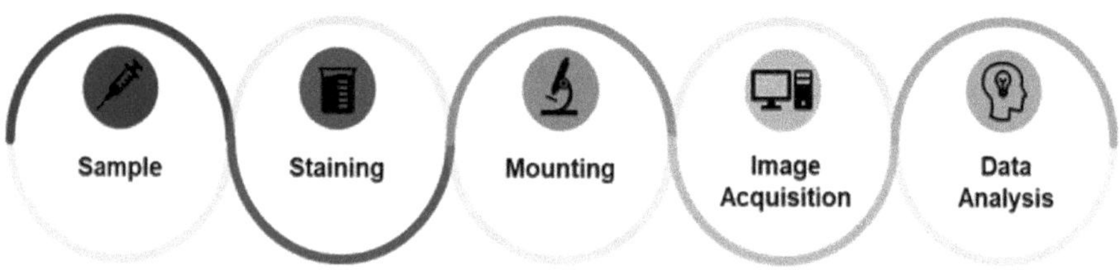

FIGURE 1.5 Workflow for sample processing and analysis for confocal microscopy.

1.4.2 SAMPLE PROCESSING

Multiple processing steps are essential to prepare samples for confocal microscopy. Sample processing can be divided into five primary phases.[29]

1. Fixation
2. Permeabilization
3. Blocking
4. Immunofluorescent (IF) Staining
5. Counterstaining and sample mounting

1.4.2.1 FIXATION

The fixation objective is to preserve cells, cellular formations, or tissue in their native state and protect the preparation by chemical reagents over an extended period. Various fixation methods are available for IF; wherein every reagent has its effect on the primary antibody's epitopes (Table 1.3). The quality of IF staining can be affected either by the masking of the antibody binding sites or damaged during fixation. As each antibody binds differently to its antigen-dependent on several fixation compounds,

TABLE 1.3 Different Types of Fixatives.

Fixatives		**Effects**	**Advantages**	**Disadvantages**
Chemical crosslinkers	Formaldehyde	Crosslink proteins through their free amino groups	Preserves cellular morphology.	May crosslink antigens
	Glutaraldehyde		Works well in the presence of fluorescent proteins.	High autofluorescence
Organic solvents	Methanol	Fixation by dehydrogenation and protein precipitation. Cells will be permeabilized.	Preserves cellular architecture. Faster procedure when compared to chemical crosslinkers.	The strong negative effect on many epitopes. Not recommended for fluorescent proteins. Soluble and lipid components will be lost.
	Acetone		Less damaging to epitopes. Faster procedure	Not recommended for fluorescent proteins. Soluble and lipid components will be lost.

it is vital to attempt several fixation techniques for a new antibody. The specifications for the appropriate fixative are available on the antibody's datasheet. Fixation should maintain the cellular and subcellular structure providing unblocked antigens for appropriate antibody binding. Fixation reagents can be classified into two categories: chemical crosslinkers and organic solvents.[28,30]

Formaldehyde is an example of chemical crosslinkers that, with the help of free amino groups, crosslink proteins; the cellular morphology is well preserved in most cases. However, the crosslinking of antigens might reduce antibody binding. Glutaraldehyde also imparts a preservative impact on cell structures, but it results in excessive autofluorescence of the specimen in microscopy.[28,30,31]

To fix cells in their native state, organic solvents like methanol or acetone are used to precipitate proteins and cause a dehydrogenating effect. But several lipid components and soluble molecules are lost during the procedure. A combination of methanol and acetone is being used; although methanol has a negative impact on many epitopes, it is still the best option for preserving cell structures. Acetone has a minimal harmful effect. With the presence of fluorescent proteins such as GFP, in cells, it will be destroyed mainly by fixation with organic solvents. Fixative like formaldehyde (4%) can be used when manufacturers do not propose the primary antibody. It can be used for 10 min at room temperature for various cell lines and antigen types.[30–32]

1.4.2.2 PERMEABILIZATION

To effectively modify cell membrane structure, the fixation with cross-linkers will not suffice. Therefore, formaldehyde fixation is followed by the cells being permeabilized to permit antibodies to access the cell's interior. Due to the large size and ionic nature of the antibodies inhibit them from entering without disrupting the cell membrane. Alcohol- and acetone-based method of fixation does not require this step. Digitonin, Leucoperm, and Saponin are moderately gentle detergents that create big enough pores for the antibodies to pass through without the cell membrane is completely dissolved. These detergents are ideal for antigens present in the cytosol or the cytoplasmic face of the plasma membrane. In the case of IF staining, Triton X-100 is most frequently used as the permeabilization agent. This detergent has the potential to dissolve the cellular membrane without interfering in the protein–protein interactions. Detergents can also be used to partially dissolve the nuclear membrane making it appropriate for nuclear antigen staining. Another detergent that is most frequently used to denature proteins is SDS, an anionic detergent that provides the protein with a negative charge for electrophoresis. It is also useful as a permeabilizing agent capable of denaturing fixed cells to expose epitopes that are usually masked from an antibody. It can be used for fixed specimens to extract small, poorly crosslinked proteins but cannot be used on samples fixed by precipitation.[26,28,30,31]

1.4.2.3 BLOCKING

To reduce non-specific binding in samples, it is imperative to "block" the sample during cellular staining of objects using antibodies. There are several reasons for non-specific binding such as crosslinking antibodies to inappropriate structures due to unreacted fixative aldehydes, trapping of antibodies within structures in the specimen, and low-affinity IgG molecules binding to inappropriate structures if using polyclonal antibodies.

To avoid non-specific binding, these samples can be treated with blocking agents like bovine serum albumin (BSA), casein (or a solution of non-fat dry milk), gelatine, or normal serum before antibody probing. Blocking is performed for 10–30 min after permeabilization using blocking agents in a 1–10% concentration.[26,28,30,31]

1.4.2.4 IMMUNOFLUORESCENT STAINING

IF staining can be carried out in two ways. First, direct or primary IF for direct visualization, specific primary antibody with a fluorochrome is precisely bound to the target molecules. Second, indirect or secondary IF is a two-step process where a specific primary antibody binds to the target molecule, followed by the binding of the secondary antibody coupled with a fluorochrome that binds to the primary antibody (Fig. 1.6).[28]

Direct IF staining is more efficient than indirect staining due to reduced steps like washing and incubation, making it faster to analyze the samples compared to the standardized IF method. The primary antibody that is fluorescence-coupled raised against its target antigen must be highly specific, making it quite expensive to develop.[26,28,30–32]

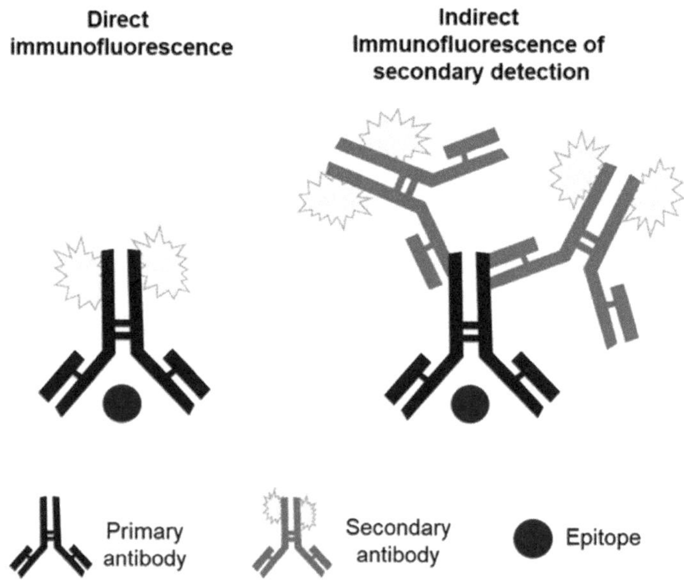

FIGURE 1.6 Direct and Indirect immunofluorescence. For direct immunofluorescence, the antibody binds to epitope that is labeled with fluorophores and in indirect immunofluorescence, the primary antibody binds to epitope and a secondary antibody specific for primary antibody binds to it.

Source: Redrawn from https://www.thermofisher.com/us/en/home/life-science/cell-analysis/cell-analysis-learning-center/molecular-probes-school-of-fluorescence/imaging-basics/labeling-your-samples/immunolabeling.html

To study various target molecules in the same sample, a distinct fluorochrome must be used in IF staining. Indirect IF staining is more flexible with the combination of fluorescence-coupled secondary antibodies with different primary antibodies. In indirect IF staining, secondary antibodies' signal amplification is achieved by multiple secondary antibody binding to the primary antibody. Indirect IF is time-consuming, but it is cost-effective as there is a possibility of using different combinations of primary and secondary antibodies[26,28,30–34] (Fig. 1.7).

1.4.2.4.1 *Choosing a Fluorescent Label*

Due to the emergence of novel fluorescent probes, the capability of the confocal instrument has substantially improved. Fluorescent probes are particles that retain light of a particular wavelength and emanate the light of a distinctive, characteristically longer wavelength, and are commonly utilized to examine biological samples. Excitation and emission of different fluorophores are now closely matched with the wavelength produced by lasers supplied by most CLSMs manufacturers. Recent advances in biochemistry have introduced new fluorophore families with excellent signal-to-noise and quantum efficiency (QE) properties. Factors like polarity of the solvent, the concentration of the metallic ion, and pH change can affect a given fluorophore's quantum yield.[24–26,28]

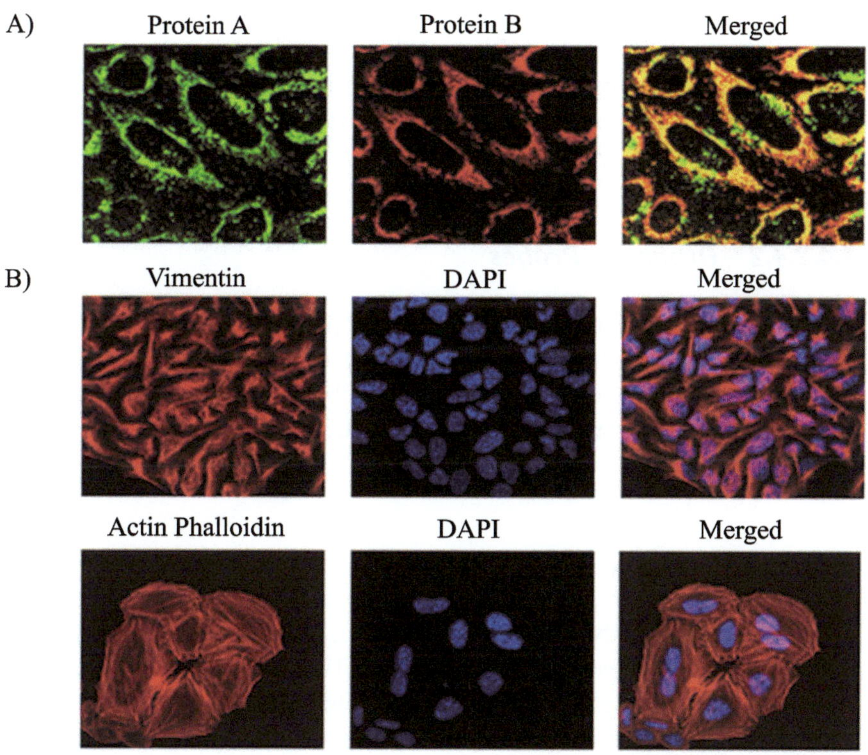

FIGURE 1.7 Colocalization refers to presence of two or more different molecules located at same physical position in the specimen. (A) A perfectly aligned protein A and B; (B) immunofluorescence showing actin phalloidin and Hoechst, and vimentin and Hoechst and their merged image.

During multicolor indirect IF staining, the primary antibodies used must infer from distinct species in arrange to distinguish the immune-complexes by examining with fluorescently labeled secondary antibodies. Secondary antibodies utilized in multicolor immunostaining ought to contrast in their wavelength range to segregate their fluorescence signal in microscopic examination. Presently, fluorescence-labeled secondary antibodies with a wavelength extend from UV to IR against primary antibodies from nearly any species are accessible. Pre-absorbed secondary antibodies' usage lowers the risk of crossreactivity with other antibodies in instances where several primary antibodies and their corresponding secondary antibodies are used simultaneously. The addition of fluorochromes in a live cell imaging system is critical, as it could be harmful to the living cells when excited by the laser. The toxic effects can be minimalized by supplementing the culture media with ascorbic acid. The cellular components that are labeled with probes can affect the cell viability during microscopic analysis. For example, nuclear stains tend to have a more adverse effect than cytosolic stains[24,26,28,30–34] (Table 1.4).

1.4.2.4.2 Organelle Probes

In live-cell imaging, to study the biological processes, fluorophores target the specific organelles of the cell such as lysosomes, mitochondria, endoplasmic reticulum, and Golgi apparatus (Table 1.2). Organelle probes are useful for studying transport, acidic compartments, membrane phenomena, respiration, mitosis, protein degradation, and apoptosis in living cells.[25,35]

1.4.2.4.3 Fluorescent Proteins

In recent years, the advancements of inherent fluorescent proteins and their derivatives have substantially developed. These non-invasive biological labels support the researcher with a stable palette for single, dual, and multispectral fluorescence analysis. Fluorescent proteins used to enhance image quality are mainly obtained from marine organisms. There are currently more than twenty-five fluorescent proteins and their derivatives.[25,36]

1.4.2.4.4 Quantum Dots

In confocal and widefield imaging, quantum dots are used in both live and fixed cell imaging as fluorescent labeling agents. Quantum dots are purified semiconductors, which are crystals of nanometer size coated with the hydrophilic polymer and conjugated with antibodies. Prolonged photostability, higher fluorescence intensity, and multiple colors with single-wavelength excitation make them advantageous over organic dyes and fluorescent proteins. While capturing the optical section in CLSM, the quantum dot has a more significant advantage as they have exceptional photostability. Unlike a quantum dot, organic fluorophores undergo excessive photobleaching after repetitive raster scanning, affecting a 3D image. With a single excitation wavelength, these fluorophores can stimulate multiple quantum dot sizes and be used for multiple labeling experiments.[25,37,38]

TABLE 1.4 Commonly Used Fluorescent Probes.

Probes	Examples	
Fluorescent probes for conjugation (to antibodies, biotin, streptavidin, etc.)	Green fluorescence	FITC, Alexa 488, Cy2
	Red fluorescence	Rhodamines, Cy3, Texas Red, Alexa 568
	Far red fluorescence	Cy5, Phycoerythrins
Probes for DNA/ RNA, Nucleus	Green fluorescence	YO-YO, Sytox green, Actimomycin
	Red fluorescence	Propidium iodide, Ethidium Bromide
	Far red fluorescence	TO-TO3, TOPRO
Viability dyes	Live cells	Calcein-AM, CFDA
	Dead cells	Propidium iodide
Organelle probes	Mitochondria	Mito Tracker, Mito Fluor, Rhodamine-123, DiOC6
	Lysosomes	Lysotracker, LysoSensor
	ER	ER Tracker
	Golgi	BODIPY-TR ceramide
Membrane markers	DiA, Dil, DiO, FM 4-64, Fluorescent Lectins	
Electrolyte/Ion concentration	Calcium	Fluo-3, Calcium Green, Calcium Crimson
	Magnesium	Mag-Fura Red, Magnesium Green
	pH	SNAFL, SNARF
Fluorescent proteins	GFP, EGFP, CFP, YFP, RFP	

1.4.2.5 COUNTER STAINING AND SAMPLE MOUNTING

The IF staining is usually followed by nuclear staining with DNA binding dyes such as Hoechst or DAPI. Other cellular structures commonly used for counterstaining are the actin cytoskeleton using fluorescently labeled phalloidin, or the plasma membrane using fluorescently tagged wheat germ agglutinin (WGA). Unlike antibodies, counterstains usually do not crossreact with one another and can stain the cellular structures even without permeabilization. These counter stains enable better orientation within the cell or tissue sections during microscopic analysis and also indicates the cellular status[28,30–33] (Fig. 1.8).

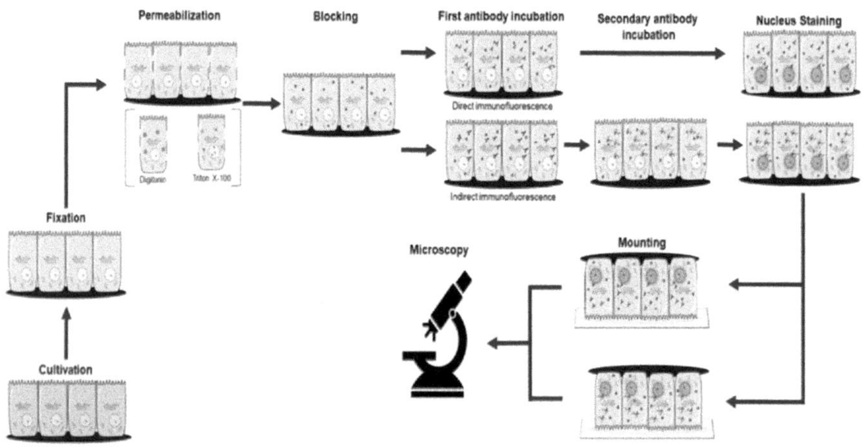

FIGURE 1.8 Schematic diagram of sample processing.

Oxidative damage and photobleaching of fluorophores are the consequences of exposure of labeled specimens to high energy light during image acquisition. Further, fixation by crosslinking is reversible in aqueous media and results in delocalization of fluorescent signals. Hence, replacing the final PBS wash with mounting media offers several benefits such as

- displaces water and thereby prevent depolymerization of crosslinkers
- stabilizes samples and fix the samples on a microscope slide
- increases refractive index which is beneficial for microscopy with oil immersion objectives
- reduces photobleaching by scavenging free radicals

Commonly used mounting media are listed in Table 1.5.

TABLE 1.5 Examples for Mounting Media.

Mounting media	Refractive index
Water	1.333
100% PBS	1.335
50% Vecta Shield + 50% PBS	1.39
50% PBS + 50% Glycerol	1.406
80% Glycerol + 20% Water	1.451
100% Vecta Shield	1.452
100% Glycerol	1.46
Mowiol	1.46
Kaiser's Glycerol Jelly	1.47
DPX	1.525
Immersion Oil	1.518
Glass	1.52

1.4.3 LIVE IMAGING

Cell biologists presently use a fluorescent microscope to obtain live cell imaging. Fluorescent agents help with live-cell staining and protein tagging to better understand the biological functions under a microscope. The significant challenges in acquiring live-cell images are maintaining a balanced signal to noise ratio with minimal photodamage and providing an appropriate condition for cells and tissue to replicate and function. Currently, there are different types of Petri dishes especially available commercially for high-resolution live-cell imaging. Petri dish featuring a circular opening cut into the center of the attachment surface and covered by a high-quality borosilicate coverslip, glass-bottom dishes (coated or uncoated imaging chambers for multi-culture assay), stage incubators, and slide warming plates (provides localized regulation of temperature) are accessible for high-resolution imaging of live cells.[39,40]

1.4.4 AUTOFLUORESCENCE

Autofluorescence complicates the analysis of IF staining. Autofluorescence is a collective term for "biological autofluorescence" and "fixative-induced autofluorescence." In general, the spectra of autofluorescence are comparatively broader than those of fluorescent probes, and hence it is challenging to differentiate between autofluorescence and the actual fluorescence of interest, leading to misinterpretation in image analysis. In order to overcome the issue of autofluorescence, it is valuable to understand its source before trying to modify the protocol or the settings of the filter[26,36] (Table 1.6).

TABLE 1.6 Sources of Autofluorescence.

Sources of autofluorescence	
Aromatic amino acid residues	
Reduced pyridine nucleotides	NADH, NADPH
Flavin and Flavin nucleotides	Riboflavin, FMN, FAD
Zinc-Protoporphyrin	
Chitin	
Chlorophyll	
Lipofuscin	
Collagen and elastin	

1.4.4.1 BIOLOGICAL AUTOFLUORESCENCE

Some cellular components show fluorescence when the appropriate wavelength is used to excite it. This phenomenon has its origin in endogenous fluorophores, and it mainly comes from lysosomes, mitochondria, and aromatic amino acids. The significant factors causing intrinsic fluorescence are flavin coenzymes (FAD, FMN) or pyridine nucleotides (NADH) (Table 1.6).

1.4.4.2 FIXATIVE INDUCED AUTOFLUORESCENCE

When aldehydes are used for fixing specimens, they react with cellular amines or proteins, leading to fluorescent products. These issues can be resolved by reducing the aldehyde group to a hydroxyl group (e.g., sodium borohydride).

1.4.5 PHOTOBLEACHING AND QUENCHING

Photobleaching and quenching are issues in all forms of fluorescence microscopies and cause a significant decrease in fluorescence emission.

These processes should be considered as crucial aspects during fluorescence imaging. Quenching is a reversible process where a collision of a fluorophore excites a non-fluorescent molecule resulting in the fluorophore deactivation bring it back to the ground state. Photobleaching is a non-reversible process where fluorescence is lost due to chemical damage and covalent modification of the fluorophore. This leads to a loss in signal-to-noise ratio comprising the image quality and the loss of data acquisition.[24,41]

1.4.5.1 TO AVOID PHOTOBLEACHING

- Use photostable dyes (Examples: Alexa, Dylight, ATTO, Cy3) whenever possible. Classical stains such as FITC and TRITC have poor photostability.
- Avoid excessive excitation or reduce excitation intensity. This lowers the frequency of fluorophore excitation and thereby reduces photobleaching.
- Use of oxygen-scavenging media, mounting media containing photo-protectants (Table 1.7).
- The addition of some vitamins to live culture dishes improve photostability.
- Reduce photosensitizers; Some components of cell culture media such as riboflavin and pyridoxal can enhance photobleaching rates.
- DO NOT use glutaraldehyde for specimen fixation. Paraformaldehyde or methanol is preferable over glutaraldehyde.

TABLE 1.7 Commonly Used Antifades.

Antifades for fixed samples	Antifades for living specimens
N-propyl gallate (NPG)	Carotenoids
Hydroquinone	Ascorbates
P-phenylenediamine (PPD)	Imidazole
DABCO	Histidine
	Reduced glutathione
	Uric acid
	Trolox (Vitamin E analogue)

1.4.6 STEPS TO SUCCESS

Following are the fundamental steps for successful confocal imaging[28]:

- Control the refraction of light between your sample and the confocal pinhole.
- Choose an appropriate sample for confocal microscopy.
- Choose the correct objective and immersion media.
- Choose the correct coverslip or sample dish.
- Avoid compression and distortion of the sample.
- Match the sample and its surrounding medium by refractive index.
- Understand the spectral profiles of the fluorophores before capturing images.

1.5 APPLICATION OF CONFOCAL MICROSCOPY

1.5.1 LOCALIZATION AND COLOCALIZATION OF MOLECULES

The most common confocal microscopy application is to determine the relative distribution of different fluorophores tagged with specific molecules. The researcher's commonly asked question is whether two molecules of interest are located in the same position. Different fluorophores can be separated with laser excitation of different wavelengths and photodetectors, provided that the excitation spectra of the fluorophores are well separated. Colocalization analysis is performed to determine if labeled proteins colocalize to specific subcellular structures. Colocalization occurs where two or more antigens are viewed in the same position by secondary antibodies labeled with fluorophores with different excitation spectra. Each dye is separately imaged from the same position and considered as colocalized when two images are overlaid. This allows researchers to investigate whether two fluorescent signals are present in the same position and different areas. Several studies are showing the importance of colocalization analysis in different disease conditions, including cancer. For example, Caroline et al. (2018) have shown the colocalization of synthetic crotamine (sCrot-Cy3) with intracellular membranes inside the tumor and non-tumor cells. Similarly, Pucci et al.[42] have shown colocalization of endothelial CD31 and calretinin markers in cardiac myxomas.[1,42]

1.5.2 PHOTOBLEACHING STUDY

1.5.2.1 FLUORESCENCE RECOVERY AFTER PHOTOBLEACHING

Axelrod and coworkers in the 1970s developed FRAP to where protein mobility of a living sample was estimated by measuring the fluorescence recovery rate at the bleached site.[43,44] FRAP was initially used to estimate the diffusion rate in cellular membranes using fluorescein.[45,46] Advancements in confocal microscopy made it capable of studying protein in the cell interior. Later, FRAP became a useful tool to study protein mobility in the cell interior with the advancement in fluorescent protein technology and confocal microscopy. Now, FRAP is utilized to analyze almost all cell biology aspects, such as nuclear protein dynamics in living cells, mRNA mobility, transcription, cytoskeletal dynamic, DNA intercalating molecules, chromatin structure, cell adhesion, mitosis, and signal transduction.[47–52] FRAP experiment uses fluorescent molecules that are irreversibly photobleached by high-intensity illumination with a focused laser beam in a small cell area. After bleaching, the fluorescence recovery rate is recorded of non-bleached fluorophores diffusing to the bleached spot at lower wavelength laser (Fig. 1.10).

The introduction of FRAP in medical research has enabled us to understand the biological system's key molecular processes. For example, Erami et al.[53] have used the FRAP technique to investigate the changes in E-cadherin mediated cell-cell junction integrity during pancreatic carcinogenesis.[53] Using human tumor xenografts, FRAP has shown the extracellular matrix (ECM) and cellular structure affecting interstitial movement and macromolecules diffusion[54] (Fig. 1.9).

1.5.2.2 FLUORESCENCE LOSS IN PHOTOBLEACHING

Fluorescence loss in photobleaching (FLIP), a complementary technique to FRAP, has been used to measure molecular mobility and living cell dynamics. In this technique, repetitive bleaching over time by a high-intensity laser beam in a specific ROI will prevent the region's fluorescence recovery. The fluorescence loss in ROI defines fluorescently labeled protein mobility, and incomplete loss of fluorescence defines the immobile part not moving into the continuously photobleached area. FLIP is considered as an ideal technique to study the exchange of molecules between cellular compartments such as the Golgi complex and the endoplasmic reticulum and the protein movement between the nucleus and cytoplasm, the nucleolus splicing factors compartment have been reported. To get active and passive transport information, FLIP, along with FRAP, is used. For example, to understand the ribosomal biogenesis using GFP-tagged proteins in living cells was made possible by FRAP and FLIP[55–59] (Fig. 1.10).

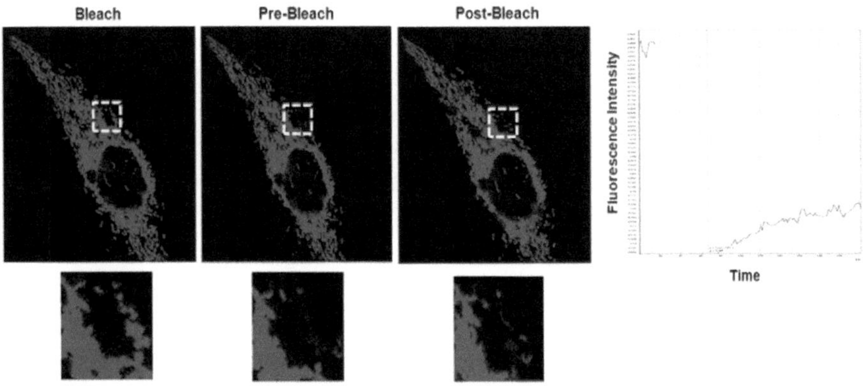

FIGURE 1.9 FRAP example illustrating mitochondrial (Mitotracker red) lateral diffusion.

1.5.2.3 *FLUORESCENCE RESONANCE ENERGY TRANSFER*

The active interaction with living cells and specific molecular species are thought to play a vital role in regulating various signaling pathways. Fluorescence resonance energy transfer (FRET) is the process where energy is transferred from an excited donor to another acceptor. It relies on the close interaction between the two molecules if the distance between donor and the receptor is less than 10 nm.[60] Theodor Forster hypothesized the resonance energy transfer between two light-sensitive molecules.[61] FRET system has an acceptor and donor fluorophore, where the donor transfers energy after excitation in an electronic state to the acceptor with a photon release. Therefore, the emission spectrum of the donor should overlap with the excitation spectrum of the acceptor. The visualization of subcellular protein activation can be achieved by FRET. For example, FRET imaging of the cell cycle was used for chemotherapy optimization in murine and patient-derived models of pancreatic cancer.[62] Mizutani et al.[63] have used FRET-based biosensor to evaluate BCR-ABL activity, and its response to drug treatment in chronic myeloid leukemia (CML) patients[63] (Fig. 1.11).

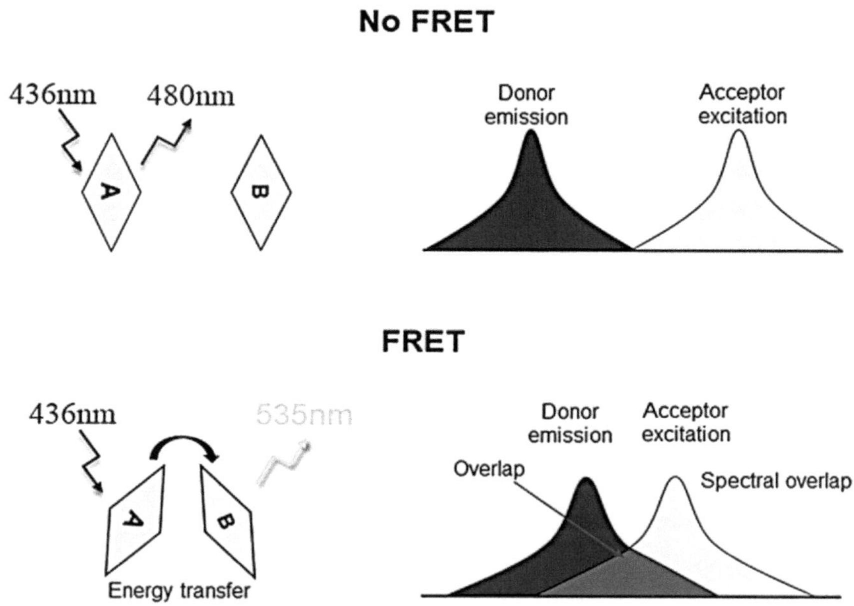

FIGURE 1.10 Schematic representation of FRAP and FLIP experiment. A region of interest (ROI) is selected and bleach with inter laser beam, and recovery is measured for ROI over a period of time. In FLIP, repetitive bleaching of ROI over a period of time and fluorescence intensity outside the selected bleached area is measured.

Source: Adapted from https://www.photometrics.com/learn/physics-and-biophysics/fret

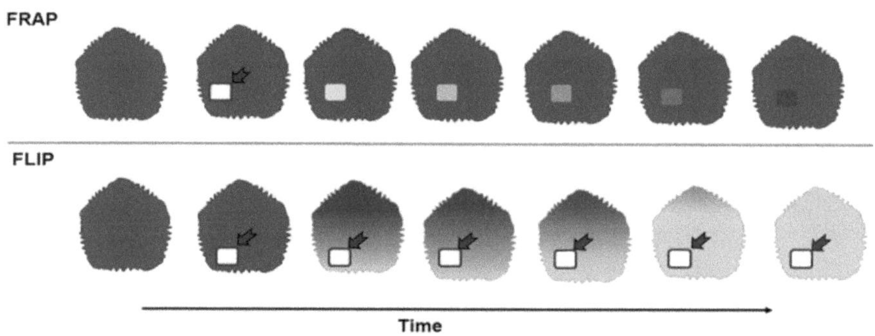

FIGURE 1.11 Schematic representation of FRAP and FLIP experiment.

1.6 ADVANTAGES AND DISADVANTAGES OF CONFOCAL MICROSCOPY

Confocal microscopes have several advantages and disadvantages over a conventional widefield optical microscope. The benefit includes the capacity to control the depth of field, and reduction and removal of background noise. There have been several advancements in confocal microscopy, creating extremely high-quality images from the specimen. The disadvantage includes the limited number of excitations wavelengths from commonly available lasers, occurring very narrow bands, and expensive to produce in ultraviolet wavelengths.

1.6.1 ADVANTAGES OF CONFOCAL MICROSCOPY

- Collects light from a single focal plane
- Has the ability to serially produce thin (0.5–1.5 μM) optical sections through fluorescent specimens
    - Eliminates artifacts
    - Improves contrast and definition through the reduction in the background and improved signal-to-noise ratio
    - Non-invasive technique
    - Enables the investigation of both fixed and living specimens under different conditions with greater clarity
    - Possible to get multi-dimensional views of living cells and tissues
    - 3D Reconstruction of Specimen
- Zoom factor—ability to adjust magnification electronically without having to change objectives
- Multicolor Imaging
- Less Crosstalk
- Improved resolution

1.6.2 DISADVANTAGES OF CONFOCAL MICROSCOPY

- A limited number of excitation wavelengths are available with common lasers
- Reduced signal strength due to detector pinhole

- Photodamage—high-intensity laser irradiation is harmful to live cells and tissues
- More labor intensive and a high cost of purchasing

1.7 CONCLUSION

Confocal microscopy is now filling the gaps between conventional wide-field microscopy and advanced transmission electron microscopy. Due to its vast applications in different research fields, it will continue as one of the major players in the scientific world. Several microscope systems manufacturers are very easy to run. There are several tools available for image processing, such as ImageJ, cell profiler/cell analyst, L-measure, and volume integration and alignment system (VIAS).

KEYWORDS

- **confocal microscope**
- **imaging**
- **live imaging**
- **fluorescence**

REFERENCES

1. Elliott, A. D. Confocal Microscopy: Principles and Modern Practices. *Curr. Protoc. Cytom.* **2020,** *92*, e68.
2. Teng, X.; Li, F.; Lu, C. Visualization of Materials Using the Confocal Laser Scanning Microscopy Technique. *Chem. Soc. Rev.* **2020,** *49*, 2408–2425.
3. Wollman, A. J. M.; Nudd, R.; Hedlund, E. G.; Leake, M. C. From Animaculum to Single Molecules: 300 Years of the Light Microscope. *Open Biol.* **2015,** *5*, 2046–2441.
4. IX. An Account of Mr. Leeuwenhoek's Microscopes. *Philos. Trans. R. Soc. London* **2006,** *41*, 503–519.
5. Anderson, D. Still Going Strong: Leeuwenhoek at Eighty. *Antonie van Leeuwenhoek, Int. J. Gen. Mol. Microbiol.* **2014,** *106*, 3–26.
6. Airy, G. B. On the Diffraction of an Object-glass with Circular Aperture. *Trans. Cambridge Philos. Soc.* **1834,** *5*, 283–291.

7. Abbe, E. X. V. The Relation of Aperture and Power in the Microscope (continued). *J. R. Microsc. Soc.* **1883,** *3*, 790–812.
8. Hell, S. W.; Lindek, S.; Cremer, C.; Stelzer, E. H. K. Confocal Microscopy with an Increased Detection Aperture: Type-B 4Pi Confocal Microscopy. *Opt. Lett.* **2008,** *19*, 222–224.
9. Minsky, M. Memoir on Inventing the Confocal Scanning Microscope. *Scanning* **1988,** *10*, 128–138.
10. Egger, M. D.; Petrăn, M. New Reflected-Light Microscope for Viewing Unstained Brain and Ganglion Cells. *Science* **1967,** *157*, 305–307.
11. Davidovits, P.; Egger, M. D. Photomicrography of Corneal Endothelial Cells in vivo. *Nature* **1973,** *244*, 366–367.
12. Shotton, D. M. Confocal Scanning Optical Microscopy and Its Applications for Biological Specimens. *J. Cell Sci.* **1989,** *94*, 175–206.
13. Prunet, N.; Duncan, K.; Wellmer, F. Imaging Flowers: A Guide to Current Microscopy and Tomography Techniques to Study Flower Development. *J. Exp. Bot.* **2020,** *71*, 2898–2909.
14. Amos, W. B.; White, J. G.; Fordham, M. Use of Confocal Imaging in the Study of Biological Structures. *Appl. Opt.* **2009,** *26*, 3239–3243.
15. Zhang, Z.; Ibrahim, M.; Fu, Y.; Wu, X.; Ren, F.; Chen, L. Application of Laser Scanning Confocal Microscopy in the Soft Tissue Exquisite Structure for 3D Scan. *Int. J. Burns Trauma* **2018,** *8*, 17–25.
16. Hao, R.; Peng, Z.; Zhang, B. Single-Molecule Fluorescence Microscopy for Probing the Electrochemical Interface. *ACS Omega* **2020,** *5*, 89–97.
17. Peterson, D. A. Confocal Microscopy. In *Encyclopedia of Biophysics*, 2012; pp 366–366.
18. Gmitro, A. F.; Aziz, D. Confocal Microscopy through a Fiber-Optic Imaging Bundle. *Opt. Lett.* **1993,** *18*, 565–567.
19. Hale, I. L.; Matsumoto, B. Resolution of Subcellular Detail in Thick Tissue Sections: Immunohistochemical Preparation and Fluorescence Confocal Microscopy. *Methods Cell Biol.* **1993,** *38*, 289–324.
20. Oreopoulos, J.; Berman, R.; Browne, M. Spinning-disk Confocal Microscopy. Present Technology and Future Trends. In *Methods in Cell Biology*, 2014; 123, pp 153–175.
21. Rai, V.; Dey, N. The Basics of Confocal Microscopy. In *Laser Scanning, Theory and Applications*; *InTech*, 2011; pp 75–96.
22. Majlof, L.; Forsgren, P.-O. Chapter 3 Confocal Microscopy: Important Considerations for Accurate Imaging. In *Methods in Cell Biology*, 2008; 38, pp 79–95.
23. Lv, J.; Zhang, K.; Chen, Q. et al. Deep Learning-Based Automated Diagnosis of Fungal Keratitis with In Vivo Confocal Microscopy Images. *Ann. Transl. Med.* **2020,** *8*, 706–706.
24. Claxton, N. S.; Fellers, T. J.; Davidson, M. W. Microscopy, Confocal. In *Encyclopedia of Medical Devices and Instrumentation*; John Wiley & Sons, Inc., 2006.
25. Wright, S. J.; Centonze, V. E.; Stricker, S. A.; DeVries, P. J.; Paddock, S. W.; Schatten, G. Introduction to Confocal Microscopy and Three-Dimensional Reconstruction. *Methods Cell Biol.* **2008,** *38*, 1–45.
26. Smith, C. L. Basic Confocal Microscopy. *Curr. Protoc. Neurosci.* **2011**.
27. Sahoo, S. K.; Crisponi, G. Recent Advances on Iron(III) Selective Fluorescent Probes with Possible Applications in Bioimaging. *Molecules* **2019,** *24*, 3267.

28. Jonkman, J.; Brown, C. M.; Wright, G. D.; Anderson, K. I.; North, A. J. Tutorial: Guidance for Quantitative Confocal Microscopy. *Nat. Protoc.* **2020,** *15*, 1585–1611.
29. Hanrahan, O.; Harris, J.; Egan, C. Advanced Microscopy: Laser Scanning Confocal Microscopy. *Methods Mol. Biol.* **2011,** *784*, 169–180.
30. Matsumoto, B. *Cell Biological Applications of Confocal Microscopy*, 2nd ed. http://www.stemcell8.cn¡ûµã»÷½øÈë (accessed Sept 7, 2020).
31. Paddock, S. W.; Eliceiri, K. W. Laser Scanning Confocal Microscopy: History, Applications, and Related Optical Sectioning Techniques. *Methods Mol. Biol.* **2014,** *1075*, 9–47.
32. Paddock, S. W. Principles and Practices of Laser Scanning Confocal Microscopy. *Appl. Biochem. Biotechnol. —Part B Mol. Biotechnol.* **2000,** *16*, 127–149.
33. Im, K.; Mareninov, S.; Diaz, M. F. P.; Yong, W. H. An Introduction to Performing Immunofluorescence Staining. In *Methods in Molecular Biology*; Humana Press Inc., 2019; 1897, pp 299–311.
34. Thompson, A. D.; Bewersdorf, J.; Toomre, D.; Schepartz, A. HIDE Probes: A New Toolkit for Visualizing Organelle Dynamics, Longer and at Super-Resolution. *Biochemistry* **2017,** *56*, 5194–5201.
35. Jensen, E. C. Use of Fluorescent Probes: Their Effect on Cell Biology and Limitations. *Anat. Rec.* **2012,** *295*, 2031–2036.
36. Younis, M. R.; He, G.; Lin, J.; Huang, P. Recent Advances on Graphene Quantum Dots for Bioimaging Applications. *Front. Chem.* **2020,** *8*, 424.
37. Deerinck, T. J. The Application of Fluorescent Quantum Dots to Confocal, Multiphoton, and Electron Microscopic Imaging. *Toxicol. Pathol.* **2008,** *36*, 112–116.
38. Ettinger, A.; Wittmann, T. Fluorescence Live Cell Imaging. *Methods Cell Biol.* **2014,** *123*, 77–94.
39. Galdeen, S. A.; North, A. J. Live Cell Fluorescence Microscopy Techniques. *Methods Mol. Biol.* **2011,** *769*, 205–222.
40. Bogdanov, A. M.; Kudryavtseva, E. I.; Lukyanov, K. A. Anti-Fading Media for Live Cell GFP Imaging. Anderson KI, ed. *PLoS One* **2012,** *7*, e53004.
41. Pucci, A.; Mattioli, C.; Matteucci, M. et al. Cell Differentiation in Cardiac Myxomas: Confocal Microscopy and Gene Expression Analysis after Laser Capture Microdissection. *Heart Vessels* **2018,** *33*, 1403–1410.
42. Axelrod, D.; Koppel, D. E.; Schlessinger, J.; Elson, E.; Webb, W. W. Mobility Measurement by Analysis of Fluorescence Photobleaching Recovery Kinetics. *Biophys. J.* **1976,** *16*, 1055–1069.
43. Koppel, D. E.; Axelrod, D.; Schlessinger, J.; Elson, E. L.; Webb, W. W. Dynamics of Fluorescence Marker Concentration as a Probe of Mobility. *Biophys. J.* **1976,** *16*, 1315–1329.
44. Liebman, P. A.; Entine, G. Lateral Diffusion of Visual Pigment in Photorecptor Disk Membranes. *Science* **1974,** *185*, 457–459.
45. Poo, M.; Cone, R. A. Lateral Diffusion of Rhodopsin in the Photoreceptor Membrane. *Nature* **1974,** *247*, 438–441.
46. Shaw, S. L.; Kamyar, R.; Ehrhardt, D. W. Sustained Microtubule Treadmilling in *Arabidopsis* Cortical Arrays. *Science* **2003,** *300*, 1715–1718.
47. Von Wichert, G.; Haimovich, B.; Feng, G. S.; Sheetz, M. P. Force-dependent Integrin-Cytoskeleton Linkage Formation Requires Downregulation of Focal Complex Dynamics by Shp2. *EMBO J.* **2003,** *22*, 5023–5035.

48. Howell, B. J.; Moree, B.; Farrar, E. M.; Stewart, S.; Fang, G.; Salmon, E. D. Spindle Checkpoint Protein Dynamics at Kinetochores in Living Cells. *Curr. Biol.* **2004,** *14*, 953–964.
49. Dundr, M.; Hoffmann-Rohrer, U.; Hu, Q. et al. A Kinetic Framework for a Mammalian RNA Polymerase In Vivo. *Science* **2002,** *298*, 1623–1626.
50. Shav-Tal, Y.; Darzacq, X.; Shenoy, S. M. et al. Dynamics of Single mRNPs in Nuclei of Living Cells. *Science* **2004,** *304*, 1797–1800.
51. Lahlou, H.; Saint-Laurent, N.; Estève, J. P. et al. sst2 Somatostatin Receptor Inhibits Cell Proliferation through Ras-, Rap1-, and B-Raf-dependent ERK2 Activation. *J. Biol. Chem.* **2003,** *278*, 39356–39371.
52. Erami, Z.; Herrmann, D.; Warren, S. C. et al. Intravital FRAP Imaging Using an E-cadherin-GFP Mouse Reveals Disease- and Drug-Dependent Dynamic Regulation of Cell-Cell Junctions in Live Tissue. *Cell Rep.* **2016,** *14*, 152–167.
53. Pluen, A.; Boucher, Y.; Ramanujan, S. et al. Role of Tumor-Host Interactions in Interstitial Diffusion of Macromolecules: Cranial vs. Subcutaneous Tumors. *Proc. Natl. Acad. Sci.* **2002,** *98*, 4628–4633.
54. Cole, N. B.; Smith, C. L.; Sciaky, N.; Terasaki, M.; Edidin, M.; Lippincott-Schwartz, J. Diffusional Mobility of Golgi Proteins in Membranes of Living Cells. *Science* **1996,** *273*, 797–801.
55. Nehls, S.; Snapp, E. L.; Cole, N. B. et al. Dynamics and Retention of Misfolded Proteins in Native ER Membranes. *Nat. Cell Biol.* **2000,** *2*, 288–295.
56. Phair, R. D.; Misteli, T. High Mobility of Proteins in the Mammalian Cell Nucleus. *Nature* **2000,** *404*, 604–609.
57. Köster, M.; Frahm, T.; Hauser, H. Nucleocytoplasmic Shuttling Revealed by FRAP and FLIP Technologies. *Curr. Opin. Biotechnol.* **2005,** *16*, 28–34.
58. Chen, D.; Huang, S. Nucleolar Components Involved in Ribosome Biogenesis Cycle between the Nucleolus and Nucleoplasm in Interphase Cells. *J. Cell Biol.* **2001,** *153*, 169–176.
59. Stryer, L. Fluorescence Energy Transfer as a Spectroscopic Ruler. *Annu. Rev. Biochem.* **2003,** *47*, 819–846.
60. Forster, T. Energiewanderung und Fluoreszenz. *Naturwissenschaften* **1946,** *33*, 166–175.
61. Vennin, C.; Chin, V. T.; Warren, S. C. et al. Transient Tissue Priming via ROCK Inhibition Uncouples Pancreatic Cancer Progression, Sensitivity to Chemotherapy, and Metastasis. *Sci. Transl. Med.* **2017,** *9*, eaai8504.
62. Mizutani, T.; Kondo, T.; Darmanin, S. et al. A Novel FRET-based Biosensor for the Measurement of BCR-ABL Activity and Its Response to Drugs in Living Cells. *Clin. Cancer Res.* **2010,** *16*, 3964–3975.

CHAPTER 2

Evaluation of Plant Extracts as Corrosion Inhibitor: A Surface Morphological Study

R. VASANTHA JOTHI and R. SARATHA*

Department of Chemistry, Avinashilingam Institute for Home Science and Higher Education for Women, Coimbatore, Tamil Nadu, India

**Corresponding author. E-mail: saratha63@gmail.com*

ABSTRACT

The advent of various surface morphological techniques has led to the study of surface structure which uses sophisticated microscopes that produce images of high spatial resolution, from the sample's atomic and molecular level. Different techniques such as optical microscopy, scanning electron microscope (SEM), transmission electron microscope (TEM), atomic force microscope (AFM), and X-ray diffraction find extensive application for the examination of surface morphology in corrosion inhibition studies using plant extracts. This chapter discusses the surface morphology and characterization of corrosion inhibitors on the mild steel surfaces, which are essential for the fruitful application to explore more of these materials.

2.1 INTRODUCTION

A major contribution in the development of Material Science can be achieved through utilizing Microscopic Techniques that facilitate the

Advanced Microscopy: A Strong Analytical Tool in Materials Science. Merin Sara Thomas, Jozef T. Haponiuk, Sabu Thomas, & Anne George (Eds.)

assessment of morphology, physical properties, composition, and the dynamism in the behavior of metals. Microscopic techniques are a type of characterization methods that explores and maps the surface and sub-surface of a material. In order to gather data about a sample's topography on a range of length scales, these techniques use photons, electrons, and ions.[1] The development of new materials that find extensive applications requires the application of this technique. The research advancements carried out, in particular, which intends to design, synthesize, and characterize, develop novel technologically sophisticated, adaptable, and multifunctional materials, are on the rapid growth in the field of material sciences. The multifunctional materials attribute with lower mass and volume but higher efficiency, affordability, and novelty. The most developing classes of new materials that pave the way to technological innovations are advanced blends, composites, and hybrid materials.[2] In order to exert a prospective safe application of these materials in the fields of Health and Environment, it is essential to acquire basic knowledge on the relationships between structure, properties, functions, and performance of the newly developed materials. The study of the physical and technical performance of the latest developments, in the areas described, will rely on microscopic techniques that find multiple industrial applications, including topographic and dynamical surface studies of many materials.[3] While many characterization techniques have been used for centuries, such as basic optical microscopy and scanning electron microscopy, newly developed surface characterization techniques are constantly emerging. The electron microscope invented in the 20th century, particularly, has helped out in bringing revolutions to the field by way of the developments in imaging and analyzing the structure and compositions, making it applicable to smaller scales than usual. This resulted in higher levels of understanding of the reason behind different materials showing different properties and behaviors.[4] The more recently invented atomic force microscopy has further contributed to the utmost possible resolution for analysis of certain samples in the last 30 years.[5]

The present study investigates the influence of plant extracts namely *Ailanthus excelsa Roxb* (AEL), *Bauhinia racemosa Lam* (BRL), and *Hardwickia binata Roxb* (HBL) as eco-friendly corrosion inhibitors for mild steel in acid medium. Plant materials are rich sources of a mixture of organic compounds (containing N, S, O atoms) and have the ability to function as effective inhibitors for metals in acid, alkaline, and neutral media The phytoconstituents present in the leaves of the selected plants were found to be flavanoids, phenolic compounds, lignin saponins sterols,[6] flavanoids coumarins,[7] saponins, tannins, terpenoids, and carbohydrates.[8]

2.2 MILD STEEL AND CORROSION

Mild steel, more specifically iron and its alloys, are renowned for their industrial applications, owing to their availability and low cost.[9] During the industrial process such as acid cleaning, pickling, and descaling, their surfaces are deployed and undergo considerable corrosion. In recent days, the study of steel corrosion phenomena has become important, especially in acid media as it plays a crucial role in uplifting the economy of a nation, and has an environmental impact.

Several corrosion protection methods were used in the industrial process. The use of corrosion inhibitors is one of the widely accepted techniques that reduces the corrosion rate and loss of base metal, when added in smaller concentrations, especially in acid solutions. Most of the inhibitors used in industries are chemical compounds composed of π electrons, aromatic rings and polar functional groups containing heteroatom with non-bonding electrons which can easily adsorb onto the metal surface by the formation of a bond between the electron pair or the π-electron cloud of the donor atoms and the metal surface, thereby reducing the corrosive attack in acid medium.[10] Many of these compounds are investigated as effective and efficient corrosion inhibitors. But some of the chemical inhibitors are very toxic to humans, have a bad environmental impact, and are highly expensive. Recently, the development of green corrosion inhibitors of plant origin for various metals in acid media, which is eco-friendly, biodegradable, cost-effective, and less toxic have been attempted. From the intensive and keen work dedicated to use plant extracts as corrosion inhibitors for mild steel in acid media, it is evident that industrial sectors deem it a boon, as its applications are growing limitless.[11–15] A number of investigations carried out by various authors could be cited as proof of the better inhibition efficiency, resulting from using different plants extracts as corrosion inhibitors in different media such as hydrochloric acid, sulfuric acid, phosphoric acid, and nitric acid.

2.3 TECHNIQUES FOR CORROSION INHIBITION MONITORING

The various techniques employed for corrosion inhibition monitoring have been classified as non-electrochemical methods and electrochemical methods.

2.3.1 NON-ELECTROCHEMICAL METHOD

These include techniques like weight loss measurement, gasometry, and thermometry. These techniques require relatively long exposure times of the corroding systems, which is a notable disadvantage. Also, the non-electrochemical methods are in general restricted to systems that do not form an adherent layer of corrosion products.

2.3.2 ELECTROCHEMICAL METHOD

The electrochemical method is one of the most widely used techniques to study the phenomenon of corrosion inhibition. Electrochemical methods are finding extensive use in corrosion research and in engineering applications. Such methods are practical because the corrosion behavior of material–electrolyte combinations is a direct function of the mechanism and kinetics. The electrochemical methods can be used, in the field or in the laboratory, to measure corrosion rates without removing the specimen from the environment or altering the sample itself.

2.4 TECHNIQUES FOR SURFACE ANALYSIS

The methods to study the corrosion inhibition process are radiotracer technique, spectroscopic methods viz. X-ray photoelectron spectroscopy, Auger electron spectroscopy, ellipsometry, hydrogen permeation, quartz crystal method, and electrochemical noise. Besides the techniques mentioned, scanning electron microscope (SEM), energy dispersive X-ray analysis (EDX), and optical laser profilometric studies are also used to analyze the corrosion inhibition process.

2.4.1 SCANNING ELECTRON MICROSCOPE

SEM is a versatile and widely used electron beam instrument. It has gained popularity by generating micrographs that are easy to interpret, to the diversity of types of information that it can produce, and to the fact that images and analytical information can readily be combined. The nature of the micrographs helps to get valuable information, particularly in the corrosion inhibition process. The corrosion inhibition ability of plant extracts namely AEL, BRL, and HBL in HCl and H_2SO_4 was investigated by SEM analysis, and the corresponding micrographs are shown in Figures 2.1 and 2.2. From the images, it is evident that the surface is severely damaged in Figures 2.1a and 2.2, a protective layer is being formed in Figures 2.1 and 2.2b–d. In this way, SEM micrographs help us study the surface morphology.[16]

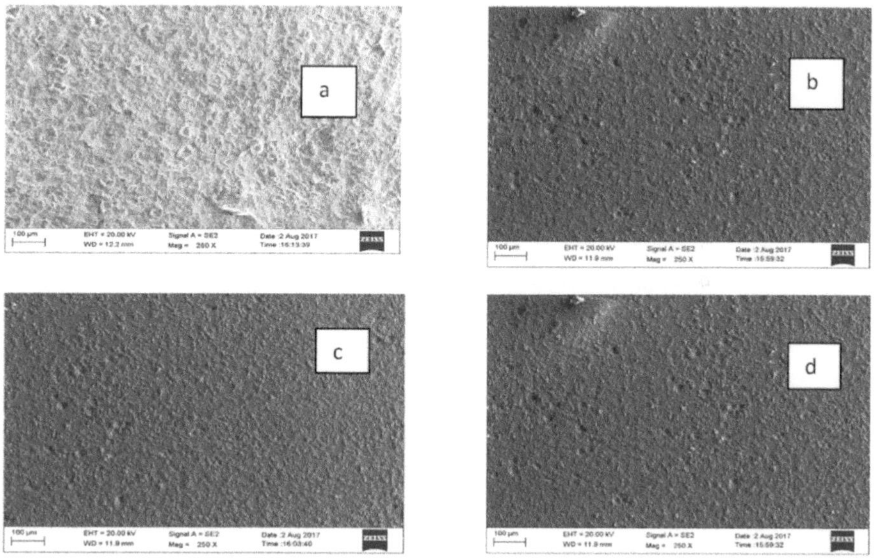

FIGURE 2.1 SEM micrographs of MS surface after immersion in 1 M HCl (a) and MS surface in 1 M HCl with extracts; AEL (b), BRL (c), and HBL (d).

2.4.2 ENERGY-DISPERSIVE X-RAY ANALYSIS

The EDX spectra were used to determine the composition of elements present on the MS surface in the absence and presence of plant extracts.

Figure 2.3a portrays the EDX spectrum of mild steel in 1M HCl for 24 h of immersion and Figure 2.3b–d in the presence of 0.5% v/v AEL, BRL, and HBL extract. Figure 2.4a–d shows the EDX spectrum of mild steel immersed in 0.5M H_2SO_4 and in the presence of 0.5% v/v AEL, BRL, and HBL extract, respectively. The surface composition of mild steel in HCl and H_2SO_4 and in the presence of plant extracts are tabulated in Tables 2.1 and 2.2.

FIGURE 2.2 SEM micrographs of MS surface after immersion in 0.5M H_2SO_4 (a) and MS surface in 0.5M H_2SO_4 with extracts; AEL (b), BRL (c), and HBL (d).

TABLE 2.1 Surface Composition (Weight %) of Mild Steel after 24 h of Immersion in 1M HCl without and with the Optimum Concentration of the Studied Extracts.

Weight (%)	Fe	C	O	Cl
Blank −1 M HCl	73.46	2.54	19.15	1.35
AEL extract	81.57	5.20	42.36	–
BRL extract	77	4.62	18.64	–
HBL extract	76	4.36	18.87	–

From the EDX spectrum of MS in 1 M HCl Figure 2.3a, it can be concluded that the MS surface corroded in aggressive medium has high chloride concentration compared to the surface in the presence of AEL/BRL/HBL extracts, chloride peak is absent as shown in Figure 2.3b–d.

TABLE 2.2 Surface Composition (Weight %) of Mild Steel after 24 h of Immersion in 0.5M H_2SO_4 solution without and with the Optimum Concentration of the Studied Extracts.

Weight (%)	**Fe**	**C**	**O**	**S**	**Ti**	**Cu**	**P**	**Si**
Blank –0.5 M H_2SO_4	60.64	4.62	29.11	1.55	1.39	1.38	1.30	–
AEL extract	77.79	7.67	13.80	–	–	–	–	0.73
BRL extract	69.87	5.15	23.50	–	–	–	–	1.48
HBL extract	76.51	4.67	19.87	–	–	–	–	–

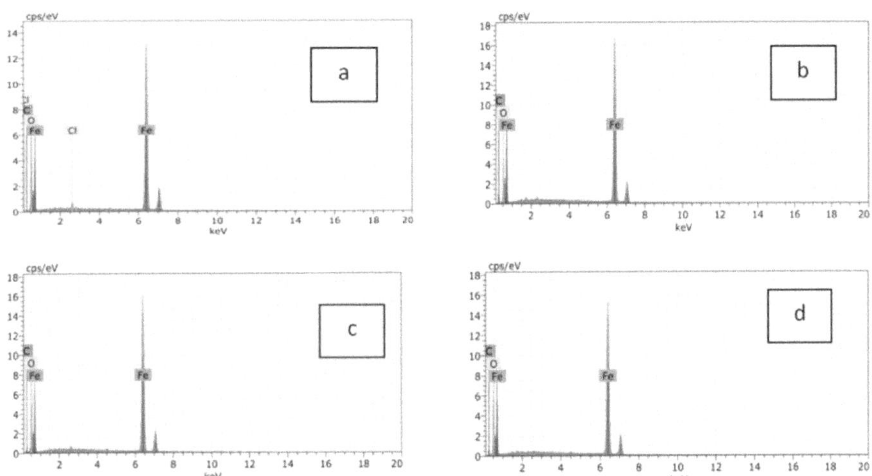

FIGURE 2.3 EDX spectra of mild steel in 1 M HCl solution after 24 h of immersion without plant extract (a); in presence of 0.5% v/v AEL (b), BRL (c), and HBL (d).

Table 2.2 shows the surface composition of MS immersed in 0.5 M H_2SO_4 without and with the optimum concentration of selected plant extracts. It can be observed from Table 2.2 that the % composition of carbon is increased in the presence of plant extracts that indicates the adsorption of a carbon atom from its constituents and the absence of sulfur peak in presence of plant extract also reveals the adsorption of plant extract.[17]

2.4.3 OPTICAL LASER PROFILOMETRIC STUDIES

It is a powerful technique to investigate the surface morphology at nano- to microscale that is being used currently to study the influence of corrosion inhibitors on metal/solution interface. From the analysis, imminent information can be gained regarding the roughness and uniformity of the study. The Roughness average value (R_a) plays an important role in identifying the efficiency of the inhibitors under study. Figures 2.5a and 2.6a show the 3D images of MS in 1 M HCl and 0.5 M H_2SO_4. Figures 2.5b–d and 2.6b–d show the 3D images of MS/1 M HCl and MS/0.5 M H_2SO_4, respectively, in the presence of investigated plant extracts.

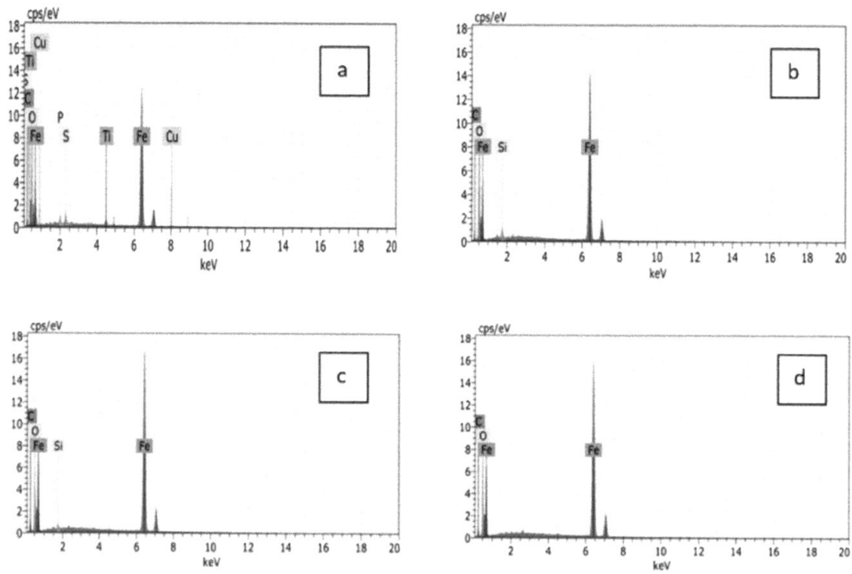

FIGURE 2.4 EDX spectra of mild steel in 0.5M H_2SO_4 solution after 24 h of immersion without plant extract (a); in presence of 0.5% v/v AEL (b), BRL (c), and HBL (d).

The Roughness average value (R_a) for MS in 1 M HCl, Figure 2.5a, was found to be 20.50 μm whereas for MS immersed in AEL, BRL, and HBL, Figure 2.5b–d, the *R*a value was found to be reduced to 10.50, 10.37, and 9.75 μm, respectively. From the difference observed in the surface morphology of the metal specimen visually, we can conclude the formation of protective film on the surface.[18]

2.5 PROBABLE MODE OF CORROSION INHIBITION MECHANISM

Adsorption is the primary step in achieving inhibition in acid solutions. This is the outcome of the fact that the corroding metal surface to be inhibited is usually oxide-free, allowing the inhibitor ready to retard the cathodic and/or the anodic electrochemical processes of corrosion. The main constituent, in aqueous acidic solutions, survives either as neutral molecules or as protonated molecules.

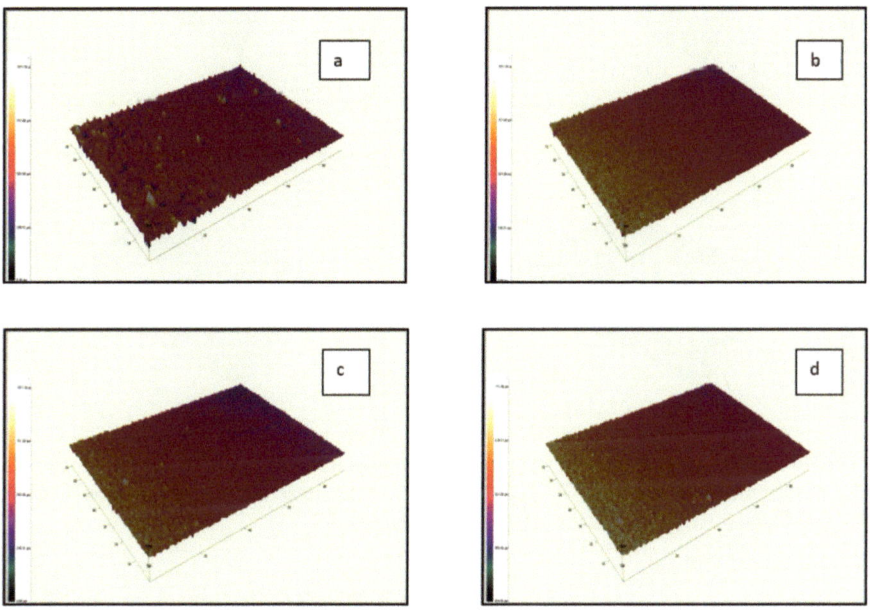

FIGURE 2.5 Laser profilometric images of mild steel in 1 M HCl and after 24 h of immersion MS in 1 M HCl (a); MS in the presence of optimum concentration of 0.5% v/v/ for; AEL (b) BRL (c), HBL and (d) extracts.

In the present situation, two modes of adsorption could be considered

i) Chemisorption mechanism: The neutral AEL/BRL/HBL may adsorb onto the metal surface, involving the displacement of water molecules from the metal surface and sharing electrons between the oxygen atom and iron. There will be adsorption of the phytochemical constituents of the selected plant extracts on the metal surface involving donor–acceptor interactions between π-electrons of the aromatic ring and vacant d-orbital of iron.

$$Fe \rightarrow Fe^{2+} + 2e^-$$

$$[AEL/BRL/HBL] + Fe^{2+} \leftrightarrow [AEL/BRL/HBL-Fe]^{2+}$$

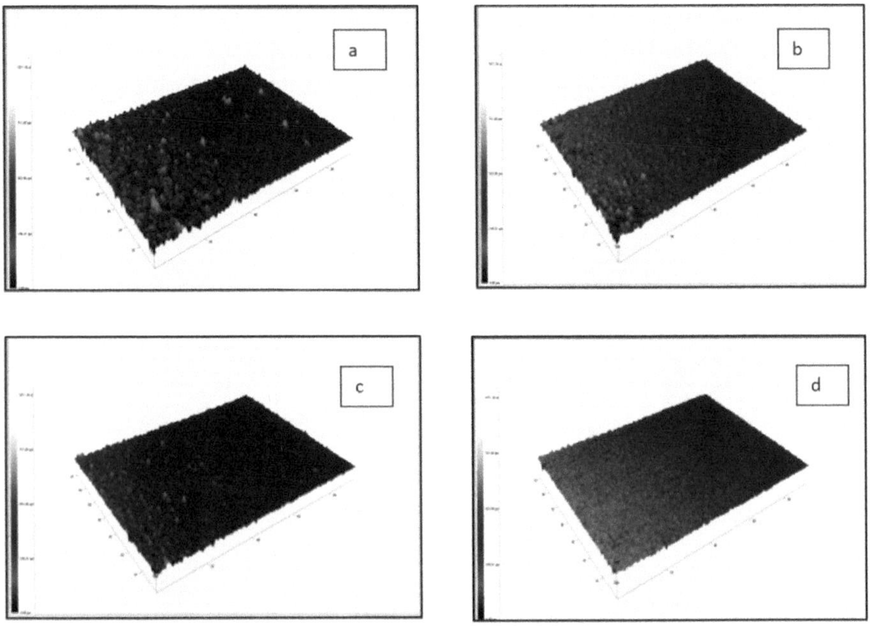

FIGURE 2.6 Laser profilometric images of mild steel in 0.5 M H_2SO_4 solution and after 24 h of immersion, MS in 0.5 M H_2SO_4 (a); MS in the presence of optimum concentration of 0.5% v/v; AEL (b), BRL (c), HBL and (d) extracts.

ii) The protonated AEL, BRL, and HBL extracts may adsorb through electrostatic interactions between the positively charged molecules and the negatively charged metal surface. In other words, there may be a synergism between Cl and AEL/BRL/HBL extract, which improves the inhibitive capability of the extract.

$$[AEL/BRL/HBLH_{\chi}]^{x+} + Fe^{2+} \leftrightarrow [AEL,BRL,HBL - Fe]^{(2+x)+}$$

$$[AEL/BRL/HBL] + Fe^{2+} \leftrightarrow [AEL/BRL/HBL - Fe]^{2+}$$

When protonated AEL, BRL, and HBL extracts are adsorbed on the metal surface, a coordinate bond may be formed by partial transference of electrons from polar atoms (oxygen atoms) to the metal surface. In addition, owing to lone-pair electrons of oxygen atom in AEL/BRL/HBL extract, protonated extracts may combine with freshly generated Fe^{2+} ions on steel surface forming metal-inhibitor complexes. These complexes are adsorbed on the MS surface by Van der Waal's force to form a protective film to keep the MS surface from corrosion. This assumption could be further confirmed by FTIR studies and SEM analysis.

2.6 CONCLUSION

The present investigation proves the extracts of AEL, BRL, and HBL (leaves extract) to be a potent green corrosion inhibitor for mild steel in acid media and showcased positive results in microscopic analysis.

KEYWORDS

- **surface morphology**
- **SEM**
- **TEM**
- **AFM**
- **corrosion inhibition**
- **plant extracts mild steel**

REFERENCES

1. Kumar, S. Z.; Lin Li, A. *Materials Characterization Techniques*; CRC Press: Boca Raton, 2009. ISBN: 1420042947.
2. Leng, Y. *Materials Characterization: Introduction to Microscopic and Spectroscopic Methods*; Wiley, 2009. ISBN: 978-0-470-82299-9.
3. Zhang, S. *Materials Characterization Techniques*; CRC Press, 2008. ISBN: 1420042947.
4. Mathys, D. Zentrumfür Mikroskopie, University of Basel: Die Entwicklung der Elektronen mikroskopievom Bildüber die AnalysezumNanolabor; p 8
5. https://mts.hindawi.com/submit/journals/amse/mtms/
6. Lavhale, M. S.; Mishra, S. H. Nutritional and Therapeutic Potential of *Ailanthus excelsa*—A Review. *Pharmacognosy Rev.* **2007,** *1*, 105–113.
7. Manohar, P.; Rajesham, V. V.; Ramesh, M.; Kiran Kumar, S.; Prasanna Kumari, J. Pharmacognostical, Phytochemical and Antimicrobial Activity of *Bauhinia racemosa* Leaves. *Int. J. Pharm. Biol.* **2011,** *1*, 10–14.
8. Gunaselvi, G.; Kulasekaren, M.; Gopal, V. Preliminary Phytochemical Studies on the Leaves of *Hardwickia binata Roxb. Int. J. Pharmacol. Biomed. Sci.* **2010,** *1*, 37–40.

9. Fouda, A. S.; Ellithy, A. S. Inhibition Effect of 4-phenylthiazole Derivatives on Corrosion of 304L Stainless Steel in HCl Solution. *Corros. Sci.* **2009,** *51*, 868–875.
10. Ji, G.; Dwivedi, P.; Sundaram, S.; Prakash, R. Aqueous Extract of *Argemone mexicana* Roots for Effective Protection of Mild Steel in an HCl Environment. *Res. Chem. Interm.* **2016,** *42*, 439–459.
11. Lee, H.; Sin, Y.; Afidah, A. R.; Gan, C. Y.; Saad, B.; Idris, M.; Umeda, S. M. *Aquilariasubtintergra* Leaves Extracts as Sustainable Mild Steel Corrosion Inhibitors in HCl. *Measurement* **2017,** *109*, 334–345.
12. Khadom, A. A.; Abd, A. N.; Arif Ahmed, N. *Xanthium strumarium* Leaves Extracts as a Friendly Corrosion Inhibitor of Low Carbon Steel in Hydrochloric Acid: Kinetics and Mathematical Studies. *South Afr. J. Chem. Eng.* **2018,** *25*, 13–21.
13. Gusti, D. R.; Emriadi, A. A.; Mai, E. Corrosion Inhibition of Ethanol Extract of Cassava (*Manihot esculenta*) Leaves on Mild Steel in Sulfuric Acid. *Int. J. Chem. Tech. Res.* **2017,** *10*, 163–171.
14. Idouhli, R.; Koumya, Y.; Khadiri, M.; Aityoub, A.; Abouelfida, A.; Benyaich, A. Inhibitory Effect of *Senecio anteuphorbium* as Green Corrosion Inhibitor for S300 Steel. *Int. J. Indust. Chem.* **2019,** *10*, 133–143.
15. Olusegun, S. J.; Joshua, T. S.; Bodunrin, M. O.; Aribo, S. Inhibition of Mild Steel Corrosion in HCl Solution by Plant Extract of *Biden pilosa. Nat. Sci.* **2018,** *16* (1).
16. Anh, H. T.; Nam, N. D. *Ficus racemosa* Leaf Extract for Inhibiting Steel Corrosion in a Hydrochloric Acid Medium, 2020. https://doi.org/10.1016/j.aej.2020.07.051.
17. Al-Bonayan, A. M. Corrosion Inhibition of Carbon Steel in Hydrochloric Acid Solution by *senna-italica* Extract. *Int. J. Recent Res. Appl. Stud.* **2015,** *22*, 49–64.
18. Laser Profile Mapping. http://www.laser-ndt.com/profilometry.html.

CHAPTER 3

CLARITY-Based Tissue Clearing Techniques and Light Sheet Microscopy: For Imaging and Investigation of Neurodegenerative Disorders

DANIA ALPHONSE JOSE and P. S. BABY CHAKRAPANI

Centre for Neuroscience, Cochin University of Science and Technology (CUSAT), Kochi 682022, Kerala, India

Department of Biotechnology, Cochin University of Science and Technology (CUSAT), Kochi 682022, Kerala, India

**Corresponding author. E-mail: chakrapani@cusat.ac.in*

ABSTRACT

Light sheet microscopy combined with state-of-the-art tissue clearing methods like Clear Lipid-exchanged Acrylamide-hybridized Rigid Imaging compatible Tissue hYdrogel (CLARITY) enables whole body antibody labeling and 3D imaging of large volumes of tissue specimens. Brain imaging modalities are not sufficient to understand the pathophysiological mechanisms of several complex brain diseases like neurodegenerative diseases. The 3D microscopy combined with CLARITY-based tissue clearing techniques promise in-vivo mapping of cyto- and myelo-brain tissue architecture. Theses histological data can also be used as a reference for the validation of imaging data to rule out the neuronal density, white matter fiber orientation, and distribution, as well as the cell-type specific properties.

Advanced Microscopy: A Strong Analytical Tool in Materials Science. Merin Sara Thomas, Jozef T. Haponiuk, Sabu Thomas, & Anne George (Eds.)

This chapter gives a general introduction about neurodegenerative diseases and the emerging role of CLARITY in neurodegenerative aspects. The chapter further discuss about the mapping of neurodegerative pathology using CLARITY-based tissue clearing techniques in combination with light sheet microscopic techniques.

3.1 INTRODUCTION

Degeneration processes within the nervous system are referred to as neurodegeneration. It can be defined as the progressive loss of neurons associated with pathologically altered proteins that deposit in the brain and in peripheral organs, which damage the nervous system over time.[1] The process of neurodegeneration includes cell death, impaired axonal regeneration, demyelination, and overall neuronal structural and functional deficits.[2] All these conditions—partially or combined, acquired or genetic, and known or unknown in origin are manifested in specific neurological disorders and collectively termed as neurodegenerative disorders (NDDs). NDD are characterized by a distinct clinical syndrome of neurological deficits including progressive decline or even the complete loss of sensory, motor, and cognitive functions.[3]

The key malefactors in the manifestation of NDD are proteins with altered physiochemical and neurotoxic properties.[4] The abnormal build up of these proteins with a conformational rearrangement endows them with a tendency to aggregate and deposits within various regions of the central nervous system (CNS) and leads to early neuronal death.[3] The pathogenetic cascades leading to neurodegeneration have not been elucidated fully, though perturbations in cellular mechanisms like handling of misfolded proteins and oxidative stress, mitochondrial and microsomal functions, and autophagy and intracellular calcium handling are known to be involved.[5]

The clinical presentations do not necessarily indicate the exact molecular pathology in neurodegenerative disorders. Also, the phenotypes differ but there are many shared clinical and biological features across NDDs.[2] The common neurodegenerative diseases are amyotrophic lateral sclerosis (ALS), Alzheimer's disease (AD), Parkinson's disease (PD), Huntington's disease (HD), front-temporal dementia (FTD), and prion disease.[6] The major proteins involved in the pathogenesis of these diseases are: aggregation of ubiquitinated proteins in ALS, β-amyloid in AD, α-synuclein accumulation in PD, misfolded Huntington protein in HD, microtubule-associated protein tau in FTD, and prion protein in the case of prion disease.[4] The molecular classification of neurodegenerative disease is protein-based; AD, PD, and HD are primarily classified as proteinopathies (Fig. 3.1). This emphasizes the role of protein-processing systems in the disease pathogenesis. These neurodegenerative diseases are also designated as conformation diseases.

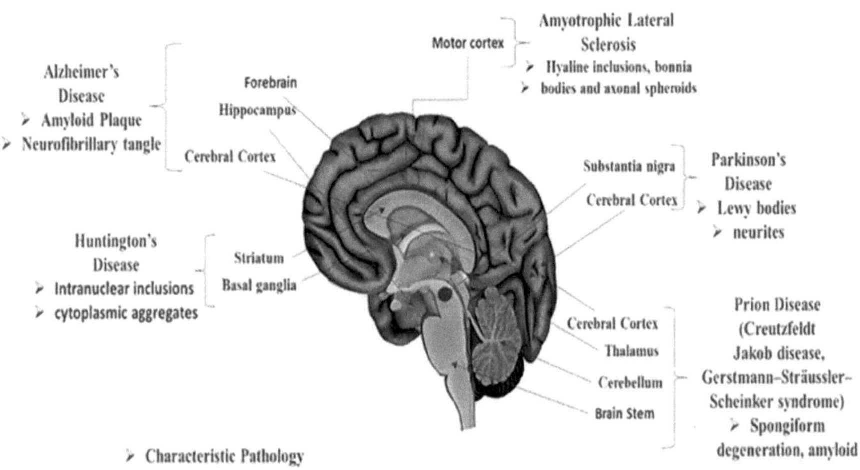

FIGURE 3.1 The image in this figure depicts the neurodegenerative disorders to their corresponding characteristic anatomical region. The brain diagram was adapted from web (http://www.innerbody.com/image/nerv02.html).

The clinical and pathological investigations of these lesions in post-mortem brain tissues routinely provides a 2D picture of the pathology associated with distinct NDDs with a limited depth of field.[7] Understanding the role of brain structural changes, which ultimately lead to functional deficits in NDDs, requires an accurate 3D representation of neuronal morphology and realistic biophysical modeling that can directly relate the pathological changes to altered neuronal firing patterns. This will help to clarify the underlying mechanisms of neurodegeneration, in neuropathological categorization of NDDs, and to translate it into a in-vivo detectable biomarkers to support better prediction of prognosis and patient stratification for therapeutic trials.

CLARITY is a novel tissue clearing technique and probably one of the most important advancement in neuroscience, which allowed researchers to see directly the optically transparent whole brains or thick blocks of intact brain tissues. Recently, various CLARITY-based tissue clearing techniques have emerged and have been utilized to visualize microscopic specimens employing the advancements of laser microscopy such as confocal/multi-photon, light sheet, volume-illumination, and Bessel-beam illumination microscope.[8] The combined approach of various CLARITY-based tissue clearing techniques coupled with light sheet microscopy will allow detection and mapping of intact brain tissues specifically the 3D visualization of neurodegenerative pathology using mouse brains and postmortem human brain tissues.

3.2 AN OVERVIEW OF MAPPING NEURODEGENERATIVE PATHOLOGY IN INTACT TISSUES

Decades of scientific advances in diagnostics, neuroimaging, and clinico-pathological correlations have permitted a better understanding of neurodegenerative diseases. Despite this considerable progress, to date, the diagnosis remains mainly clinical and only confirmed by autopsy.[9] Nevertheless, the overlap in symptomatology and clinical heterogeneity in disease presentation sometimes cause clinical misdiagnosis in the early stages of the disease, even in the setting of a movement disorder clinic.[10] The extraction of detailed structural and molecular information from intact brain tissues has long been a fundamental challenge in the field of neuroscience and now it has spurred considerable technological innovation. The main goal of modern neuroscience is to study the brain structure–function relationships at cellular, circuit, and organ-wide scale. Current pioneering methods, CLARITY and CLARITY-based tissue clearing methods are suitable for full structural analysis of intact mammalian brain and this requires, in particular, the 3D anatomical and phenotypical mapping.[11] Tissue clearing methods with the depth of light sheet microscopy plays a pivotal role in this regard by providing an in-vitro opportunity to visualize the neuroanatomical and functional signatures of these disorders, as well as identifying disease-specific biomarkers of the underlying neurodegenerative processes. These biomarkers have the potential to eventually serve as reliable neuropathologic indicators to improve the sensitivity and specificity of clinical diagnoses.

3.3 TISSUE CLEARING TECHNIQUES AND LIGHT SHEET MICROSCOPY

What Is Tissue Clearing?

Tissue clearing is a technique that renders a large biological sample transparent using a series of chemical treatment. It can be a whole organ, entire organoids, or even an entire young rat, ranging in thickness from around 100 μm to several centimeters.[12] The typical classification of tissue clearing techniques is based only on the chemicals used and all tissue clearing methods share the same goal. This new method allows researchers to see directly the optically transparent whole brains or thick blocks of brain tissue instead of thick sections of tissue.

Why is tissue clearing required? Typically, biological tissues are complex with a 3D structure; most biological tissues are not transparent, including human tissues except for the cornea, vitreous, lens, and retina of the eye.[13] These tissues contain a mix of components of small size with different refractive indices (RIs) and, thus, the interactions of light with these heterogeneous components do not allow light to pass.[12] This leads to a lack of transparency and it is difficult to analyze using visible wavelengths of light with a microscope. Optical tissue clearing (OTC) has been developed to address this issue. Clearing of tissue tries to homogenize the RI of a sample by removing, replacing, and modifying some of its components.

In the last decade, the tissue clearing techniques along with light sheet microscopy opened up new research avenues to view unprecedented 3D views of enormous volumes of tissue, including the central nervous system and the large networks of neurons, with unprecedented ease and accuracy. Chemical clearing provides an ideal way for researchers to transform intact biological tissues to achieve optical transparency and it preserves both the molecular and structural information with minimal disassembly of the sample.[14]

In short, if your object of interest are microscopic but extend more than a few tens of microns inside biological tissue, tissue clearing can be very beneficial.

3.4 CLARITY-BASED TISSUE CLEARING TECHNIQUES

Recent years have witnessed a proliferation of tissue clearing methods based on CLARITY technique, CUBIC, PACT/PARS, iDISCO, and ACT-PRESTO.

The basic difference between these methods is the chemistry used for the clearing of tissues and the tradeoffs between speed, complexity, cost, conservation of protein-based fluorescence, compatibility with immunostaining, and other criteria.

3.4.1 CLARITY

CLARITY is a novel tissue clearing technique, in principle applicable to any tissue. It was first devised by Kwanghun Chung and Karl Deisseroth at Stanford University in California and was selected as one of ten breakthroughs in 2013 by Science.[15] CLARITY tissue clearing method transforms intact biological tissues into a nanoporous hydrogel-tissue hybrid.

The basic CLARITY technique, also called as clarification, includes (1) fixation of tissues, (2) hydrogel monomer infusion, (3) hydrogel-tissue hybridization, (4) lipid extraction, (5) molecular labeling (if required), (6) and refractive index matching and imaging.[16] The process requires the use of specialized electrophoresis devices that must be operated within a narrow range of parameters to yield meaningful results.The electrophoretic chamber (EC) is not essential but when it is not used, it takes a long time to clear tissues and this approach is called passive CLARITY whereas CLARITY with EC is called active CLARITY.[17]

The protocol includes infusing hydrogel monomers (acrylamide and bisacrylamide), formaldehyde, and thermally triggered initiators into tissue at 4°C. In this step, formaldehyde not only cross links the tissue, but also covalently links the hydrogel monomers to biomolecules including proteins, nucleic acids, and small molecules, followed by polymerization of the hydrogel at 37°C for 3 h, at this point, the tissue and hydrogel gets converted to a hybrid construct. This hydrogel–tissue hybridization physically supports tissue structure and chemically incorporates biomolecules into the hydrogel mesh, except the lipids. The lipids are removed, using electrophoretic tissue clearing (ETC). The clearing solution contains boric acid, sodium dodecyl sulfate, sodium hydroxide, and distilled water. During this process, the proteins and nucleic acids are preserved and the unbound lipids are removed. To extract lipids, either ionic detergent-based solution or hydrophobic organic solubilization can be used.[18] Although organic solvents can extract lipids, ionic extraction technique is more efficient. Organic solvents can reduce refractive-index variations, but these solvents quench fluorescence, thereby limiting the imaging time. The resulting lipid-extracted and structurally stable tissue-hydrogel hybrid can be immersed in refractive-index-specified matching homogenization solution (for example, 85% glycerol or Focus Clear) for 24 h at 37°C (Fig. 3.2). The resulting sample is highly suitable for LSFM imaging.[19]

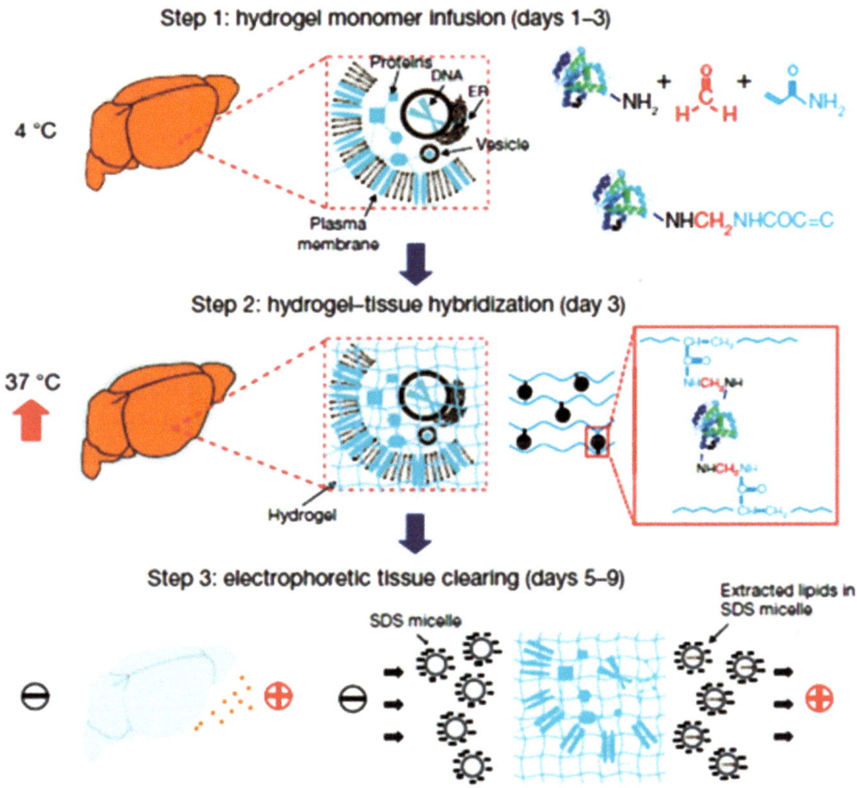

FIGURE 3.2 A schematic of the CLARITY process on a molecular level.

Source: Adapted from web ([http://wiki.claritytechniques.org).

CLARITY has shown to be successful in comparison with previous clearing method like Scale where the protein loss during tissue clearing is 41% but CLARITY causes only 8% protein loss from tissues during clearing.[13] However, the original CLARITY technique has several technical limitations: time-consuming, expensive, and rely on electrophores technique to remove lipids to make the tissue transparent, which impede its application in wider fields. Therefore, several modified tissue clearing methods based on CLARITY have emerged.[16] An another caveat is the application of lipophilic fluorescent dyes, for example, carbocyanine dyes like DiD and DiI, which are commonly used in histology. These dyes bind to the lipids, and therefore are unintentionally washed out during in the clearing process. Alternative dyes that are also aldehyde fixable have been identified.[20]

3.4.2 CUBIC

In 2014, Susaki et al. developed CUBIC—clear, unobstructed brain imaging cocktail—method to clear mouse brains.[21] But its advancement in the upcoming years made it scalable from a primate brain to the sub-cellular structures. CUBIC clearing method enables rapid, high-performance, whole-brain imaging with light sheet fluorescence microscopy (LSFM). The basic mechanism involves the removal of lipids using detergents, high concentrations of urea, and a final immersion in a refractive-index-specified matching homogenization solution in the range 1.38–1.48.[14] The main advantages of CUBIC are its simplicity, that is, it does not require any special equipment. It contains only nontoxic water-soluble chemicals, so it is easy to handle, good clearing performance, lasts approximately 2 weeks, and the preservation of fluorescent proteins.[21] The applications of CUBICs include mapping progenitors during heart morphogenesis,[22] studying interneurons during brain development,[23] and visualizing amyloid beta plaques in 18-month old mice brain.[24]

3.4.3 PACT/PARS

Passive CLARITY Technique (PACT) and PerfusionAssisted agent Release in Situ (PARS), are another two techniques to clear thick tissues developed by Yang et al. in 2014.[25] These two approaches were proposed to facilitate fast, whole-body, and whole-brain clearing using the circulatory system or the cerebrospinal fluid route to directly deliver clarifying agents. PACT is for passive tissue clearing and immunostaining of intact organs and PARS is a method for whole-body clearing and immune-labeling. Both these techniques are the improvements of the original CLARITY method that takes a long time for tissue clearing. To image PACT-cleared tissues, a solution termed refractive index matching solution (RIMS) is used, which is much more economical and customized recipe, with outcome similar to Focus Clear™.[8] These techniques can be used to clear mouse brain, kidney, heart, and lung, as well as human biopsy tissues of basal cell carcinoma.[13]

3.4.4 iDISCO

iDISCO is a modification of existing whole-mount immunolabeling (a combination of whole-mount immunolabeling using Alexa Fluor dyes and 3DISCO) method that results in transparency of various organs by a better removal of lipids.[26] Its advantages include: it enables deep antibody diffusion in to tissues, shown to be compatible with many antibodies.[17] The reagents for sample processing are common and inexpensive and do not require special equipments. Compared to other similar clearing techniques, iDISCO clearing takes only a day. This relatively fast turnover rate in iDISCO allows multiple rounds of sequential analysis in a reasonable time frame. These features together with light sheet microscopy makes it a reliable and consistent analytic platform to scale volume imaging of large biological samples.[27] One disadvantage of this method is it quenches the fluorescence signal.

3.4.5 ACT-PRESTO

Active Clarity Technique-Pressure Related Efficient and Stable Transfer of macromolecules into Organs (ACT-PRESTO) is a highly reproducible and rapid clearing technique.[13] This technique dramatically accelerates antibody penetration into large samples and clears the tissues or the whole body within 1 day. In addition, it preserves the tissue architecture and protein-based signals derived from endogenous fluorescent proteins. Moreover, it is compatible with conventional immunolabeling techniques. ACT-PRESTO allows high-content mapping and analysis of normal and pathological features in intact organs and bodies.[8]

3.5 LIGHT SHEET FLUORESCENCE MICROSCOPY

Selective plane illumination microscopy (SPIM) or light sheet flurescence microscopy comes under the family of planar illumination techniques, which revolutionized optical imaging of biological specimens. LSM uses a focused light sheet to illuminate the specimen from the side. The emitted fluorescent light is collected by the objective and projected to fast capturing image sensors like electron multiplication CCD (EMCCD) camera systems. It is an imaging method that is capable of imaging an optically cleared entire brain with sufficient resolution to capture single cells and projections and thus enables to discriminate between cells with different functions. In the past decade, LSFM has evolved as the imaging method of choice in biomedical research as it overcomes the issues of poor spatial resolution and limited penetration depth combined with the retained integrity of the specimen.

In LSFM, the sample is illuminated by a thin sheet of light perpendicular to the direction of objective. Moreover, it is not focused to a point, but shaped into a flat, extended "light sheet" that coincides with the focal plane of the detection objective. Thus, only one plane of the specimen is illuminated. SPIM illuminates only the region surrounding the focal plane of the detection objective and hence this method is also called as selective plane illumination microscopy or SPIM.[28] So, to image the entire tissue volume, the specimen needs to move along the direction of the optical axis. This way of laser configuration significantly reduces out-of-focus fluorescence and enables imaging deep inside intact specimens and avoids any loss of information. In LSFM, for 3D imaging, either the specimen needs to be moved or the light sheet and detection focal plane are moved across the sample volume. Therefore, LSFM images the entire plane at once from different angles, collecting a 3D image stack, which requires scanning in only one dimension.[29] Another major advantage of light sheet microscopy technique is that hundreds of 3D image stacks can be acquired with minimal photo bleaching and phototoxicity, this allows long-term imaging over many days or even weeks.

3.6 TISSUE CLEARING AND LIGHT SHEET IMAGING IN NEURODEGENERATION

The central nervous system communicates through dense synaptic, electric, and modulatory connections and develops through concurrent growth and interlinking of neurons, processes, glia, and blood vessels. In the brain, each constituent cell works independently, creating an exceedingly complex dynamic system. Understanding the brain function is challenging; these factors drive the development of techniques capable of imaging function of living biological systems like anatomy of various organs, neural signaling, and developmental processes at microscopic scales. The earlier imaging approaches were invaluable in the advancement of understanding the nervous system but their limitations put greater impact on the design of imaging experiments.

Recent innovations in tissue clearing, combined with advances in light sheet microscopy, resolved the limitations of the existing conventional techniques and effectively addressed previous limitations and made it possible to investigate exceptionally large volumes at micrometer resolution. Neurodegenerative diseases affect various regions of the brain because of the improper folding and aggregation of disease-specific proteins in various regions of the brain.

Combining tissue clearing and LSFM imaging can delineate the underlying pathology of the protein aggregation in the case of various neurodegenerative diseases. The initial optical clearing studies were focused on the analysis of anatomy and connectivity within the brain using mouse models. Currently, light sheet microscopes' live-imaging approach have been used to image the dynamic processes over a wide range of spatial and temporal scales in the whole-embryo imaging of *Caenorhabditis elegans* and *Drosophila melanogaster* nervous system development with sub-cellular resolution.[19] In addition, some groups developed strategies for structural whole brain imaging, with specific focus on rapid imaging of fixed and cleared brain tissues.[19]

In addition to this, LSFM provided comprehensive reconstruction of cell morphologies in intact, centimeter-sized brains with cellular resolution. A study using light sheet microscopy recorded and reported genetically encoded calcium indicator GCaMP5G, from the entire volume of brain of the larval zebrafish in vivo at 0.8 Hz. It could capture more than 80% of the neurons at single-cell resolution, demonstrating how LSFM could be used to reveal functionally defined circuits across the brain.

These experimental demonstrations illustrate the capabilities of light sheet microscopy in neuroscience application. The rapid imaging ability and low photo damage of light sheet microscopy is useful not only for noninvasive observation of dynamic processes in living specimens but also for whole brain imaging even for fairly large, mammalian brains.[30]

A study by Susaki et al. showed that the CUBIC treatment suited well with fluorescent proteins, and demonstrated multicolor imaging with various kinds of fluorescent proteins. They found that subcellular structures, including axons and dendritic spines, are well-preserved in the CUBIC-cleared tissues.[31] In addition, in another study, they were able to image fixed brains with single-cell resolution and perform reconstructions of dendritic trees and spines in populations of CA1 neurons in isolated hippocampus. In addition, Tomer et al. used CLARITY combined with light sheet microscopy and developed a rapid imaging strategy for entire mouse brain based on optical tiling of the fairly large tissue volume for high-resolution imaging at the whole-brain level.[19] Optical imaging methods using small vertebrate and invertebrate brains demonstrated that they can capture anatomy, neural activity, and information about neuron locations and morphology. So, this can be a very useful technique for imaging brain slices and areas of larger vertebrate brain and also to study processes like neurodegeneration where individual neuron are affected.

3.7 APPLICATIONS OF LIGHT SHEET MICROSCOPY

LSFM is a non-destructive, versatile imaging tool, which operates as a microscope and microtome, and has been present in cell biology laboratories for quite some time. It uses a plane of light to optically section and view tissues mainly as custom-made systems, with imaging applications ranging from single cells (in the micrometer scale) to small organisms (in the millimeter scale). The strengths of the LSFM over the other fluorescence microscopy techniques are higher speed, lower photobleaching and phototoxicity, and high tissue penetration even in scattering tissues.[32] Currently, few fields like material macrography, life sciences, photomicrography, and developmental biology make use of the wide range of imaging applications of light sheet fluorescence microscopy.

In developmental biology, this technology enables prolonged observations of embryonal development in small organisms mostly in embryos and larvae. In life sciences, it is used in applications such as live imaging of fast dynamic processes in spheroid cultures and in single cells. In brain morphological and activity studies, it has been used to image at different levels from nervous systems of small organisms to single neurons.[33] When advanced camera technology is combined with the LSFM method, the complete development of the nervous system can be viewed with excellent temporal resolution. Current experimental setups make extensive use of light sheet techniques combined with micro devices (bioMEMS-bio-microelectro mechanical systems) for the 3D characterization of active cells, to investigate mechanisms underlying cellular responses to drugs.[32]

3.8 CONCLUSION

Neurodegenerative disorders also called as conformational disorders with limited light penetration and/or insufficient resolution limits the usual conventional imaging of brain samples due to the thick plaque formation or protein aggregation in various regions of the brain. The advanced clearing techniques like CLARITY combined with light sheet florescence microscopy can delineate the underlying pathology using autopsy samples. In addition, in NDDs, the histopathological analysis is limited mostly to postmortem samples and brain biopsies; this again delays the diagnosis of individual patient.

Understanding the cellular and molecular mechanism and getting 3D pathological feature behind the progressive neuronal death may possibly able to target a biological entity that can control the disease progression. Conformational alteration and aggregation of protein is considered the foremost causative link in neurodegenerative disorders. Therefore, the mechanism of aggregation demands an in-depth exploration in the near future. Neurodegenerative disorders are progressive and it affects various regions of the brain; light sheet microscopy can easily image over large areas. In addition, LSFM can acquire image with multiple view obtained from different angles and, therefore, can reconstruct the related lesions such as the aggregations of protein plaques in samples. This will shed light to their 3D structure that appeared either dense and focal or diffused. Volume imaging coupled with automated detection can successfully find the disease hallmark of this debilitating disorder. Future researches should focus on the interaction of multiple factors, like dysregulation of protein aggregate formation and clearance of neurofibrillary degeneration, genetic predisposition, stressors, inflammatory mechanisms, vascular insufficiency, and other neurochemical anomalies. In neurodegenerative diseases, this system-level approach may be able to answer how the state of one brain region affects the dynamics of other regions and how do sensory systems represent the world at a larger network level.

KEYWORDS

- **neurodegenerative diseases**
- **CLARITY techniques**
- **tissue clearing**
- **microscopy**
- **3D imaging**

BIBLIOGRAPHY

1. Kovacs, G. G. Concepts and Classification of Neurodegenerative Diseases. *Handb. Clin. Neurol.* **2017,** *145*, 301–307.
2. Dugger, B. N.; Dickson, D. W. Pathology of Neurodegenerative Diseases. *Cold Spring Harb. Perspect. Biol.* **2017,** *9*, 7.
3. Hussain, R. et al. Neurodegenerative Diseases: Regenerative Mechanisms and Novel Therapeutic Approaches. *Brain Sci.* **2018,** *8*, 9.
4. Shamsi, T. N. et al., A Review on Protein Misfolding, Aggregation and Strategies to Prevent Related Ailments. *Int. J. Biol. Macromol.* **2017,** *105*, 993–1000.
5. Ganguly, G. et al. Proteinopathy, Oxidative Stress and Mitochondrial Dysfunction: Cross Talk in Alzheimer's Disease and Parkinson's Disease. *Drug Des. Devel. Ther.* **2017,** *11*, 797–810.
6. Lee, S. J. et al. Protein Aggregate Spreading in Neurodegenerative Diseases: Problems and Perspectives. *Neurosci. Res.* **2011,** *70*, 339–348.
7. Ando, K. et al. Inside Alzheimer Brain with CLARITY: Senile Plaques, Neurofibrillary Tangles and Axons in 3-D. *Acta Neuropathol.* **2014,** *128*, 457–459.
8. Du, H. et al. Advances in CLARITY-Based Tissue Clearing and Imaging. *Exp. Ther. Med.* **2018,** *16*, 1567–1576.
9. Beach, T. G. A Review of Biomarkers for Neurodegenerative Disease: Will They Swing Us Across the Valley? *Neurol. Ther.* **2017,** *6*, 5–13.
10. Saeed, U. et al. Imaging Biomarkers in Parkinson's Disease and Parkinsonian Syndromes: Current and Emerging Concepts. *Transl. Neurodegener.* **2017,** *6*, 8.
11. Jensen, K. H. R.; Berg, R. W. Advances and Perspectives in Tissue Clearing Using CLARITY. *J. Chem. Neuroanat.* **2017,** *86*, 19–34.
12. Mano, T. et al. Whole-Brain Analysis of Cells and Circuits by Tissue Clearing and Light-Sheet Microscopy. *J. Neurosci.* **2018,** *38*, 9330–9337.
13. Azaripour, A. et al. A Survey of Clearing Techniques for 3D Imaging of Tissues with Special Reference to Connective Tissue. *Prog. Histochem. Cytochem.* **2016,** *51*, 9–23.
14. Richardson, D. S.; Lichtman, J. W. Clarifying Tissue Clearing. *Cell* **2015,** *162*, 246–257.
15. Chung, K. et al. Structural and Molecular Interrogation of Intact Biological Systems. *Nature* **2013,** *497*, 332–337.

16. Jensen, K. H. R.; Berg, R. W. Advances and Perspectives in Tissue Clearing Using CLARITY. *J. Chem. Neuroanat.* **2017,** *86*, 19–34.
17. Ariel, P. A Beginner's Guide to Tissue Clearing. *Int. J. Biochem. Cell Biol.* **2017,** *84*, 35–39.
18. Chung, K.; Deisseroth, K. CLARITY for Mapping the Nervous System. *Nat. Methods.* **2013,** *10*, 508–513.
19. Tomer, R. et al. Advanced CLARITY for Rapid and High-Resolution Imaging of Intact Tissues. *Nat. Proto.* **2014,** *9*, 1682–1697.
20. Jensen, K. H. R.; Berg, R. W. CLARITY-Compatible Lipophilic Dyes for Electrode Marking and Neuronal Tracing. *Sci. Rep.* **2016,** *6*, 32674.
21. Susaki, E. A. et al. Whole-brain Imaging with Single-Cell Resolution Using Chemical Cocktails and Computational Analysis. *Cell* **2014,** *157*, 726–739.
22. Chabab, S. et al., Uncovering the Number and Clonal Dynamics of Mesp1 Progenitors during Heart Morphogenesis. *Cell Rep.* **2016,** *14*, 1–10.
23. Calvigioni, D. et al. Functional Differentiation of Cholecystokinin-Containing Interneurons Destined for the Cerebral Cortex. *Cereb. Cortex.* **2017,** *27*, 2453–2468.
24. Hama, H. et al. ScaleS: An Optical Clearing Palette for Biological Imaging. *Nat. Neurosci.* **2015,** *18*, 1518–1529.
25. Yang, B. et al. Single-cell Phenotyping within Transparent Intact Tissue through Whole-Body Clearing. *Cell* **2014,** *158*, 945–958.
26. Ariel, P. A Beginner's Guide to Tissue Clearing. *Int. J. Biochem. Cell Biol.* **2017,** *84*, 35–39.
27. Bode, J.; Krüwel, T.; Tews, B. Light Sheet Fluorescence Microscopy Combined with Optical Clearing Methods as a Novel Imaging Tool in Biomedical Research. *Eur. Med. J.* **2017,** *1*, 67–74.
28. Funane, T. et al. Selective Plane Illumination Microscopy (SPIM) with Time-Domain Fluorescence Lifetime Imaging Microscopy (FLIM) for Volumetric Measurement of Cleared Mouse Brain Samples. *Rev. Sci. Instrum.* **2018,** *89*, 053705–053705.
29. Gómez-Gaviro, M. V. et al. Biomedical Applications of Tissue Clearing and Three-Dimensional Imaging in Health and Disease. *iScience* **2020,** *23*, 101432.
30. Ueda, H. R. et al. Tissue Clearing and Its Applications in Neuroscience. *Nat. Rev. Neurosci.* **2020,** *21*, 61–79.
31. Xu, Y. et al. Imaging the Brain in 3D Using a Combination of CUBIC and Immunofluorescence Staining. *Biomed. Opt. Express* **2019,** *10*, 2141–2149.
32. Albert-Smet, I. et al. Applications of Light-Sheet Microscopy in Microdevices. *Front. Neuroanat.* **2019,** *13*, 1.
33. Corsetti, S.; Gunn-Moore, F.; Dholakia, K. Light Sheet Fluorescence Microscopy for Neuroscience. *J. Neurosci. Methods.* **2019,** *319*, 16–27.

CHAPTER 4

Characterization of 2D-Poly(Glycerol Sebacate) Bioelastomer Films through Spectroscopic Analysis

RAJU B. MALIGER*, PETER J. HALLEY, and JUSTIN J. COOPER-WHITE

Australian Institute of Bioengineering & Nanotechnology, School of Chemical Engineering, University of Queensland, St. Lucia, Brisbane 4072, Queensland, Australia

Corresponding author. E-mail: r.maliger@gmail.com

ABSTRACT

Poly(glycerol-sebacate) (PGS) is a bioelastomer, which finds potential applications in tissue engineering. It is known that the average functionality of a polymer system has a profound effect on crosslinking density, mechanical properties, and the reaction kinetics. PGS films synthesized from systems with varying average functionalities should have differences in their percentage crystallinity, crosslinking density, stiffness, thermal transitions, and contact angle measurements. The 2D-PGS films, prepared from non-stoichiometric monomer mixtures (glycerol: sebacic acid), were characterized using Fourier transform-infrared spectroscopy (FTIR), differential scanning calorimetry (DSC), wide-angle X-ray scattering (WAXS), and X-ray photoelectron spectroscopy (XPS). FTIR spectra of various PGS films were used for calculating the percentage of esterification and degree of polymerization. DSC thermograms of the PGS films were used to identify crystallization exotherms and melting endotherms, and to determine the effect of average functionality of the system on the

Advanced Microscopy: A Strong Analytical Tool in Materials Science. Merin Sara Thomas, Jozef T. Haponiuk, Sabu Thomas, & Anne George (Eds.)

density of crosslinking, degree of crystallinity, and the network structure of bioelastomers. The WAXS patterns of the PGS oligomers were key to understanding the relationship between the average functionality and the degree of crystallinity. XPS was instrumental in proposing the composition and the structure of PGS bioelastomers. Spectroscopic study, along with kinetics parameters, is vital in designing 2D-films and 3D-scaffolds for targeted tissue engineering applications. The study also revealed that average functionality, degree of crystallinity, chemical constitution of the surface, and post-polymerization techniques would help control the network structure of such bioelastomers.

4.1 INTRODUCTION

Network structure of Poly(glycerol-sebacate) (PGS) bioelastomers can be controlled by choosing appropriate average functionality, post-polymerization techniques, chemical constitution of the surface, and the degree of crystallinity. Novelty of this work lies in characterizing 2D PGS films of varying stiffness (defined by average functionality) through various spectroscopic techniques. By tailoring the properties of such PGS films, the desired cell phenotypes, their attachment, and differentiation can be studied.

4.1.1 KINETICS

Bioelastomers have rubber-like elasticity and good mechanical properties, which make them ideal candidates for potential applications in soft tissue engineering. For example, in PGS, glycerol is the basic building block for lipids, whereas sebacic acid is the natural metabolic intermediate in β-oxidation of medium-to-long-chain fatty acids. The molecular weight distribution of such polyesters synthesized through step-growth polymerization is completely defined by the molar ratio of the reactants and the extent of the reaction. When reactants are present in non-stoichiometric amounts, such as in case of glycerol and sebacic acid, the concept of non-equal reactivity of functional groups,[1] Pinner's equation,[2] kinetics, and gelation of the system need to be considered. Many reactive polymer systems employ an overall formulation such that the average functionality is 2 or slightly greater than 2. This means that the density of crosslinking can be controlled to obtain flexible-to-tough bioelastomers, which can then be used for numerous tissue engineering applications.[3,4] In fact, for a given concentration and temperature, a unique value of the reaction rate may be defined. Therefore, PGS films synthesized from systems with varying average functionalities should have differences in their percentage crystallinity, crosslinking density, stiffness, thermal transitions, and contact angle measurements. Tailoring the properties of 2D films and 3D scaffolds is essential for every tissue engineering application. The following section deals with characterization of 2D PGS films prepared from non-stoichiometric monomer mixtures (glycerol: sebacic acid).

4.1.2 SYNTHESIS OF POLY(GLYCEROL-SEBACATE) PREPOLYMERS AND FILMS

PGS prepolymer is synthesized following a step-growth reaction scheme. In the presence of a solvent-like toluene, this is generally a very slow reaction requiring 2–3 days to reach gelation. The process is detailed in our earlier publication.[5] Glycerol (G) and sebacic acid (SA) were chosen in three non-stoichiometric quantities (G:SA = 0.6, 0.8, 1.0) and charged into three round-bottomed flasks. Glycerol was dried by bubbling nitrogen gas through it in the round-bottomed flask at 120°C before the start of the experiment. Sebacic acid was dried in a vacuum oven for 2 h at 50°C under a stream of nitrogen gas before transferring it into the round bottomed flask. Toluene was used as a solvent (40% of combined monomer weight) as it forms a heterogeneous azeotrope at 85°C with water formed during polyesterification. The round-bottomed flask was connected to a Dean-Stark trap, which in turn was connected to a reflux condenser. The volume of water collected in the Dean-Stark trap was used for calculating the extent of esterification (p). The temperature of the round-bottomed flask was maintained at 130°C using a silicone oil bath. Table 4.1 shows the parameters associated with the reaction. A diagram of the experimental set-up is shown in Figure 4.1.

A BUCHI R-210 Rotavapor, operated at 77 mbar and 40°C, was used for separating toluene from the reaction mixture. Remaining traces of toluene were removed by keeping the prepolymer in a Salvis-LAB vacuum oven (Extech Equipment Pty Ltd, Boronia, Australia) for 12 h at 10 mbar and 80°C. The prepolymer was allowed to cool and stored at room temperature in a desiccate environment until further use. The purified prepolymers were dissolved in 1,4-dioxane to make up 25% (w/v) solution. Requisite volume of the solution was transferred into a petrie-dish and subjected to post-polymerization in an oven for 30 h at 130°C and 600 mbar, which produced firm, highly crosslinked, PGS elastomers of approximately 1 mm. The films were shaken in 70% v/v ethanol solution for 5 min at 100 rpm to extract unreacted monomers or oligomers. Then, 70% v/v ethanol solution was replaced with solutions containing decreasing levels of ethanol with 10 min shaking periods after each addition. Then, the films were washed thoroughly several times with distilled water to remove any traces of ethanol and 1,4-dioxane and finally they were statically incubated in ddH_2O for 4 h. The films were then cut into 15 mm diameter shapes and dehydrated in an oven at 110°C for 12 h and stored in a desiccator at room temperature until further use. The bioelastomer oligomers and films thus synthesized from three different prepolymers will be designated as PGS 0.6, PGS 0.8, and PGS 1.0, respectively, throughout the remainder of the discussion.

TABLE 4.1 Parameters Associated with the Step-Growth Polymerization.

G:SA	m_{gl}	m_{SA}	n_{gl}	n_{SA}	eq_{gl}	eq_{SA}	f_{av}	DP	p	p_{gel}	m_{tol}
0.6	8.28	30.3	0.09	0.15	0.27	0.3	2.25	2	0.444	0.889	15.432
0.8	11.04	30.3	0.12	0.15	0.36	0.3	2.22	2	0.451	0.901	16.536
1.0	13.8	30.3	0.15	0.15	0.45	0.3	2.00	2	0.5	1.00	17.64

G:SA, molar ratio of glycerol: sebacic acid; f_{av}, average functionality of the system; n_{gl}, number of moles of glycerol; m_{gl}, mass of glycerol (gm); n_{SA}, number of moles of sebacic acid; m_{SA}, mass of sebacic acid (gm); eq_{gl}, equivalence of glycerol (= 3 n_{gl}); eq_{SA}-equivalence of sebacic acid (=2 n_{SA}); p_{gel}, gel point (%); *DP*, degree of polymerization; p, extent of reaction; m_{tol}, mass of toluene (g).

The degree of polymerization (DP) was calculated using Carothers' eq 4.1

$$DP = 2/(2-pf_{av}) \tag{4.1}$$

4.2 CHARACTERIZATION OF PGS OLIGOMERS AND FILMS

The PGS oligomers and films were thoroughly characterized using Fourier transform-infrared spectroscopy (FTIR), differential scanning calorimetry (DSC), wide angle X-ray scattering (WAXS), X-ray photoelectron spectroscopy (XPS), and contact angle measurements.

FIGURE 4.1 Experimental set-up for synthesizing PGS prepolymers. 1. Round bottom flask; 2. Silicone oil bath; 3. Dean-Stark trap; 4. Reflux condenser.

4.2.1 FOURIER TRANSFORM-INFRARED SPECTROSCOPY

The evidence of esterification in the prepolymers of PGS 0.6, PGS 0.8, and PGS 1.0 was confirmed by using FTIR spectroscopy. FTIR spectra were recorded on a Nicolet 5700 spectrometer from 4500 to 525 cm^{-1} with an average of 32 scans at 4 cm^{-1} resolution and 1 cm^{-1} sensitivity.

4.2.2 DIFFERENTIAL SCANNING CALORIMETRY

A TA instrument, modulated DSC (TA 2920), was used to study the thermal transitions of the PGS prepolymers. The samples were dried in a vacuum oven at 80°C and 200 mbar for 12 h. To eliminate thermal history, the sample was equilibrated at 25°C, heated to 200°C at 10°C/min, maintained under isothermal conditions for 5 min, cooled to −150°C at 10°C/min, maintained under isothermal conditions for 5 min, and heated to 200°C min^{-1}at 10°C min^{-1}.

4.2.3 *WIDE-ANGLE X-RAY SCATTERING*

Wide-angle X-ray patterns of various oligomers were taken at room temperature using a Rigaku Miniflex XRD unit, where the cobalt radiation is generated with a tube tower within a steel cabinet that houses the goniometer. The X-ray generator typically operates at set power of 30 kV/15 mA. The scanning regions of the diffraction angle 2θ were 1.5°–149° with a step size of 0.04° and a step time of 2.4 s.

4.2.4 *X-RAY PHOTOELECTRON SPECTROSCOPY*

Three types of PGS films were subjected to XPS using a PHI model 560 XPS/SAM/SIMS I multitechnique surface analysis system incorporating a model 25–270 AR electron energy analyzer. Mg K_ X-rays (1253.6 eV) generated using 400 W (15 kV, 27 mA) was used to produce photoelectrons from the sample surface. Survey (wide) scans, at analyzer pass energy of 100 eV, over 1000 eV were taken, followed by multiplex (narrow) scans of C 1s, O 1s, and Si 2p energy levels at high resolution using a pass energy of 25 eV. Curve fittings of C (1s) and O (1s) were carved out using Resident PHI V6.0 curve-fitting software to establish the relative percentage of different functional groups.

4.2.5 *CONTACT ANGLE MEASUREMENTS*

For contact angle experiments, requisite quantities of oligomers (PGS 0.6, PGS 0.8, and PGS 1.0) were cured on 12 mm diameter glass coverslips. Another set of cured films were subjected to aminolysis. For this, the films were immersed for 30 min in AEPDA (N-Aminoethyl-1,3-propanediamine) diluted at 0.05M with water and adjusted to a pH of 9.8 by adding hydrochloric acid. The films were washed with cold water and further soaked for 1 h. Finally, the films were dried in an oven (40°C) for a few hours.

Equilibrium static air/water contact angles were measured with a contact angle goniometer (OCA 20) and drop shape analysis software (First Ten Angstroms, Inc.) using the captured sessile drop method on aminolyzed and non-aminolyzed PGS surfaces at a temperature of 22.5°C. The size of the water-drop was maintained at 5 μL throughout the experiment. The drop was allowed to reach equilibrium for 15 min before measuring the contact angle. Three drops were imaged for each surface. The angle between the baseline of the drop and the tangent at the drop boundary was measured. Values reported are the average contact angles ± standard deviation.

4.3 RESULTS AND DISCUSSION

4.3.1 FOURIER TRANSFORM-INFRARED SPECTROSCOPY

FTIR spectra of oligomers of PGS 0.6, PGS 0.8, and PGS 1.0 are shown in Figure 4.2. The spectra are offset by an increment of 1.5 to allow for easier discrimination of peak positions.

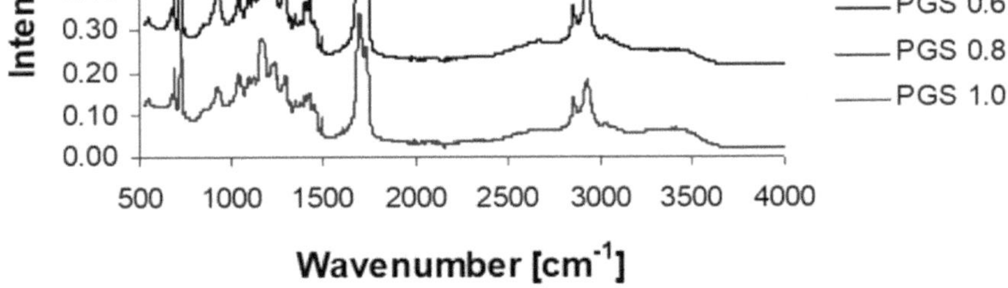

FIGURE 4.2 FTIR spectra of PGS oligomers (PGS 0.6, PGS 0.8, and PGS 1.0).

The band assignments were made according to standard spectroscopy literature.[6–8] The infrared absorption by the carboxylic acid groups of sebacic acid is shown by a peak at 1690 cm^{-1}, whereas the appearance of a sharp peak at 1740 cm^{1}, which is due to the stretching of the ester carbonyl group, confirms the formation of ester groups (see Figure 4.3). The peak at 1460 cm^{-1} is due to the deformation of methylene groups (CH_2) in the sebacic acid. The C–O–H bend in the glycerol is indicated by a band at 1415 cm^{-1}. A cluster of bands in the region 1150–1300 cm^{-1} confirms the stretching of the (O=)C–O group in the ester. The in-phase and the out-of-phase C–C–O stretchings are indicated by bands at 1051 and 1174 cm^{-1}, respectively.

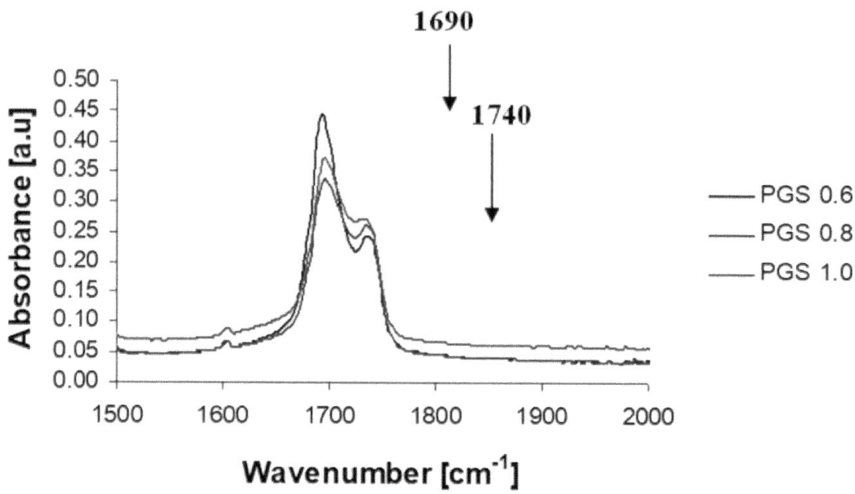

FIGURE 4.3 FTIR spectra of the region showing acid and ester peaks.

The intensity of peak at 1305 cm^{-1} was used for all spectra as an internal normalization standard. The normalized acid and ester peak areas and the calculated extent of esterification and degree polymerization (DP) are shown in Table 4.2.

TABLE 4.2 Normalized Acid and Ester Peak Areas of PGS Oligomers.

PGS system	Acid (1690 cm^{-1}) Normalized	Ester (1740 cm^{-1}) Normalized	% esterification	DP
PGS 0.6	5.048	0.957	15.936	1.218
PGS 0.8	4.514	1.454	24.363	1.371
PGS 1.0	4.589	1.172	20.344	1.255

DP, Degree of polymerization.

4.3.2 DIFFERENTIAL SCANNING CALORIMETRY

Thermal transitions such as melting and crystallization temperatures are quite important in polymer processing techniques. A thermogram showing the crystallization exotherms of the oligomers PGS 0.6, PGS 0.8, and PGS 1.0 is shown in Figure 4.4.

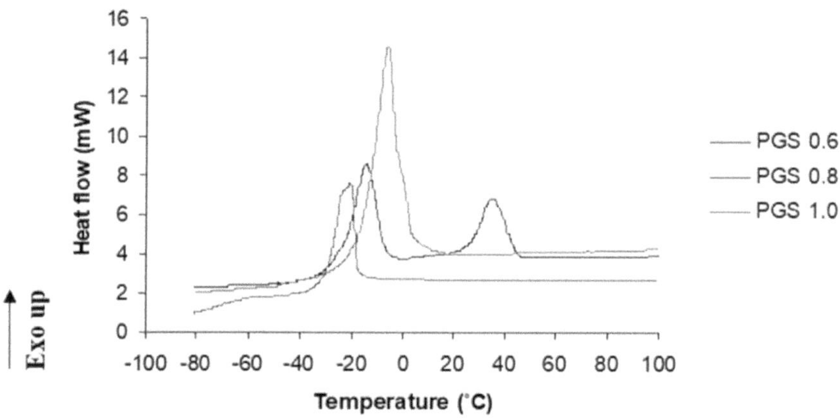

FIGURE 4.4 Crystallization exotherms of PGS oligomers.

The oligomer, PGS 0.6, displayed two crystallization exotherms at −12°C and 33°C, whereas PGS 0.8 and PGS 1.0 exhibited single crystallization temperature at −21°C and −6°C, respectively. In a polymer network, the amorphous regions impart elasticity and network-structure to the elastomer, whereas crystalline regions provide strength, stiffness, and thermoplasticity to the system.[9] The presence of extra crystalline domains in PGS 0.6 makes it tougher and stiffer than the other two oligomers. This is in agreement with the WAXS results discussed in Section 4.3.3. The melting endotherms are shown in Figure 4.5.

The thermogram showed four melting endotherms at 2.8°C, 25.3°C, 48°C, and 58°C for PGS 0.6. The other two oligomers, PGS 0.8 and PGS 1.0, displayed three melting temperatures, each at 2.5°C, 19.6°C, 39°C, and 15°C, 44C, 54°C, respectively. A positive shift in the melting endotherms is observed as the molar ratio (G:SA) is increased from 0.6 to 1. From crystallization exotherms and melting endotherms, it can be concluded that variation in the average functionality affects the density of crosslinking, degree of crystallinity, and the network structure of bioelastomers.

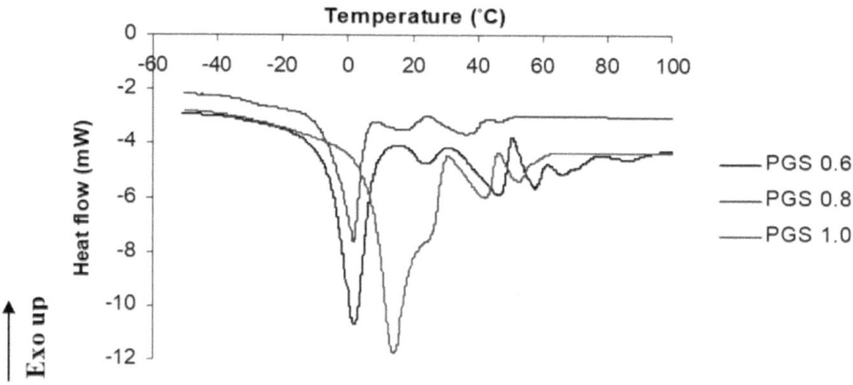

FIGURE 4.5 Melting endotherms of PGS oligomers.

4.3.3 *WIDE-ANGLE X-RAY SCATTERING*

The WAXS patterns of the PGS oligomers are shown in Figure 4.6.

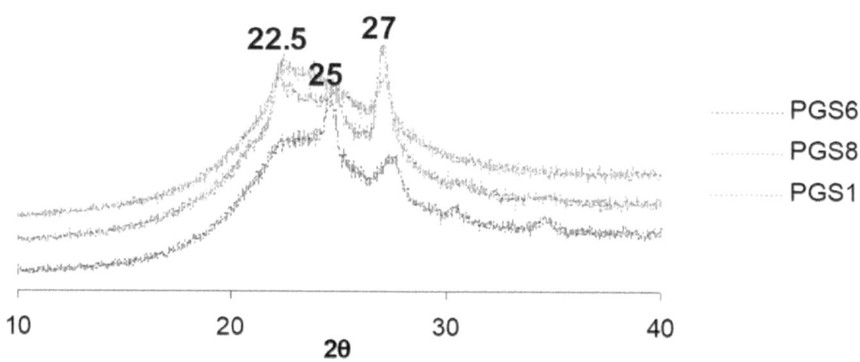

FIGURE 4.6 WAXS patterns of PGS 0.6, PGS 0.8, and PGS 1.0 oligomers.

A distinct single diffraction peak appears at 22.5° in all oligomers, which indicates the formation of some ordered structure owing to the regular network of symmetric structure of the multifunctional monomers.[10] A sharper diffraction peak is observed at 25° in PGS 0.6 and PGS 0.8 suggesting a higher degree of crystallization in these oligomers. Further, the presence of a sharp diffraction peak at 27° in all three oligomers indicates the presence of multiple crystalline domains. In addition to this, the presence of smaller peaks in PGS 0.6 in the region 2θ = 30°–49° confirms that it (PGS 0.6) has more crystalline domains than the other two oligomers. The percentage crystallinity of the oligomers was calculated using the software Traces, version 3.1. The results are in agreement with the WAXS pattern shown in Figure 4.6 (see also Table 4.3).

TABLE 4.3 % Crystallinity in PGS Oligoesters.

PGS	% Crystallinity
PGS 0.6	12.72
PGS 0.8	7.43
PGS 1.0	3.31

It can be observed from Table 4.3 that the percentage crystallinity decreased in the following order: PGS 0.6 > PGS 0.8 > PGS 1.0. The higher the average functionality of a system, the greater the degree of polymerization and the density crosslinking of the system. Therefore, it is not surprising to find PGS 0.6 with the highest degree of crystallinity.

4.3.4 X-RAY PHOTOELECTRON SPECTROSCOPY

Although XPS is a surface analysis technique, it still provides valuable information about the structure of crosslinked polymers.[11] PGS films were washed with hexane to remove surface impurities. In the survey scans of all graphs, 2p peaks were observed in the range 100–102 eV, with the atomic concentration varying between 0.43% and 0.62%. Distinct peaks corresponding to C (1s) and O (1s) were observed in the range of 280–290 eV and 530–540 eV, respectively.

Thereafter, high-resolution multiplex scans of C (1s) and O (1s) energy levels of the films were carried out at a pass energy of 25 eV. All the data was charge-corrected at 282.24 eV. The peaks corresponding to COO, C–O, C–COO, C–C, O–C, and O=C of PGS 0.6, PGS 0.8, and PGS 1.0 were estimated. The proposed structure of PGS is shown in Figure 4.7.

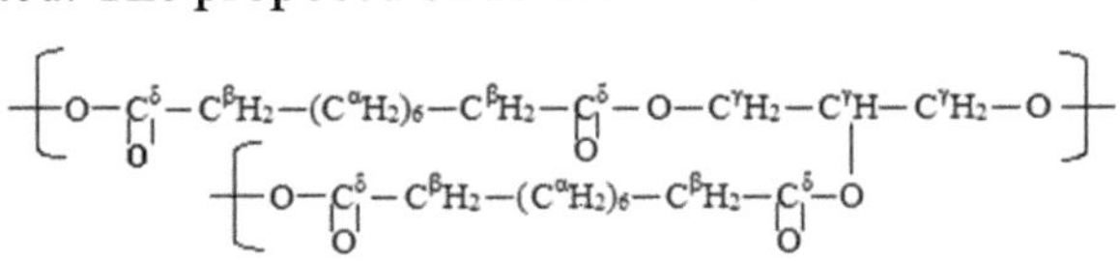

FIGURE 4.7 Proposed structure of PGS.

The proposed structure is compared with the C (1s) energy levels of PGS films and the results are summarized in Table 4.4a–c.

TABLE 4.4(a–c) Proposed Composition and Structure of PGS Using XPS.

PGS 0.6	No. of groups	Ratio (theoretical)	XPS (experimental)
(a)			
C^{α}–C^{α}	12	6.67	4.49
C^{β}–COO	4	2.22	1.22
C^{γ}–O	3	1.67	1.18
C^{δ}OO	4	2.22	1
PGS 0.8			
(b)			
C^{α}-C^{α}	12	5	4.29
C^{β}-COO	4	1.67	1.12
C^{γ}-O	3	1.25	1.11
C^{δ}OO	4	1.67	1
PGS 1.0			
(c)			
C^{α}-C^{α}	12	4	4.29
C^{β}-COO	4	1.33	1.04
C^{γ}-O	3	1	1.42
C^{δ}OO	4	1.33	1

The theoretical ratios of groups change with the feed composition (molar ratio). From Table 4.4, it can be observed that the ratio of various C groups according to XPS results vary with the theoretical results. In cases of PGS 0.6 and PGS 0.8, the theoretical values of all moieties are found to decrease when compared with the respective values in the experimental results. In case of PGS 1.0, the experimental value of C–O group is higher than the theoretical value. The shuffle in the values of C–COO, O–C=O, C–C, and C–O ratios in all PGS films indicates the surface reorientation of the moieties to minimize the surface energies. As a result, the all-PGS films are hydrophobic, which is confirmed by contact angle measurements in Section 4.3.5.

4.3.5 CONTACT ANGLE MEASUREMENTS

Figure 4.8 shows the pictures taken on the PGS 0.6 film at the beginning of the test and after 5 min using a contact angle goniometer (OCA 20).

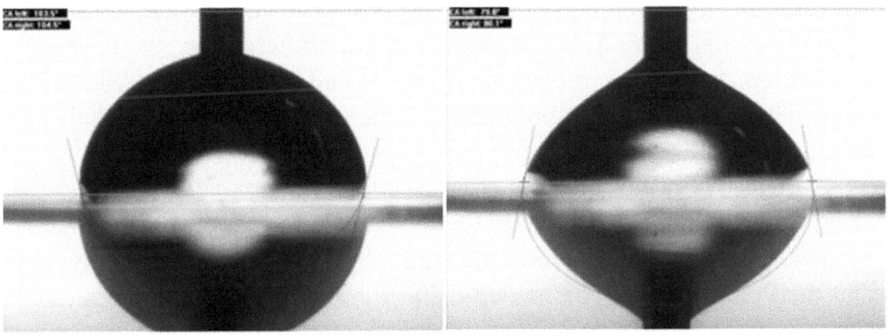

FIGURE 4.8 Contact angle measurements on PGS 0.6 at different times.

It can be noticed that the contact angle decreased with time due to the spreading of water drop on the film. Contact angles were measured after 15 min so as to allow the drop to reach equilibrium. The contact angle results of PGS 0.6, PGS 0.8, and PGS 1.0 for both non-aminolyzed and aminolyzed films are shown in Table 4.5.

TABLE 4.5 Contact Angles Before and After Aminolysis.

	Average contact angle	Standard deviation
Before aminolysis		
PGS 0.6	72.8	1.53
PGS 0.8	76.2	1.57
PGS 1.0	81.6	1.30
After aminolysis		
PGS 0.6	60.8	3.19
PGS 0.8	64.8	0.43
PGS 1.0	67.7	1.54

Generally, a contact angle is indicative of the hydrophobicity of a given solid surface and is seen as physical manifestation of the fundamental concepts of surface energy and surface tension. High-contact angle suggests the surface has relatively low surface energy.[12] It can be observed from Table 4.5 that for non-aminolyzed PGS films, the contact angle increased with an increase in the molar ratio (G:SA), and at all ratios, the surfaces are hydrophobic (contact angle > 65°). Therefore, the surface energy of the films and hydrophilicity should be in the following order: PGS 1.0 < PGS 0.8 < PGS 0.6. The decrease in the surface energy of the films is essentially due to the reduction in the polar components, wherein the surface hydroxy groups are replaced by the non-polar aliphatic chains as a result of esterification.[13]

It can also be noticed that the average contact angle of the films decreased after aminolysis by approximately 12°. The introduction of active amino groups on the surface rendered the films hydrophilic. These results show that the PGS films may be treated using surface treatment techniques such as aminolysis (and hydrolysis) to increase hydrophilicity and incorporate active amino groups, and therefore, affect better cell attachment and growth.

4.4 CONCLUSIONS

PGS films prepared from non-stoichiometric amounts of monomers exhibit different physical and mechanical characteristics. Therefore, consideration of kinetics parameters is of vital importance for designing 2D-films and 3D-scaffolds for specific tissue engineering applications. Not only can the network structure of such bioelastomers be controlled by choosing the appropriate average functionality and post-polymerization techniques, but also can the chemical constitution of the surface and the degree of crystallinity (and number of crystalline forms) be possible in the bulk, which will effect degradation timeframes. Since cells, such as mesenchymal stem cells, have been shown to be sensitive to substrate stiffness.[14, 15] Tailoring the properties of bioelastomers is a key feature for obtaining desired cell phenotypes and encouraging attachment and differentiation. A number of recent studies on bioelastomer synthesis and characterization have yielded promising results. Dopant-free conductive polyurethane elastomer (DCPU) is a potential candidate for soft tissue repair and soft electronics.[16] Poly(itaconate-co-citrate-co-octanediol) is reported to exhibit tunable elastomeric networks, which provides controllable, rapid gelation required in a range of healthcare applications.[17] Synthesis of sustainable bioelastomer from soybean oil-based monomers *via* ring-opening polymerization has been reported.[18] Photocrosslinkable bioelastomer prepolymers synthesized from dimethyl itaconate, 1,8-octanediol, and triethyl citrate are reported to be good materials for 3D printing of vascular tubes.[19] PGS and other new-age bioelastomers can be functionalized to develop tailored 3D scaffolds with unique mechanical properties, which influences cell proliferation and lineage.

KEYWORDS

- **biomaterials**
- **PGS bioelastomers**
- **spectroscopy**
- **Fourier transforms infrared spectroscopy (FTIR)**
- **wide angle x-ray scattering (WAXS)**
- **x-ray photoelectron spectroscopy (XPS)**
- **differential scanning calorimetry (DSC)**
- **thermogram**

REFERENCES

1. Solomon, D. H. Ed. *Step-Growth Polymerizations*; Marcel Dekker, Inc.: New York, 1972; pp 345–355.
2. Pinner, S. H. *J. Polym. Sci.* **1956,** *21*, 153–157.
3. Mark, J. E.; Erman, B.; Eirich, F. R. Ed. *Science and Technology of Rubber*; Academic Press: San Diego, 1994; pp 189–210
4. Odian, G. *Principles of Polymerization*; McGraw-Hill, Inc.: New York, 1970.
5. Maliger, R. B.; Halley, P. J.; Cooper-white, J. J. *J. Appl. Polym. Sci.* **2013,** *127* (5), 3980–3986.
6. Daimay, L. V. *The Handbook of Infrared and Raman Characteristic Frequencies of Organic Molecules*; Academic Press: Boston, 1991.
7. Lambert, J. B. *Organic Structural Analysis*; Macmillan: New York, 1976.
8. Socrates, G. *Infrared and Raman Characteristic Group Frequencies: Tables and Charts*; Wiley: New York, 2001.
9. Liu, Q. Y.; Tian, M.; Ding, T.; Shi, R.; Feng, Y. X.; Zhang, L. Q.; Chen, D. F.; Tian, W. *J. Appl. Polym. Sci.* **2007,** *103*, 1412–1419.
10. Nagata, M.; Ibuki, H.; Sakai, W.; Tsutsumi, N. *Macromolecules* **1997,** *30*, 6525–6530.

11. Maliger, R. B.; McGlashan, S. A.; Halley, P. J.; Matthew, L. G. *Polym. Eng. Sci.* **2006,** *46*, 248–263.
12. Kim, Y.; Yang, J. C.; Kim, D. E. *Int. J. Precis. Eng. Manuf.* **2004,** *5*, 13–18.
13. Freire, C. S. R.; Silvestre, A. J. D.; Neto, C. P.; Gandini, A.; Fardim, P.; Holmbom, B. *J. Colloid Interface Sci.* **2006,** *301*, 205–209.
14. Discher, D. E.; Janmey, P.; Wang, Y. L. *Science* **2005,** *310*, 1139–1143.
15. Engler, A. J.; Sen, S.; Sweeney, H. L.; Discher, D. E. *Cell* **2006,** *126*, 677–689.
16. Xu, C. et al. *Sci. Rep.* **2016,** *6*, 1–13
17. Huyer, L. D et al. *Adv. Healthcare Mater.* **2019,** *8*, 1900245–1900255.
18. Wang, Z. et al. *Chem. Sci. Biomol. Eng.* **2020,** *1*, 103–110.
19. Savoji, H. et al. *ACS Biomater. Sci. Eng.* **2020,** *6*, 1333–1343.

CHAPTER 5

Nanoparticles in Drinking Water Treatment: An Overview of the Significance of Analytical Instrumentation

T. C. PRATHNA

Department of Irrigation and Flood Control, Government of NCT of Delhi, New Delhi 110002, India (formerly with Department of Environmental Engineering and Water Technology, UNESCO-IHE, The Netherlands)

Corresponding author. E-mail: prathna.tc@gmail.com

ABSTRACT

Access to clean drinking water is a right to be exercised by citizens and yet many people in particularly developing countries lack access to safe water. Technologies such as physical-based methods of filtration, chemical-based methods, such as coagulation and flocculation, and biological-based methods have been used traditionally since ancient times. However, the increase in population and urbanization of towns have increased water demand and the need for more advanced technological interventions. Advances in applications of nanotechnology over the last two decades show great promise in the field of water treatment. Knowledge of analytical instrumentation techniques for nanoparticle characterization is critical in ensuring the precise use of nanoparticles for a specific application. The chapter provides an overview of the recent applications of nanoparticles in

Advanced Microscopy: A Strong Analytical Tool in Materials Science. Merin Sara Thomas, Jozef T. Haponiuk, Sabu Thomas, & Anne George (Eds.)

the field of drinking water treatment together with advances in analytical characterization techniques used.

5.1 INTRODUCTION

One of the 17 sustainable development goals (SDG) calls for ensuring safe drinking water to all (SDG 6) by the year 2030[1] and reports suggest that around 780 million people around the world lack access to safe drinking water[2] and millions of people die every year due to diseases spread by unsafe drinking water. Most of the major cities in China, for example, suffer from water shortages as the major rivers supplying water are highly polluted.[3] Safety of drinking water sources is often classified as either safe or unsafe by national standards of the country or by international guidelines. Guidelines put forth by the World Health Organization (WHO) regarding drinking water quality are the most commonly used guidelines in developing countries[4], and national standards in many countries are more often framed taking into consideration WHO guidelines. These reports point to an urgent need to adopt techniques to ensure safe drinking water in addition to taking measures aimed toward sustainable use of available water resources. Adoption of advanced technologies can play a critical role in achieving SDG 6.[1] Many emergent and persistent contaminants enter the municipal water supply of both developing and developed countries mainly due to anthropogenic activities leading to public health concerns.[5] Pathogenic microorganisms causing water-borne diseases are a major cause of fatality in developing countries.[6] Advanced water treatment with efficient removal of contaminants cannot only drastically reduce healthcare costs but also significantly help in economic development.[7]

Water treatment has traditionally been carried out using techniques such as filtration, precipitation (coagulation and flocculation), adsorption, distillation, and oxidation to name a few.[8] Some of the oldest water treatment techniques have included filtration, that is, slow sand filtration to provide safe drinking water using the concept of size exclusion. However, slow sand filtration is ineffective when the influent water is rich in chemical contaminants. Coagulation and flocculation using chemicals such as alum are still used extensively for traditional water treatment.[9] Use of natural coagulants for the removal of suspended solids from water is being extensively studied in recent times and their potential for large-scale deployment for water treatment is being explored. However, natural coagulants are observed to generate more voluminous sludge after treatment due to mucilage, and sufficient technologies to handle the sludge post-treatment need to be considered. Recently, large-scale water treatment is being carried out using membranes, particularly in Middle Eastern and water-scarce nations. Approximately 75% of the world population is expected to face freshwater scarcity by 2050[10] and therefore need for robust, cost-effective, and environmentally safe technologies to treat water is very critical. Over the last decade, there has an increased interest in applying nanotechnology for environmental applications, more specifically in water treatment. Nanotechnology has shown promise to play a critical role in addressing problems related to drinking and wastewater treatment, desalination, and water reuse.[11]

5.2 NANOTECHNOLOGY IN CLEAN WATER APPLICATIONS

Nanomaterials (nanoparticles, nanofibers, nanotubes, etc.) are being widely tested as adsorbents to remove emerging and persistent contaminants in both water as well as wastewater treatment. Nanotechnology can help in reducing treatment costs since treatment systems can be designed to remove specific problematic contaminants and also avoid wastage of treatment systems for nonharmful constituents.[12] An extensive search on the search engine Google Scholar for application of nanoparticles in drinking water treatment reveals 87,700 entries till date with 12,100 entries since 2020 alone. The search terms in Google Scholar were "nanoparticles" and "drinking water treatment." This clearly indicates the increased interest in exploring the potential of nanoparticles for drinking water treatment.

Silver nanoparticles and iron-based magnetic nanoparticles are some of the major types of nanoparticles that dominate the publication space with regard to use in water treatment. Some of the advantages that nanoparticles offer in water treatment include the further development of compact water purification devices due to the smaller size and increased reactivity of nanoparticles. The nanomaterial used for the water treatment process can have different sizes and shapes depending on the process of synthesis. For example, nanoparticles, nanoclusters, nanodots, nanotubes, nanofibers, nanowires, and nanomembranes can be prepared by fine-tuning of the process parameters[13] describe the synthesis of silver nanoparticles with different shapes using various methods and the properties of the nanoparticles have been studied to greatly depend on their size and shape.

Nanomaterials are usually prepared by approaches such as laser ablation, chemical reduction in addition to biological-based methods using microorganisms, plants, etc., as reductants. In brief, the process is either a top-down or bottom-up approach for the synthesis. In addition to higher reactivity, nanomaterials also provide additional advantages such as being applied to remove multiple hazardous contaminants simultaneously.[14] This can be achieved by optimizing and functionalizing the surface of the nanomaterials with specific surface-active agents. Nanofiltration membranes on the other hand are usually produced by electrospinning process and effectively filter out contaminants in water and are extensively applied in water, wastewater treatment, and desalination.[15] Nanofibers, for example, are highly porous, small pore size with well-connected pores, and high permeability. Nanofibers are also cost-effective and promising filtration membranes in terms of low fouling[5,16] synthesized electrospun nanofibers with high flux and rejection rates as compared to traditionally used membranes. In addition, the synthesized nanofibers were also cost-effective and had lower energy consumption. Nanofiltration membranes have shown immense potential in the removal of common contaminants in drinking water such as total dissolved solids, pathogens, and minerals in lab-scale studies.[17]

Some of the common contaminants found in water are arsenic, fluoride, nitrates, lead, selenium, chlorides, iron, heavy metals, chromium, pathogens, pharmaceuticals, dyes, as well as other emerging contaminants. Nanoparticles while being used in drinking water treatment can either function as nanoadsorbents and aid in the removal of contaminants, nanosensors, and aid in the detection of micro- and nano-molar concentrations of contaminants, nanomembranes to filter out contaminants, and also help in increasing the life span of conventional membranes by preventing biofouling. Some of the major nanomaterials used in drinking water treatment are noble metal nanoparticles, metal oxide nanoparticles, magnetic nanoparticles, and polymeric nanoparticles.

5.2.1 ANALYTICAL INSTRUMENTATION

The significance of analytical instrumentation in these applications lies in the accurate characterization of (1) nanomaterials used for contaminant removal and purification as well as (2) trace detection of the treated water for the presence of persistent contaminants.

5.2.1.1 COMMONLY USED NANOMATERIAL DETECTION TECHNIQUES

The techniques can roughly be grouped into preliminary characterization techniques (used to confirm the identity of the nanomaterial), surface analytical techniques (used to determine the surface-active groups as well as identify the surface charge), stability tests (used to determine the heat stability of the materials), and microscopic techniques (used to determine the surface topography, particle size distribution of the material).[18]

5.2.1.1.1 Preliminary Characterization

Techniques such as UV–vis spectroscopy are used as a preliminary test to confirm the type of nanoparticle and are based on the Beer–Lambert law. Noble metal nanoparticles such as silver and gold have their characteristic peak due to surface plasmon resonance phenomena. A preliminary idea of the size and size distribution of the nanoparticle can be gauged from the position of the peak. For example, a broad peak can signify a polydisperse sample, whereas a red-shifted peak (shift toward higher wavelength) may relate to higher particle size.[19] Similarly shifting of a peak toward the blue end of the spectrum indicates a narrow and smaller size range while redshift of the peak on the variation of process parameters usually indicates aggregation of particles and a larger size range. Programs such as MiePlot can be utilized to estimate the size of the nanoparticle synthesized based on the UV-Vis peak.

Structural identification techniques such as X-ray diffraction (XRD) can also be used to confirm the structural identity and also determine the phase variation of the nanomaterial. Each material has its own signature diffraction peak which can be confirmed with the JCPDS datasheet. For example, zinc oxide nanoparticles with hexagonal structure show three peaks with 2θ values at 31.76°, 34.58°, and 36.67° corresponding to (100), (002), and (101) planes. Sharp and narrow intensity of the XRD peak can confirm the high-quality nature of the nanoparticle indicating fine grain size and good crystallinity.[20] The average size of the nanocrystals in the sample can be deduced using the Debye Scherrer formula from the XRD data.[21]

X-ray photoelectron spectroscopy (XPS) is another quantitative elemental detection technique that is widely used for nanoparticle characterization.

The technique operates under high vacuum and involves bombarding of the sample surface with X-rays. XPS spectra are constructed from the kinetic energy and the number of electrons that escape from the surface. Kinetic energy provides information on the binding energy of electrons that gives an idea on the oxidation state of the nanoparticle.[22,23]

5.2.1.1.2 Surface Analytical Techniques

Surface analytical techniques such as Fourier transform infrared spectroscopy can be used to identify the surface functional groups that are active on the surface on the material used. Any surface interactions with the material of interest can also be identified based on the shifts in the peak vibrations in the spectra.[24] Phases and phase transitions of nanomaterials upon adsorption of various contaminants together with the nature of the nanoparticle (amorphous or crystalline) can be identified using Raman spectroscopy. The sample preparation for Raman spectroscopy is simpler than the traditional IR spectroscopy.[25] Samples are usually coated onto a glass slide and sputter coated with gold.[26] Advanced forms of Raman spectroscopy include surface-enhanced Raman spectroscopy, stimulated Raman spectroscopy, and resonance Raman spectroscopy.

Techniques such as zeta potential analysis can be used to identify the potential overall charge of the system which can be used to predict the stability of the system. For example, if the zeta potential of the system ranges between +20 and −20 mV, it signifies that the system is highly unstable.[27] The nanoparticle is least unstable at the isoelectric point and precipitates out of solution. Surface charge characterization is a very critical parameter to measure for nanoparticles being used in water treatment.[28] The surface charge is the first parameter that determines the stability of the nanoparticle when it comes in contact with the water sample. Processes like functionalization of the nanoparticle surface with capping agents—both chemical and biological can provide stability to the system and prevent aggregation.

An important function required for nanomembranes is its permeability. Nanomembrane permeance can be determined using dead-end stirred cells[29], determined the membrane permeance of cationic cellulose nanofibrils by forcing deionized water through the nanopapers, and measuring the permeate volume to determine the active filtration area.

5.2.1.1.3 Thermal Stability Techniques

Heat stability techniques such as differential scanning calorimetry and thermogravimetric analysis (TGA) can be used to analyze the thermal stability of the material. Weighed quantities of samples are placed in either platinum or graphite crucibles and heated from room temperature to ranges around 700°C. Certain applications may require the nanomaterial to be stable over a very high temperature and these techniques can help in their identification. In these techniques, the nanomaterial is heated over a specific temperature range under controlled conditions and the change in weight due to loss of moisture along with measurements such as glass transition temperature is observed. For example, in point of use water filters constructed using fired clay and impregnated silver nanoparticles, the silver nanoparticles are required to be stable over the firing temperature and heat stability studies are critical to evaluate their potential application. Recent studies have shown the application of TGA techniques to determine the amount of coating on the surface of nanoparticles.[30]

5.2.1.1.4 Microscopic Techniques

Electron microscopic techniques are often used to determine the topography of the particles (shape, size, and distribution) under high magnification and resolution. The energy dispersive analysis X-ray spectroscopy (which works in tandem with scanning electron microscopy (SEM)) is often used to confirm the percentage composition of the sample analyzed. The technique is usually used to determine the composition of the nanoparticles near or at the surface of the sample and is not very sensitive. Nanoparticles of higher atomic number can be easily identified using the technique.

Nanoparticle samples for SEM are sputter coated with gold or platinum (in the case of nonmetals) while samples for transmission electron microscopy (TEM) are placed on carbon-coated copper TEM grids and examined under different magnifications.[26] As compared to TEM, SEM images are of lower resolution. Very thin samples are preferred for TEM imaging as they should be able to withstand the high vacuum conditions operating inside the instrument. Statistical software such as Sigmaplot can be used to determine the mean diameter, histogram, and standard deviation by sampling nanoparticles at a specific magnification.[31] Selected area electron diffraction is a crystallographic technique that can be performed inside a TEM to confirm the crystallinity of the particle analyzed. For example, the crystallographic pattern of rings of silver nanoparticles identified in a SAED pattern correspond to (111), (200), (220), (311), and (331).[31]

Force microscopic techniques such as atomic force microscopy are often used as a confirmatory tool to measure the surface roughness, as well as depth of the sample. This technique can be operated in contact, noncontact, and tapping mode[32] and can produce three-dimensional images of the sample. The technique is of immense interest to both scientists and engineers since it is nondestructive and can operate under nonvacuum conditions. In the contact mode, the cantilever tip moves in close contact with the sample surface and the deflections at the tip are measured. The attracting forces between the tip and sample surface are measured and used to construct images in the noncontact mode.

Other advanced microscopic techniques such as scanning tunneling microscopy which works on quantum tunneling are being extensively used nowadays in addition to the conventional microscopic techniques. The microscope can operate both under air and vacuum but requires the presence of a sharp tip and clean surface to ensure its operation.[22] Wettability is an important parameter to determine the efficiency of membranes since greater wettability helps in getting the surface of the membrane wet further enhancing the effectiveness of water purification. Wettability of membranes can be determined using contact angle measurements and image processing software can also be applied to assess contact angle through tangent means.[5]

5.2.1.2 COMMONLY USED CONTAMINANT DETECTION TECHNIQUES

A number of techniques are often used to determine the concentration and presence of persistent contaminants in the treated water. These techniques are often quantitative while some may be qualitative too. Qualitative techniques are often methods to determine the presence of any possible presence of nanoparticle which may have entered the treated water. Examples are XRD methods, nanoparticle sensors, particle size analyzer (used to determine particle size distribution of the treated water), etc.

Techniques such as atomic absorption spectroscopy and optical emission spectroscopy are used to quantitatively determine the concentration of contaminants left untreated or post-treatment. They can also be used to determine the loading of nanoparticles onto a specific system for treatment. Contaminants such as arsenic, selenium can be effectively determined using these techniques. The quantity of samples required for these techniques is very less and presample treatment is not complicated.[33] Ion-selective electrodes on the other hand are more simpler and user-friendly and can be set up in small laboratories. Pen model of electrodes can be carried to the field for easier measurements. This electrode is usually used to determine concentration of inorganic contaminants, such as fluoride.

5.3 REGENERATION AND RE-USE OF NANOPARTICLES

Leaching of ions from nanoparticles into the environment is a concern with regard to its potential toxicity to humans. There are many studies reporting the toxicity of different types of nanoparticles using microcosm, mesocosm, and animal studies. However, there are very few studies reporting the toxicity of nanoparticles at environmentally relevant concentrations. Currently, there are no guidelines to determine the maximum permissible limits of nanoparticles in drinking water[34] since the behavior of nanoparticles depends on the type, size, and stability in a specific system. Until there are clear guidelines from environmental guidelines, it is very critical that measures are undertaken to limit the release of nanoparticles into water. Design and use of supported nanoparticles can dramatically reduce the leaching of ions into the water by increased stability[26,35] synthesized composites of cellulose and goethite nanoparticles which demonstrated greater stability and reduced nanoparticle leaching[31] synthesized silver nanoparticles in porous cellulose fibers with strong interactions between the silver ions and the hydroxyl and ether groups of cellulose macromolecules. Recycling and regeneration of nanoparticles is another sustainable approach that can tremendously limit the impact of nanoparticles on the ecosystem.

5.4 IMPLEMENTATION BARRIERS

Nanotechnology can be considered as a sustainable alternative technology when other conventionally available techniques are ineffective in treating water to desired standards. The use of nanotechnology in such cases can prove to be cost-effective in removing multiple contaminants and has a smaller footprint as compared to other techniques. For example, conventional techniques such as sand filtration is ineffective in removing dissolved toxic heavy metals, such as arsenic.[36] Coating of sand with magnetic nanoparticles can prove to be a cost-effective solution in removing arsenic from water.[37] Although rapid advances have taken place in the field of nanotechnology for water treatment, a detailed knowledge of advantages and limitations of various engineered nanoparticles with regard to its cost-effectiveness, environmental impact, pollutant removal capacity, upscaling, and regeneration capacity is needed.[36] An increase in surface area to volume ratio for example might not necessarily translate to increased pollutant removal efficiency in all cases.[38] Currently, there are nanotechnology-enabled products in the market for water treatment and many are still in the stage of scalability and impact assessment.[39] Thorough knowledge of the pros and cons of each engineered nanoparticle is thus critical before its availability in the market. In addition, even though each country, which produces nanomaterials, is fully aware of potential risks that they may pose to the environment, there are no specific regulations being implemented so far.[3] This is another major reason which has slowed down the upscaling of nano-based products in the commercial arena.

5.5 CONCLUSION

Nanomaterials in the form of nanoparticles, nanospheres, nanocylinders, nanofibers, nanomembranes, nanotubes, and nanodots find tremendous application in drinking water treatment. They are either used as adsorbents or as membranes to filter contaminants. Modification of process parameters can produce nanomaterials of varying size ranges and shapes. The accurate characterization of the nanomaterial used for drinking water treatment applications is as important as the characterization data of the contaminants/nanomaterials present in the treated water. Critical characterization data of nanomaterials can often be used to determine the specifications and suitability of the nanomaterial for a particular application. Though there are many characterization technologies available, a thorough knowledge of the pros and cons of each technology and its data analysis is critical to ensure its precise use for nanoparticle characterization.

KEYWORDS

- **nanoparticles**
- **drinking water**
- **analytical**
- **techniques**
- **instrumentation**
- **characterization**

REFERENCES

1. United Nations. Report of the Secretary General the Sustainable Goals Report 2018, 2018.
2. World Health Organization (WHO). Progress on Drinking Water and Sanitation 2012 Update, 2012.
3. Gehrke, I.; Geiser, A.; Somborn-Schulz, A. Innovations in Nanotechnology for Water Treatment. *Nanotechnol. Sci. Appl.* **2015,** *8*, 1–17.
4. World Health Organization (WHO). *Guidelines for Drinking Water Quality*, 4th ed.; IWA Publishing: London, 2011.
5. Tlili, I.; Alkanhal, T. A. Nanotechnology for Water Purification: Electrospun Nanofibrous Membrane in Water and Wastewater Treatment. *Water Reuse Desalin.* **2019,** *9* (3), 232–248.
6. Lima, A. A. M.; Moore, S. R.; Barboza, M. S. Jr.; Soares, A. M.; Schleupner, M. A.; Newman, R. D.; Sears, C. L.; Nataro, J. P.; Fedorko, D. P.; Wuhib, T.; Schorling, J. B.; Guerrant, R. L. Persistent Diarrhea Signals a Critical Period of Increased Diarrhea Burdens and Nutritional Shortfall: A Prospective Cohort Study among Children in Northeast Brazil. *J. Infect. Dis.* **2000,** *181* (5), 1643–1651.
7. Van Minh, H.; Hung, N. V. Economic Aspects of Sanitation in Developing Countries. *Environ. Health Insights* **2011,** *5*, S8199.
8. Behbahani, M.; Moghaddam, M. R. A.; Arami, M. Techno-economical Evaluation of Fluoride Removal by Electrocoagulation Process: Optimization through Response Surface Methodology. *Desalination* **2011,** *271* (1–3), 209–218.
9. Jadhav, S. V.; Bringas, E.; Yadav, G. D.; Rathod, V. K.; Ortiz, I.; Marathe, K. V. Arsenic and Fluoride Contaminated Groundwaters: A Review of Current Technologies for Contaminants Removal. *J. Environ. Manage.* **2015,** *162*, 306–325.
10. Kargari, A.; Shirazi, M. M. A. Water Desalination: Solar-assisted Membrane Distillation. In *Encyclopedia of Energy Engineering and Technology*, 2nd ed.; Capehart, B. L., Ed.; CRC Press: Boca Raton, FL, USA, 2014; pp 1–15.

11. Khan, M. N.; Khan, W. A.; Tlili, I. Forced Convection of Nanofluid Flow across Horizontal Elliptical Cylinder with Constant Heat Flux Boundary Condition. *J. Nanofluids* **2019,** *8* (2), 386–393.
12. Zodrow, K. R.; Li, Q.; Buono, R. M.; Chen, W.; Daigger, G.; Duenas-Osorio, L.; Elimelech, M.; Huang, X.; Jiang, G.; Kim, J-H.; Logan, B. E.; Sedlak, D. L.; Westerhoff, P.; Alvarez, P. J. J. Advanced Materials, Technologies, and Complex Systems Analyses: Emerging Opportunities to Enhance Urban Water Security. *Environ. Sci. Technol.* **2017,** *51*, 10274–10281.
13. Khodashenas, B.; Ghorbani, H. R. Synthesis of Silver Nanoparticles with Different Shapes. *Arab. J. Chem.* **2019,** *12* (8), 1823–1838.
14. Qiao, J.; Cui, Z.; Sun, Y.; Hu, Q.; Guan, X. Simultaneous Removal of Arsenate and Fluoride from Water by Al-Fe hydr(oxides). *Front. Environ. Sci. Eng.* **2014,** *8* (2), 169–179.
15. Mohammad, A. W.; Teow, Y. H.; Ang, W. L.; Chung, Y. T.; Oatley-Radcliffe, D. L.; Hilal, N. Nanofiltration Membranes Review: Recent Advances and Future Prospects. *Desalination* **2015,** *356*, 226–254.
16. Timoumi, Y.; Tlili, I.; Ben; Nasrallah, S. Performance Optimization of Stirling Engines. *Renew. Energy* **2008,** *33*, 2134–2144.
17. Asif, M.; Ul Haq, S.; Islam, S.; Khan, I.; Tlili, I. Exact Solution of Non-Newtonian Fluid Motion between Side Walls. *Results Phys.* **2018,** *11*, 534–539.
18. Ali, I. New Generation Adsorbents for Water Treatment. *Chem. Rev.* **2012,** *112*, 5073–5091.
19. Mulvaney, P. Surface Plasmon Spectroscopy of Nanosized Metal Particles. *Langmuir* **1996,** *12*, 788–800.
20. Moharram, A. H.; Mansour, S. A.; Hussein, M. A.; Rashad, M. Direct Precipitation and Characterization of ZnO Nanoparticles. *J. Nanomater.* **2014,** Article ID 716210, 5 pages.
21. Chen, C.; Yu, B.; Liu, J.; Dai, Q.; Zhu, Y. Investigation of ZnO Films on Si<111> Substrate Grown by Low Energy O^+ Assisted Pulse Laser Deposited Technology. *Mater. Lett.* **2007,** *61* (14–15), 2961–2964.
22. Titus, D.; Samuel, E. J. J.; Roopan, S. M. Nanoparticle Characterization Techniques. In *Green Synthesis, Characterization and Applications of Nanoparticles*; Elsevier, 2019; pp 303–319.
23. Baer, D. R.; Engelhard, M. H. XPS Analysis of Nanostructured Materials and Biological Surfaces. *J. Electron Spectrosc. Rel. Phenom.* **2010,** *178–179*, 415–432.
24. Eastman, J. A.; Thompson, L. J.; Kestel, B. J. Narrowing of the Palladium-Hydrogen Miscibility Gap in Nanocrystalline Palladium. *Phys. Rev. B* **1993,** *48*, 84–92.
25. Israelsen, N. D.; Hanson, C.; Vargis, E. Nanoparticle Properties and Synthesis Effects on Surface-Enhanced Raman Scattering Enhancement Factor: An Introduction. *Sci. World J.* **2015,** Article ID, 124582.
26. Kong, D.; Wilson, L. D. Synthesis and Characterization of Cellulose-Goethite Composites and Their Adsorption Properties with Roxarsone. *Carbohydr. Polym.* **2017,** *169*, 282–294.
27. Savage, R. N.; Hieftje, G. M. Development and Characterization of a Miniature Inductively Coupled Plasma Source for Atomic Emission Spectrometry. *Anal. Chem.* **1979,** *51*, 408–413.

28. Carvalho, P. M.; Felicio, M. R.; Santos, N. C.; Goncalves, S.; Domingues, M. M. Application of Light Scattering Techniques to Nanoparticle Characterization and Development. *Front. Chem.* **2018,** *6*, 237.
29. Mautner, A.; Kobkeatthawin, T.; Bismarck, A. Efficient Continuous Removal of Nitrates from Water with Cationic Cellulose Nanopaper Membranes. *Resour. Efficient Technol.* **2017,** *3*, 22–28.
30. Dongargaonkar, A.; Clogston, J. D. Quantitation of Surface Coating on Nanoparticles Using Thermogravimetric Analysis. *Methods Mol. Biol.* **2018,** *1682*, 57–63.
31. He, J.; Kunitake, T.; Nakao, A. Facile In Situ Synthesis of Noble Metal Nanoparticles in Porous Cellulose Fibers. *Chem. Mater.* **2003,** *15*, 4401–4406.
32. Vanhecke, D.; Rodrigues-Lorenzo, L.; Clift, M. J. D.; Blank, F.; Petri-Fink, A.; RothenRutihauser, B. Quantification of Nanoparticles at the Single-Cell Level: An Overview about State of the Art Techniques and Their Limitations. *Nanomedicine* **2014,** *9* (12), 1885–1900.
33. Prathna, T. C.; Chandrasekaran, N.; Raichur, A. M.; Mukherjee, A. Recent Developments on Biosynthesis of Noble Metal Nanoparticles: Synthesis, Characterization and Potential Applications. *Rev. Adv. Sci. Eng.* **2014,** *3* (3), 239–249.
34. Li, Q.; Mahendra, S.; Lyon, D. Y.; Brunet, L.; Liga, M. V.; Li, D.; Alvarez, P. J. J. Antimicrobial Nanomaterials for Water Disinfection and Microbial Control: Potential Applications and Implications. *Water Res.* **2008,** *42* (18), 4591–4602.
35. Kwon, J. H.; Wilson, L. D.; Sammynaiken, R. Synthesis and Characterization of Magnetite and Activated Carbon Binary Composites. *Synth. Met.* **2014,** *197*, 8–17.
36. Alvarez, P. J. J.; Chan, C. K.; Elimelech, M.; Halas, N. J.; Villagran, D. Emerging Opportunities for Nanotechnology to Enhance Water Security. *Nat. Nanotechnol.* **2018,** *13*, 634–641.
37. Farrell, J. W.; Fortner, J.; Work, S.; Avendano, C.; Gonzalez-Pech, N. I.; Araiza, R. Z.; Li, Q.; Alvarez, P. J. J.; Colvin, V.; Kan, A.; Tomson, M. Arsenic Removal by Nanoscale Magnetite in Guanajuato, Mexico. *Environ. Eng. Sci.* **2014,** *31*, 393–402.
38. Waychunas, G. A.; Kim, C. S.; Banfeld, J. F. Nanoparticulate Iron Oxide Minerals in Soils and Sediments: Unique Properties and Contaminant Scavenging Mechanisms. *J. Nanopart. Res.* **2005,** *7*, 409–433.
39. Prathna, T. C.; Sharma, S.; Kennedy, M. Nanoparticles in Household Level Water Treatment: An Overview. *Sep. Purificat.Technol.* **2018,** *199*, 260–270.

CHAPTER 6

Assessment of ZnO Thin Films for Their Suitability as a Transparent Conducting Layer in DSSC Fabrication

MOHAMMAD RAMZAN PARRA[1,2], HAFSA SIDDIQUI[1], PADMINI PANDEY[1,3], JYOTI BAMNE, and FOZIA Z. HAQUE[1,*]

[1]*Optical Nanomaterials Lab, Department of Physics, Maulana Azad National Institute of Technology, Bhopal 462003, India*

[2]*Department of Physics, Government Degree College Boys Sopore, Jammu and Kashmir, 193201, India*

[3]*Department of Physics, Savitribai Phule Pune University, Pune, Maharashtra 411007, India*

[*]*Corresponding author. E-mail: foziazia@rediffmail.com*

ABSTRACT

This chapter embarks on the fabrication and characterization of spin-coated ZnO transparent layers prepared using different deposition cycles at the constant substrate post-heat temperature of 400°C. The ZnO nuclei could be formed by the hydrolysis and thermal decomposition of zinc acetate during the preheating process. Confirmed wurtzite lattice with Zn–O vibrational modes implied the hexagonal phase formation of ZnO. Smooth surface and the uniform distribution of a surface grain without cracks were observed through scanning electron microscopic and atomic force microscopic results, established the suitability of the nanocrystalline ZnO thin films as a transparent layer. Optical transmittance and strong photoluminescence

Advanced Microscopy: A Strong Analytical Tool in Materials Science. Merin Sara Thomas, Jozef T. Haponiuk, Sabu Thomas, & Anne George (Eds.)

peaks preferred the easy charge transfer and the light-transmitting capacity of the prepared films. Around 350 nm thick (five layers) ZnO film obtained 92% transparency without any doping. Probably, a combination of optical transmittance with smooth surface ZnO nanostructured thin film will be helpful for future developments of substrates for solar cells with ZnO as a transparent layer.

6.1 INTRODUCTION

Application of nanostructured ZnO photoanodes in dye-sensitized solar cell (DSSC) significantly enhanced efficiency (η), compared to simple ZnO nanoparticle mesoporous films. To the best of our knowledge, the highest η (8.3%) was reported by hierarchical assembled ZnO nanocrystallites, which offer a large specific surface area for dye loading while poly-dispersed aggregates act as efficient light scatters, enhancing the probability of photon absorption.[1–8] Another strategy to boost η in ZnO DSSCs is the application of a ZnO buffer layer in between the FTO and the ZnO active layer, to inhibit the electron back reaction from the FTO to electrolyte and to enhance the chemical capacitance during the transport and collection processes.[9–15] Various methods have been used for nanostructured ZnO thin film preparation, including laser molecular-beam epitaxy (L-MBE), pulsed laser deposition, sputtering, cathodic magnetron sputtering and reactive electron beam evaporation, spray pyrolysis, chemical vapor deposition, and electrodeposition. Direct deposition techniques in aqueous solution, based on heterogeneous nucleation and subsequent crystal growth on a specific substrate, have attracted much attention to prepare ZnO nanorod films.[16–25] Soft-chemistry routes, in particular sol–gel methods, feature advantages such as the possibility of obtaining metastable materials, achieving superior purity, and compositional homogeneity of the products at moderate temperatures with simple laboratory equipment, and influencing the particle morphology during the chemical transformation of the molecular precursor to the final oxidic network.[26–28]

Although aqueous sol–gel methods were highly successful to synthesize bulk metal oxides they brought some major limitations, when it came to the preparation of their nanoscale counterparts.[29] Aqueous sol–gel chemistry is rather complex, mainly due to the high reactivity of the metal oxide precursors and the double role of water as solvent and ligand.[30,31] In many cases, the three reaction types (hydrolysis, condensation, or aggregation) occur simultaneously (and are difficult to control individually), so slight variation in experimental conditions results in altered particle morphologies and is a serious concern regarding the reproducibility of a synthesis.[33] Furthermore, the as-synthesized metal oxides are often amorphous, and it is difficult to retain full control over the crystallization process. However, nonaqueous (or nonhydrolytic) sol–gel processes in organic solvents under the exclusion of water molecules are able to beat some of the major limitations of aqueous systems.[34–40]

The advantages are closely related to the diverse role of the organic solvent (ethers, alcohols, ketones, or aldehydes) and organic constituents of the precursor (alkoxides or acetylacetonates) involved in the reaction mixture.[41–45] They act as oxygen-supplying agent for the formation of metal oxide and strongly influence particle size, shape, surface, and assembly properties, and, in selected cases, even composition and crystal structure as well. Moreover, stabilizer plays vital role in sol–gel process.[46,47] It prevents the uncontrolled growth of particles and also prevents particles from aggregation. Amino ethanols such as monoethanol, diethanol, and triethanol amines act as bidentate ligands to stabilize the solution against any precipitation thus producing clear solutions for coating procedures.[48] Diethanol amine (DEA) contains two functional segments which play significant role as a complex agent to govern the hydrolysis of zinc acetate dihydrate.[49–51] DEA extends into particle surface to loss their conformational entropy and to inhibit the particles agglomeration in polymer matrix by steric repulsion.[52] It retards the condensation process to promote the formation of ZnO because of the presence of amine which increases the solution pH. In alcoholic solutions, zinc acetate dihydrate dissolve to form Zn^{2+}, CH_3COO^-, H^+, and OH^-.

The slow reaction rates, mainly a consequence of the moderate reactivity of the C–O bond, in combination with the stabilizing effect of the organic species lead to the formation of well-aligned single-crystalline nanorods of ZnO that can be grown along the C-axis direction in a perpendicular fashion onto the substrates.[53] In general, films are prepared by dip- or spin-coating of substrate from solutions or sols freshly prepared. Singh et al.[54] reported the preparation of ZnO nanorod arrays using the dip-coating technique and explained the effect of heat treatments.

Here, in this chapter, we have demonstrated a simple method for spontaneous formation of nanostructured ZnO thin films fabricated using a sol–gel process. Although the nanostructured ZnO delivered some enhancement of light absorption via light scattering, current homogeneity was significantly disturbed at the interface between the photoactive and ZnO layers.

6.2 EXPERIMENTAL DETAILS

Zinc acetate dihydrate [$Zn(CH_3COO)_2 \cdot 2H_2O$], 2-methoxyethanol [$H_3C$–O–$C_2H_5$–OH] (2MOE), and DEA [$HN(CH_2CH_2OH)_2$] of AR grade having 99.99% purity procured from Sigma Aldrich, used as precursor material, solvent, and stabilizer, respectively. A sol was prepared by dissolving zinc acetate dihydrate in a 2-methoxyethanol-DEA solution. The molar ratio of DEA to zinc acetate was kept at 1.0 and the concentration of zinc acetate was 0.5 mol/L. The resultant solution was stirred at 50°C for 60 min to yield a homogeneous solution that served as coating material on glass substrates. Single and multilayered nanocrystalline ZnO thin films were prepared by varying the number of deposition cycles, to observe the effect of temperature, and film thickness on their properties. Spin-coating technique was adopted for deposition on microscopic glass slide (75 mm × 25 mm × 1 mm: dimensions), which were rinsed ultrasonically in acetone, ethanol, and distilled water 15 min for each.

Single-layer nanocrystalline ZnO thin film was obtained through spin coating of the sol onto the glass substrate using Holmarc spin coater (Model No: HO-TH-05), operated at the speed of about 6000 rpm for 60 s (single step), and directly heated at 400°C for 120 min lead to the formation of nanocrystalline ZnO thin film. To understand the effect of post-heating treatment, some changes have been made in the above procedure, as shown in Figure 6.1. The coating procedure was repeated twice (two steps) at the speed as mentioned above, after each step the coated substrate was introduced to a preheated oven at 150°C for 30 min. Finally, after the third coating, the substrate was directly put into the furnace for post-heating treatment at 400°C for 120 min. The process for coating and preheating to prepare five-layer deposited thin film was adopted as the above-mentioned procedure. The deposition cycle was repeated four times for preheat treatment and finally after the fifth coating was post-heated.

The growth and deposition of ZnO thin film are referred to the method of chosen atomic species at a target, moving through the surrounding intermediate, and condensation of the deposited material at the substrate. The surface phenomenon of as-prepared nanocrystalline ZnO films is highly dependent on the film thickness, in our case which was measured through Ambios XP-Plus 200 Stylus thin film analyzer. The thickness of as-prepared films was found to be 50, 190, and 350 mm for "single layer," "triple layer" and "five layers" coated ZnO thin film, respectively.

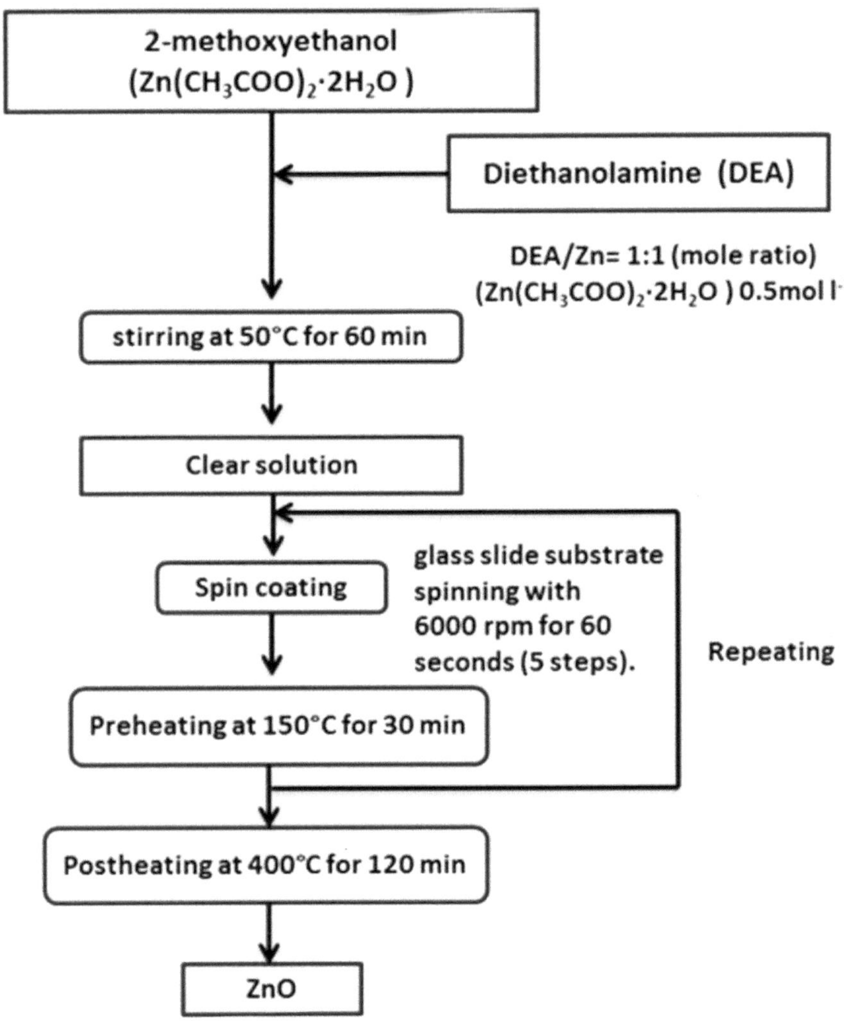

FIGURE 6.1 Flowchart of ZnO thin films preparation.

6.3 RESULTS AND DISCUSSION

6.3.1 STRUCTURAL ANALYSIS

The formation of nanocrystalline ZnO films was confirmed by using X-ray diffraction. XRD patterns of the simple glass substrate and ZnO-coated glass substrates were shown in Figure 6.2. The amorphous behavior of glass substrate with a continuous curve ~20°–35° in 2θ range was depicted in Figure 6.2a and was clearly evidenced in ZnO-coated substrates. A comparatively low intensity and broadness in diffraction peaks were observed in single layer deposited ZnO thin film (Fig. 6.2b), which revealed that film consists of coarse fine grains with nanocrystallinity. It was observed that the peak intensities gradually increased from the triple layer (Fig. 6.2c) to five layers (Fig. 6.2d) deposited ZnO thin film and hence confirmed the high crystalline nature of as-prepared films. The strongest reflection along (101) compared to (002), (102), and (103) peaks, indicating a texture effect in accordance with the standard JCPDS profile ZnO (card # 80-0075). The average crystallite size of ZnO was determined by using Scherrer's equation.[55,56] The average crystallite size is increased from 25 to 28 nm with the increase in deposition cycles. The complete structural and microstructural parameters of the as-prepared ZnO films were obtained by the Rietveld refinement procedure. The whole pattern fitting was carried out with the help of the FullProf software.[57–60]

The characteristic XRD plots of a single layer, triple layer, and five layers deposited on nanocrystalline ZnO films along with the crystal structure were extracted by Rietveld analysis and is represented in Figure 6.3a–c. The values of the average crystallite size obtained using Rietveld analysis are closer to those obtained by the Debye Scherrer formula. The slight difference (~4–5 nm) in crystallite size was observed due to the internal strain (ε) that was not considered in Scherer equation.[57]

However, in Rietveld refinement, the peaks profile is fitted using Pseudo-Voight functions, which is linear combination of Gaussian and Lorentzian functions.[58] The internal strain (ε) may arise from the chemical reaction involved in the growth process of ZnO films and having a negligible effect on peak broadening.[61] The values of the crystallite size, strain, and lattice parameter obtained from the Rietveld refinement and Scherrer methods are shown in Table 6.1.

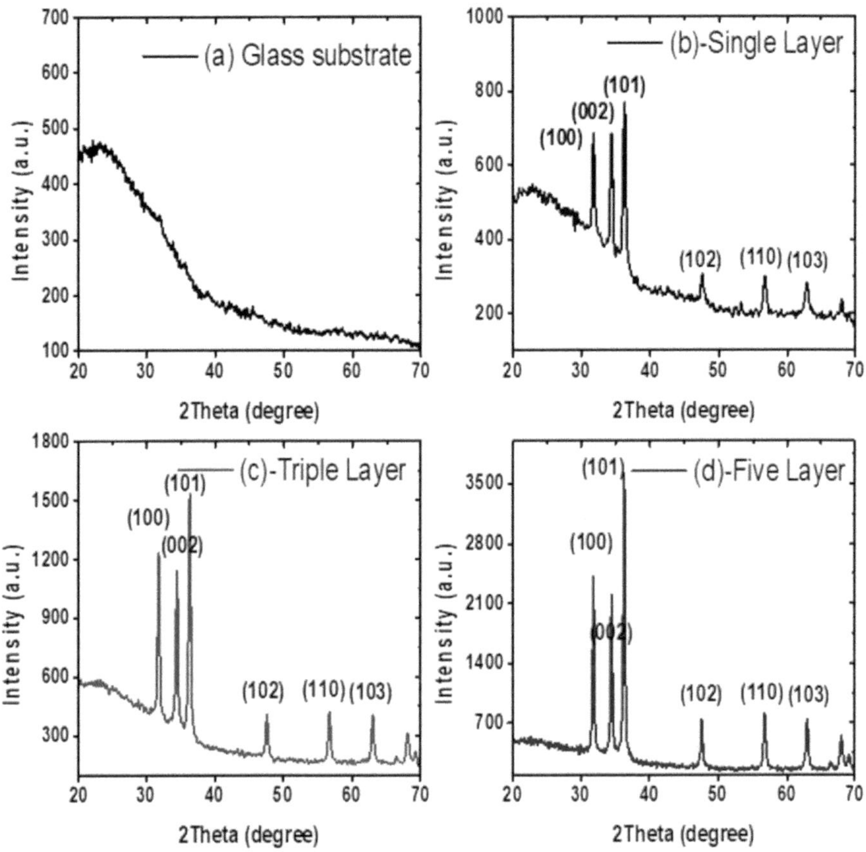

FIGURE 6.2 XRD pattern of (a) glass substrate, (b) single layer, (c) triple layer, and (d) five layers deposited nanocrystalline ZnO thin film.

In Table 6.1, the microstructural analysis shows the lattice parameters are well-matched lattice constants for the standard values of the hexagonal wurtzite structure of ZnO. Furthermore, the variation of single- to five-layer-coated ZnO film lattice parameters "a" and "c" is not remarkable when compared with the standard ZnO, due to the complexity of the hexagonal wurtzite crystal structure of ZnO and anisotropic variation in the crystal lattice. These results are consistent with the previously reported result, in which film thickness has no significant changes with the change of the film thickness.[55] As seen in Table 6.1, the ratio of the lattice parameters a/c is slightly affected by deposition cycles. However, the unit cell volume of as-prepared ZnO film is decreased from 47.598 to 47.501 Å^3 with increasing the deposition cycle from single to five layers. A similar performance was observed for the Zn–O bond length (l) that decreases from 2.09° to 2.03° with increasing deposition cycles from a single to five layers.

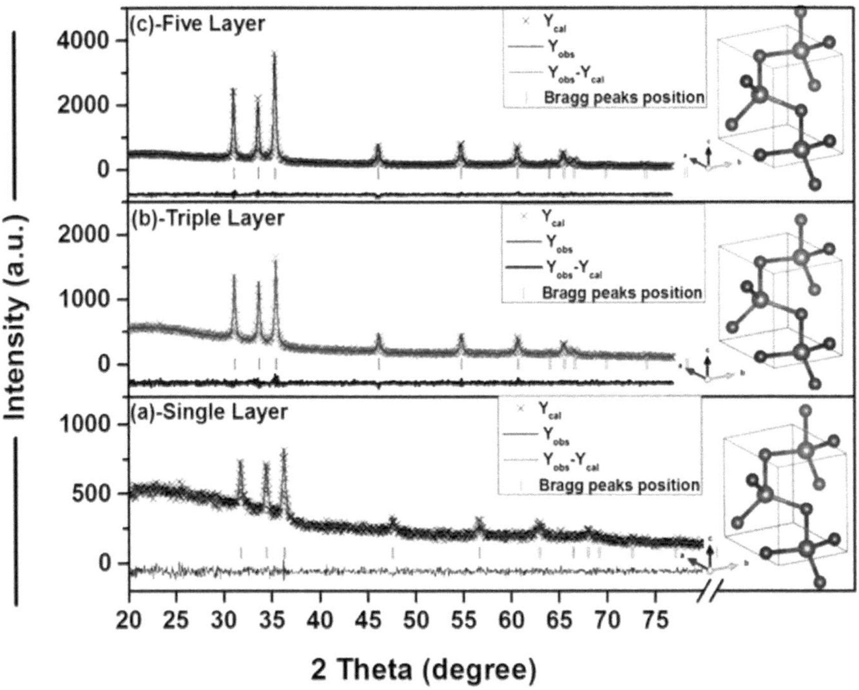

FIGURE 6.3 Rietveld fitted XRD pattern of (a) single layer, (b) triple layer, and (c) five layers deposited nanocrystalline ZnO thin film. The insert crystal structure of hexagonal ZnO showing the measured bond angles and bond length, extracted from the crystallographic information file (.cif) by Rietveld analysis. The green balls represent Zn^{2+} and the red ones O^{2-}.

The term "dislocation" states a crystallographic defect in material sciences.[62] It has opposite relation to crystallite size and is calculated by using the relation as mentioned by Siddiqui et al.[63,64] and it was found that the crystallite size increases with the increasing film thickness and as a result, both the strain and dislocation density decreased, indicated a reduction in the concentration of lattice imperfections and led to the

TABLE 6.1 Structural Parameters of Different Layers of ZnO Thin Films on Glass Substrates.

Sample coding	X-ray diffraction analysis											
	Crystallite size (nm)		a(Å)	c(Å)	c/a	(V)Å^3	bond length (l) Å	Strain (ε)10^{-3}	Dislocation density (δ) 10^{-15} (nm^{-2})	Density g/cm^3	χ^2	Goodness of fit (S)
	Scherrer (D)	Rietveld refinement (D)										
Standard ZnO	–	–	3.253	5.215	1.601	47.81	2.21	–	–	5.610		
One-layer	25	29.0	3.248	5.207	1.603	47.598	2.09	3.9	1.260	5.703	1.09	1.4
Three-layer	27	31.8	3.247	5.203	1.602	47.521	2.07	4.0	1.353	5.715	1.08	1.1
Five-layer	28	36.0	3.246	5.202	1.602	47.501	2.03	4.3	1.544	5.713	1.17	1.1

better crystallization of ZnO films. The film with five layers had better crystallinity that was well matched with UV-visible spectrum results as reported by Z. Shaban et al.[65]

Endorsement of as-prepared ZnO thin films was accomplished by the FT-IR technique. Figure 6.4 shows the FT-IR spectra of ZnO thin films with various deposition cycles and corresponding bands information are summarized in Table 6.2. From Figure 6.4, bands located in one-circle at ~505–484 cm^{-1} corresponds to the characteristic stretching vibration of Zn–O.[66] The band position in two-circle ranges in between 1554 and 1591 cm^{-1} represents the asymmetric stretching vibration of C=O mode of zinc monoacetate superimposed with the effect of N–H bonding present in DEA. The peaks located at 2343, 2344, and 2341 cm^{-1} allocated to CO_2 molecules.[67–69]

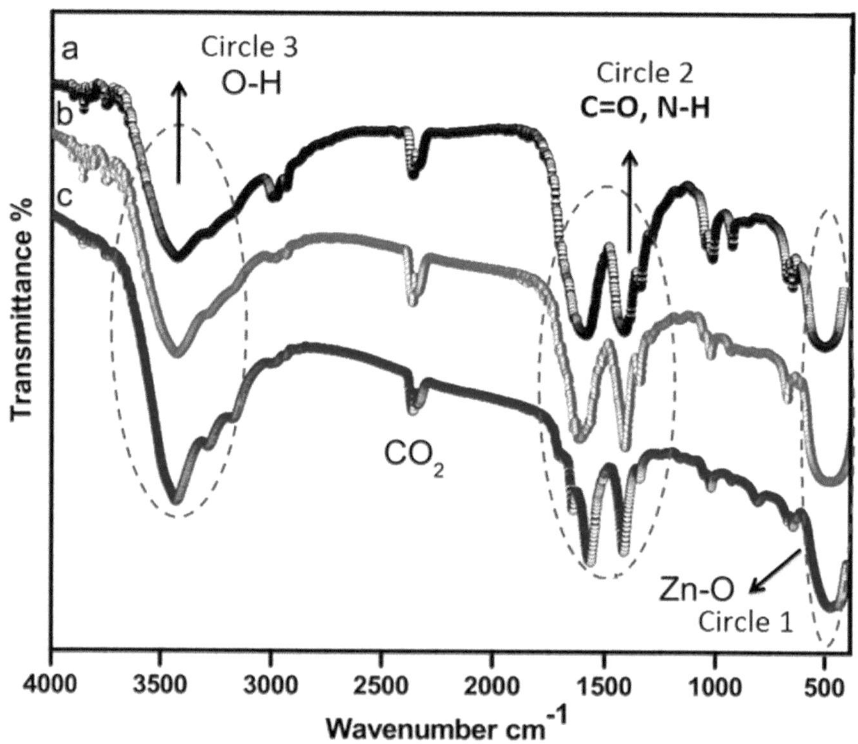

FIGURE 6.4 FTIR spectra of (a) single layer, (b) triple layer, and (c) five layers-deposited nanocrystalline ZnO film.

TABLE 6.2 Assignment of FT-IR Bands of Nanocrystalline ZnO Thin Films with Different Deposition Cycles. All values are in (cm^{-1}).

Functional group	**Wavenumber (cm^{-1})**		
	a—Single layer	**b—Triple layer**	**c—Five layers**
O–H bond	3490	3501	3522
C–H bond stretching	2927	2922	2926
CO_2 molecules in air	2343	2344	2341
Asymmetric stretching of C=O	1554	1564	1591
Asymmetric stretching of C–O	1385	1381	1391
Zn–O bond (tetrahedral)	505	496	484

Moreover, the broadband position in three-circle at 3540–3490 cm^{-1} ascribed the hydroxyl vibrations corroborated the presence of acetates, probably as a surface species chelated to Zn^{2+}ions. After five-layer depositions, the intensity of peaks increases, this implied that the deposition cycles were efficiency involved the ZnO crystal, rather a quantitative analysis is not feasible by FTIR analysis. Apart from the above-mentioned FTIR peaks, no other prominent peak was associated with any other functional group, which clearly reflected that the synthesized films were without any major impurity.

Raman spectroscopy is one the most popular techniques to study the vibrational properties, crystal structure, and defects of the nanomaterials.[70,71] ZnO has eight sets of optical phonon modes and two formula units in the primitive cell, at the point "Γ" of the Brillouin zone, ordered as $A_1 + E_1 + 2E_2$ modes (Raman active), $2B_1$ modes (Raman silent), and A_1+E_1 modes (infrared active).[72] The nonpolar E_2 (high) modes are Raman active while the B_1 modes are silent in Raman scattering.[73] The high-frequency E_2 mode rises from the vibration of oxygen sublattice, which is known as the strongest mode and characteristic of the wurtzite structure that of heavy Zn sublattice gives rise to the low-frequency E_2 mode.[74] The Raman spectra of the single layer, triple layer, and five layers deposited ZnO thin films were obtained in the backscattering geometry and shown in Figure 6.5. The Raman spectra are fitted to Gaussian line, shape, and the components derived from fitting give the frequencies of the observed Raman modes. The positions of phonon vibrational frequencies and their assigned Raman modes were summarized in Table 6.3. The micro Raman spectra of the different layer coated ZnO film show nonpolar and polar phonon modes, such as E^2_2- $E^1_{2,}$ A_1-TO, E_1-TO, E_2(high), A_1-LO, E_1-LO, multiphonon Raman peaks, and few weak peaks. All the spectra predicted by group theory are observed in the Raman spectra and exhibit prominent peak with sharp and narrow E_2 (high) mode. However, most of the Raman modes become relatively less intense (weak) with appreciable shifting in frequency for triple and five layers compared to single-layer ZnO thin film. This is due to the fact that an increase in film thickness did change the crystallinity of ZnO lattice. In Figure 6.5a, the observed B_1 (high) silent mode in the Raman spectrum is due to breakdown of the translational symmetry that occurred as result of disorder-activated Raman scattering by lattice defects or impurities.[75] This mode has been observed only in single-layered deposited ZnO film and their intensity is found in negligible on increasing the deposition cycles which strengthens the arguments that the defects on the surface are responsible for the appearance of this mode. The presence of optical phonon confinement was observed by blue shifting in E_2H mode. The significant peak intensities of Raman modes reflect the high crystallinity of ZnO thin films, consequently from the XRD and Raman analysis; the crystallinity of ZnO thin films was improved by increasing the film width. To quantify these deviations, the ratios of A_1TO/E_2H of the Raman peak-intensities were calculated and given in Table 6.3. From the Raman study, it was endorsed the growth of ZnO film with respect to deposition cycles.

TABLE 6.3 Comparison of Raman Active Modes of with Standard ZnO Material (in cm^{-1}).

Symmetry	Theory	a—Single layer	b—Triple layer	c—Five layers
E^2_2–E^1_2	323	323.61	332.15	332.77
A_1TO	380	383.50	383.02	381.8
E_1TO	407	413.1	411.2	409.6
E_2H	437	438.03	438.9	439.1
B_1(high)	540	542.9	–	–
A_1LO	574	576.04	577.69	577.90
E_1LO	583	584.0	585.06	587.82
TO+LO	650	661.3	660.8	655.67
$A1TO/E_2H$	–	0.308	0.222	0.102

6.3.2 GROWTH MECHANISM OF NANOCRYSTALLINE ZnO THIN FILMS

For blocking layer, it is necessary that the prepared film is compact without cracks and voids.[76] Our work showed the importance of the solvent and the

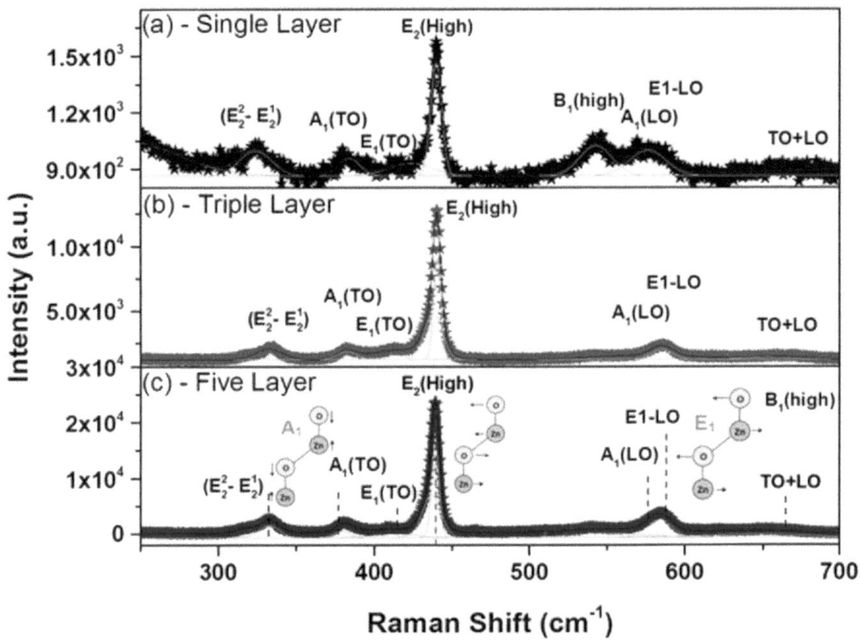

FIGURE 6.5 Raman spectra of (a) single layer, (b) triple layer, and (c) five layers-deposited nanocrystalline ZnO film with the Gaussian peak fitted curves. The Raman active vibrational modes are also shown in Figure 1.5c in which the gray solid spheres represent Zn^{2+} and the yellow ones O^{2-}.

stabilizer for the formation of smooth surface and fine structure without cracks and voids on the films. When $Zn(CH_3COO)_2{\cdot}2H_2O$ transforms to mono-acetate $Zn(C_2H_3O_2)$ in 2-methoxyethanol, it can obtain the following[77] molecular image likely on a parallel plane for zinc acetate in 2-methoxyethanol as follows:

$$\underset{\text{O-H-OH}}{\overset{\text{O-H-OH}}{}}\quad H_3C\text{-}\overset{\|}{C}\text{-O-Zn-O-}\underset{\|}{C}\text{-HC}_3 \longrightarrow [CH_3\text{-}\underset{\underset{O}{\|}}{C}\text{-O-Zn}]^+ + [O\text{-}\underset{\underset{O}{\|}}{C}\text{-}CH_3]^- + 2H^+ + 2(OH)^-$$

When the glass substrate is heated, it is supposed that the zinc monoacetate was formed. The hydrolysis process may produce acetic acid as the reaction between:

$$H^+ \text{ and } [O\text{-}\underset{\underset{O}{\|}}{C}\text{-}CH_3]^-$$

In this work, hydrolysis and condensation of Zn^{2+} are relatively slow, due to the low quantities of water. DEA acts as a complexing agent, also retarding Zn^{2+} condensation.[78] Moreover, the amine group also helps in the formation of ZnO. The reaction between mono-acetate [$Zn(C_2H_3O_2)$] and DEA [$HN(CH_2CH_2OH)_2$] are listed below, which results in the formation of ZnO.

$$[CH_3\text{-}\overset{\overset{\displaystyle O}{\|}}{C}\text{-}O\text{-}Zn]^{+} + \begin{matrix} HO\text{-}CH_2\text{-}CH_2 \diagdown \\ \qquad\qquad\qquad NH \\ HO\text{-}CH_2\text{-}CH_2 \diagup \end{matrix} \rightarrow CH_3COOH + Zn(OH)_2$$

$$Zn(OH)_2 \longrightarrow ZnO + H_2O$$

6.3.3 MORPHOLOGY AND SURFACE ANALYSIS

Scanning electron microscopic images of ZnO thin films with different coating layers are shown in Figure 6.6 with 10,000 magnifications. Figure 6.6a "single layer coated ZnO thin film" consists of spherical particles with diameter ~20 nm distributed over the entire surface of the substrate. Figure 6.6b "triple layer coated ZnO thin film" also consists of spherical particles but smaller in diameter ~20–15 nm. Defects like voids, pinholes, or peeling free surface are observed in five-layer coated ZnO thin film, as shown in Figure 6.6c. This kind of dense and compact morphology is essential for the application of dye-sensitized solar cells, which might be helpful in preventing the leakage of photocurrent.[79–82] Furthermore, the immediate effect of the annealing process leads to the reduction of porosity and increment particle size.

The porosity of thin-film becomes lesser at higher annealing temperatures; this observation correlated with the density of ZnO as mentioned in XRD results. Figure 6.6 also represents the EDX patterns (left panel of Figure 6.6) which indicate that the as-prepared ZnO thin films were composed of only Zn and O. No evidence of other impurities was found and the ZnO thin films were nearly stoichiometric. The EDX data show almost the same peaks for all three films, which confirmed the phase purity of nanocrystalline ZnO thin films. Additionally, atomic force microscopy

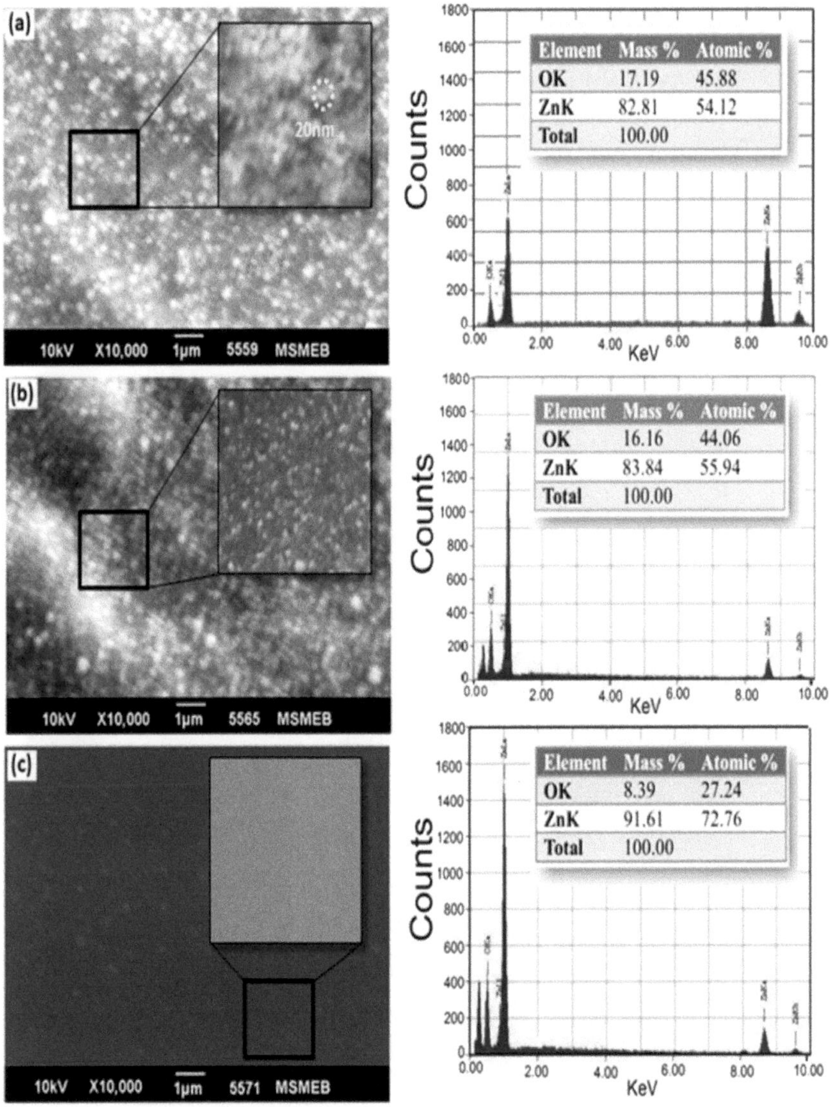

FIGURE 6.6 SEM images with energy-dispersive analysis X-ray (EDAX) patterns of (a) single layer, (b) triple layer, and (c) five layers deposited nanocrystalline ZnO thin film.

was used to examine the particle size distribution and surface topography of ZnO nanocrystalline thin films on selected places with 1.0 × 1.0 μm^2 area. Two-dimensional phase (2D) diagrams and three-dimensional (3D) images with surface roughness as a function of number of deposited layers are shown in Figure 6.7. The smoothness over the surface of the thin films can be observed clearly with lesser and uniform grain distributions. It was clearly observed from Figure 6.7a that the 2D film comprise of a few dots in comparison with other films due to the lack of ZnO atoms and nucleation sites to form dots but for ZnO thickness of 350 nm, dots merged together and form a smooth film, thus it can be considered due to excess ZnO atoms.

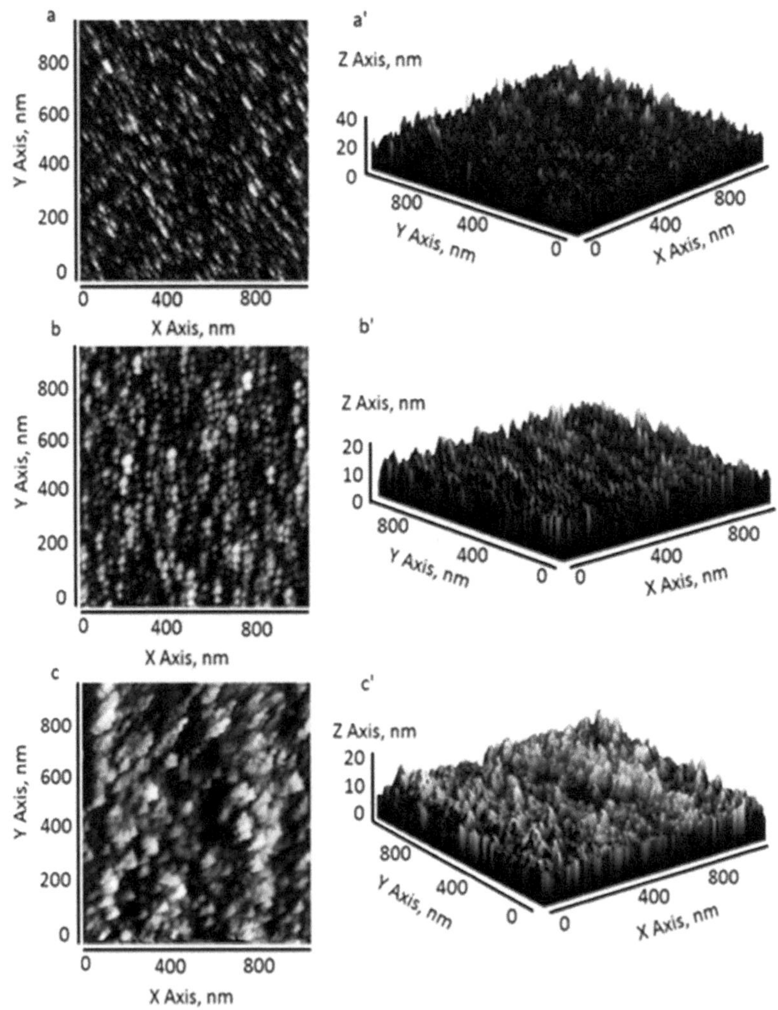

FIGURE 6.7 AFM two-dimensional phase diagrams and three-dimensional images of nanocrystalline ZnO thin films for (a–a') single layer, (b–b') triple layer, and (c–c') five layers' deposition.

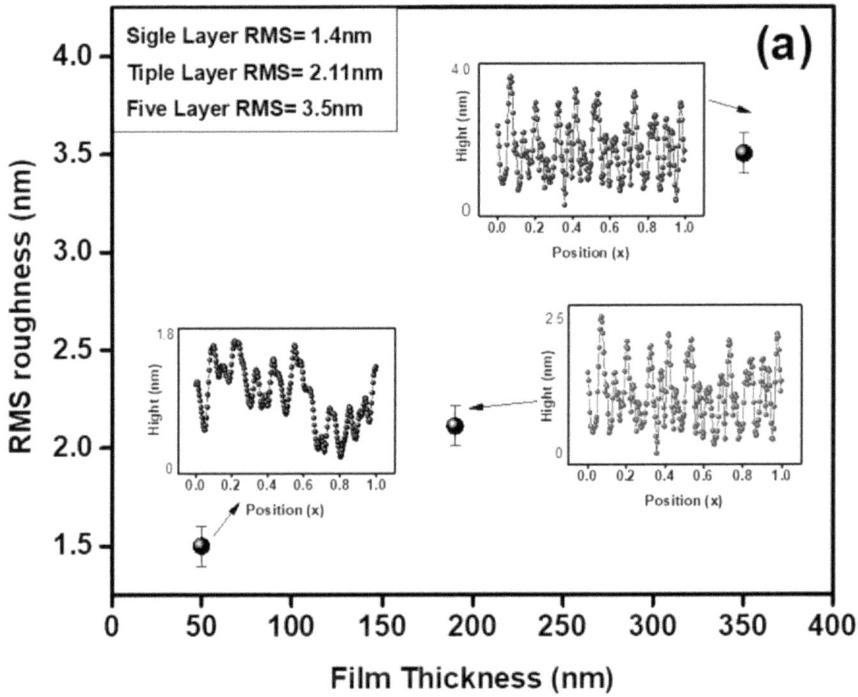

FIGURE 6.8(a) Variation of RMS roughness as a function of as-deposition film thickness along with the line scan along the x-axis.

The formation of such particles can be explained based on nucleation and coalescence process.[83] Smooth surface and the uniform distribution of a surface grain without cracks were established the suitability of the nano-crystalline ZnO thin films as blocking layers.[84] Increasing the thickness to 350 nm, origins to increase the area engaged by grains, and increases the RMS roughness, as shown in Figure 6.8a, the line profile and the deviation of RMS roughness of the films calculated from AFM data clearly indicates the role of film thickness on the height and width of the grains. Also calculated the height–height correlation function $H(r)$ by given formula[85] along the x-direction is shown in Figure 6.8b.

$$H(r) = 2w^2\left[1-\exp\{-\left(\frac{r}{\xi}\right)^{2x}\}\right] \tag{6.1}$$

where w is the interface width of RMS roughness, ξ is the correlation length, a is the roughness exponent.[86]

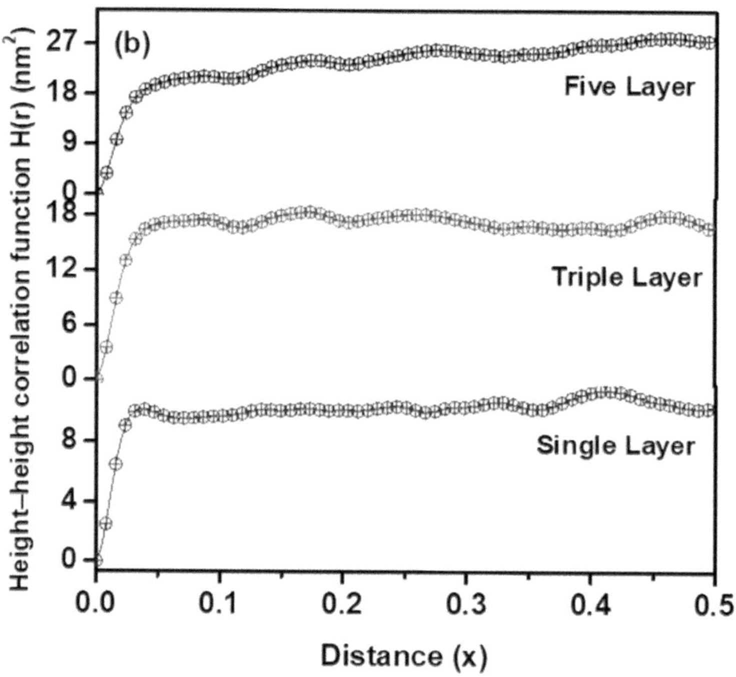

FIGURE 6.8(b) Height–height correlation function *H*(*r*) versus position (*x*) of the single layer, triple layer, and five layers' deposition ZnO thin films.

The correlation lengths calculated by fitting the height–height correlation function with eq 6.1 are given as 9, 24, and 29 nm for the single layer, triple layer, and five layers coated ZnO films, respectively. Meanwhile the correlation length is proportional to the grain size, thus it was concluded that film-coated with five has larger grain size. The values of the roughness exponent *(x)* are 0.80–0.84 as 0.30, 0.34, and 0.33 for single layer, triple layer, and five layers coated ZnO films, respectively, which indicates that all the as-prepared films have similar type of morphology. AFM results clearly indicate that the thickness to 350 nm, film has better crystallization in terms of the height and ordering which is required for the blocking layer.[87]

6.3.4 OPTICAL ANALYSIS

According to Shaban et al.[65] and Amala Rani et al.[88] "The ZnO blocking layer should be highly transparent, so that it will help in the absorbance of maximum light in visible rang." Therefore, owing to the importance of the blocking layer transmission in DSSC, the optical transmission analysis was carried out and shown in Figure 6.9. The average transmittance of thin films in the visible range was about ~77%, 87%, and 92% respectively, with sharp absorption edges observed in the ultraviolet region, which is significant for applications as blocking layer in DSSC.[89] Results showed that the transmission of five layers was more uniform than the other. Moreover, the transmission spectra results support all the above mention characterization analysis. Film deposited in five layers, shows 92% transparency it is also able for TCO application.[90,91]

The inset Figure 6.9a–c determined the energy band gap value by the extrapolation of the linear portion of the $(\alpha h\nu)^2$ curve versus the photon energy $h\nu$. Figure 6.9d shows the absorption edge is almost observed in the visible region. It was noticed that the position of absorption edge depends on thickness of thin films, where it was slightly shifted toward the higher energy value by increasing deposition cycles. The absorbance spectra of as-deposited ZnO thin films were calculated using the following equation [92]:

$$A = 2 - \log(\%T) \tag{6.2}$$

Optical band gap energy values were found in the range from 3.03 to 3.05 eV. Owing to the bandgap absorption, the transmission decreases suddenly near the short wavelength region (UV-region). These findings are in good agreement with the findings of S. Sharma et al.[92] It is also noticed that the cut-off wavelength shifts toward higher wavelength and optical bandgap changes toward a smaller energy value when the thickness increases. Further the obtained optical absorption coefficient (α) was in the range 10^4 cm^{-1} reveals the stoichiometric compositions of as-prepared films.[93,94] The absorption coefficient (α) is associated with photon energy ($h\nu$) by the recognized equation as:

$$\text{Absorption coefficient } (\alpha) = 2.303A/t \tag{6.3}$$

where (A) is absorbance and (t) is thickness of thin film[95] and shown in Figure 6.10. The analysis of inserted Figure 6.10 shows that the value of absorption coefficient (α) decreases with the increase in the incident wavelength for all the films up to a certain wavelength depending on the film sample, after that wavelength (about 400 nm), α almost becomes invariant.

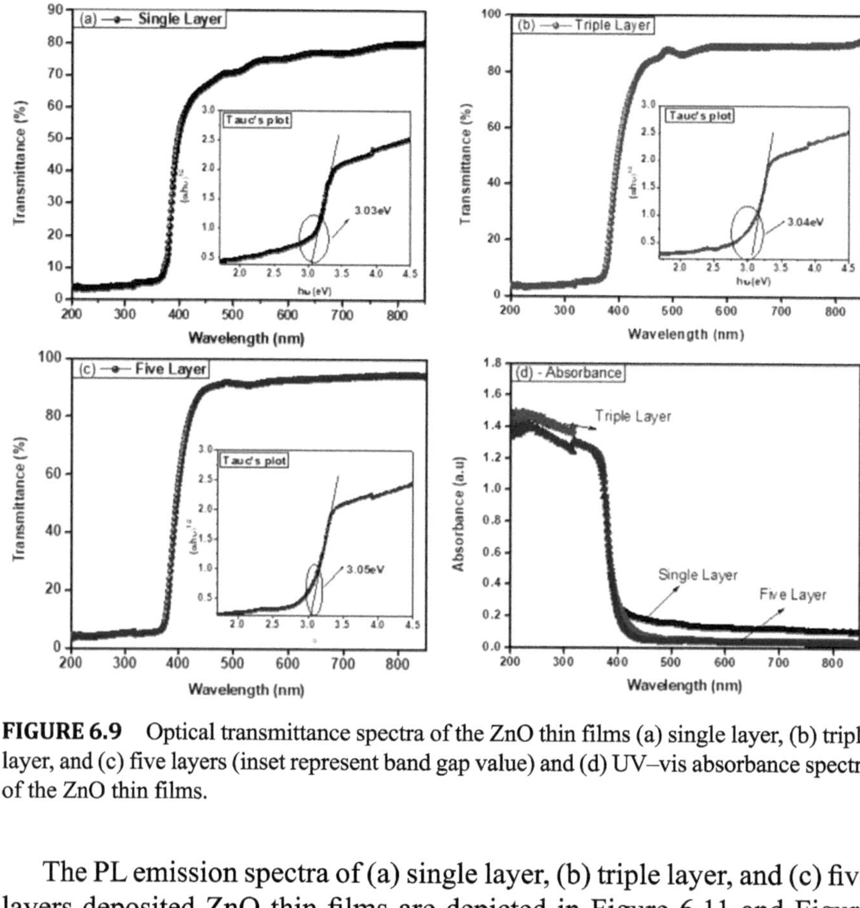

FIGURE 6.9 Optical transmittance spectra of the ZnO thin films (a) single layer, (b) triple layer, and (c) five layers (inset represent band gap value) and (d) UV–vis absorbance spectra of the ZnO thin films.

The PL emission spectra of (a) single layer, (b) triple layer, and (c) five layers deposited ZnO thin films are depicted in Figure 6.11 and Figure 6.11d shows the proposed band structure of the ZnO with different defect levels within the range of 410–530 nm. With the increase in deposition cycles, the PL peak position shifts toward higher wavelengths, which might be attributed to the variation in the optical band gaps of the films. All the films exhibit blue–green emission peaks between 428 and 550 nm with the absented UV-emission peak corresponding to ZnO. The ZnO film vanished UV emission in many cases as reported previously.[95–97] Therefore, the origin of these peaks may result from the emission through variation of the intrinsic defects in ZnO film,[98–101] such as zinc vacancy (VZn), oxygen vacancy (V_O), interstitial zinc (Zni), interstitial oxygen (Oi), and antisite oxygen (OZn) which is shown in Figure 6.11d. In the present study, the violet–blue emission at 428 nm is due to the vacant V_{Zn}

and interstitial O_i, violet emission corresponds to Z_{ni}, and the transition involving V_{Zn} would result in blue emission. The broad emission peak in the blue emission peak at 465 nm might be related to the surface defect in ZnO thin films. The emission peak at ~ 480–525 nm attributed to the presence of V_O states and emission peak at ~528–550 nm ascend due to ionized O_i-induced transitions as reported by Pandey et al.[102] The intensity of the peaks increased with the increase in the layer coating, which may be attributed to the improvement in the crystallinity. The obtained optical bandgap and high intense photoluminescence peaks favored, respectively, the easy charge transfer and the light-transmitting capacity of the ZnO films.

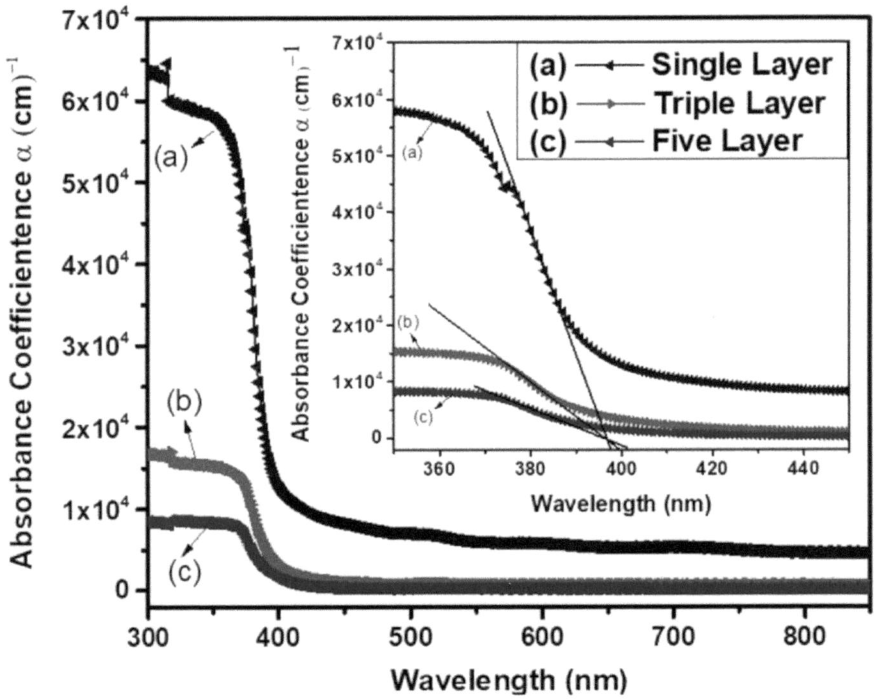

FIGURE 6.10 The variation of the obtained optical absorption coefficient (α) with the incident photon energy *hv* (eV).

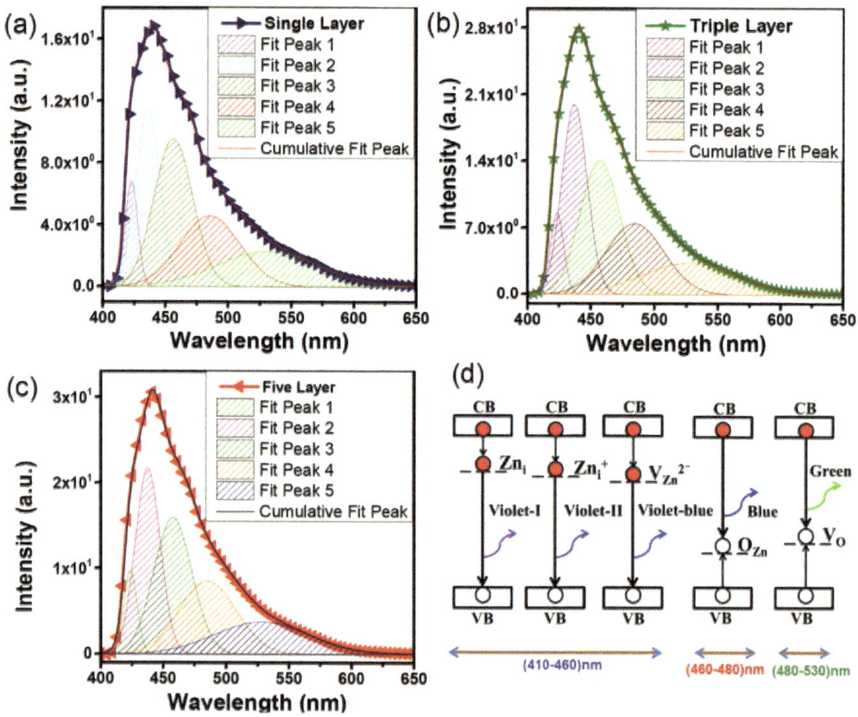

FIGURE 6.11 Room temperature photoluminescence spectra of (a) single layer, (b) triple layer, (c) five layers deposited ZnO film, and (d) schematic band diagram based on PL data.

6.4 CONCLUSION

The nanocrystalline ZnO thin films were successfully fabricated via the sol–gel method. As opposed to traditional depositions via physical methods, sol–gel process offers a simple, low-cost, and highly controlled way for ZnO film deposition. Results indicate that the organic ligand and pre- and post-heat treatment affected the growth of as-prepared ZnO films. As the wetting films were preheated at 150°C, the hydrolysis and thermal decomposition of zinc acetate simultaneously occurred to form ZnO nuclei. Additionally, the immediate effect of the annealing process leads to the reduction of porosity and increment in particle size. SEM and AFM results revealed that the as-prepared ZnO films had smooth surface without having cracks and voids. The grains of films become larger and denser with increasing deposition cycles, which was consistent with the

XRD results and optical transmittance results. The photoluminescence spectra revealed that the concentration of imperfections and surface defects decreased after each deposition cycle. For comparing to recently reports, our report has advantages such as fine and smooth surface of ZnO thin film within the combination of high (92%) optical transmittance that will be highly suitable for DSSCs as blocking layer.

ACKNOWLEDGMENTS

Authors are pleased to acknowledge the Director, MANIT, Bhopal for providing necessary facilities for this work, and the Director, UGC-DAE-Consortium for Scientific Research, Indore Centre (M.P.), India, for XRD, FTIR, Micro-Raman, Thin film analyzer and UV-Vis measurements. The appreciation also goes to Mr. K. K. S. Gautam for helping in SEM and AFM characterization. M.R.P. is thankful to CSIR, New Delhi, for financial support as CSIR-SRF [Grant 09/0953(0011)/2015; EMR-I (Ack. no. 163320/2K14/1)].

COMPETING INTERESTS

The authors declare that they have no competing interests and are alone responsible for the content and the writing of this chapter.

KEYWORDS

- **dye-sensitized solar cell**
- **transparent conductor**
- **FT-IR of nanocrystalline ZnO**
- **AFM and SEM**
- **photoluminescence**

REFERENCES

1. Sarcheshmeh, H. M.; Arazi, R.; Ardakani, M. M. Application of Bifunctional Photoanode Materials in DSSCs: A Review. *Renew. Sustain. Energy Rev.* **2020,** *34*, 110249.
2. Ye, M.; Wen, X.; Wang, M.; Iocozzia, J.; Zhang, N.; Lin, C.; Lin, Z. Recent Advances in Dye-Sensitized Solar Cells: From Photoanodes, Sensitizers and Electrolytes to Counter Electrodes. *Mater. Today* **2015,** *18* (3), 155–162.
3. Ghann, W.; Kang, H.; Sheikh, T.; Yadav, S.; Chavez-Gil, T.; Nesbitt, F.; Uddin, J. Fabrication, Optimization and Characterization of Natural Dye Sensitized Solar Cell. *Sci. Rep.* **2017,** *7* (1), 1–12.
4. Chen, H.; Yang, S. Hierarchical Nanostructures of Metal Oxides for Enhancing Charge Separation and Transport in Photoelectrochemical Solar Energy Conversion Systems. *Nanosc. Horiz.* **2016,** *1* (2), 96–108.
5. Hezam, A.; Namratha, K.; Drmosh, Q. A.; Chandrashekar, B. N.; Jayaprakash, G. K.; Cheng, C.; Swamy, S. S.; Byrappa, K. Electronically Semitransparent ZnO Nanorods with Superior Electron Transport Ability for DSSCs and Solar Photocatalysis. *Ceram. Int.* **2018,** *44* (6), 7202–7208.
6. Umar, A.; Akhtar, M. S.; Al-Assiri, M. S.; Al-Salami, A.E; and Kim, S. H. Composite CdO-ZnO Hexagonal Nanocones: Efficient Materials for Photovoltaic and Sensing Applications. *Ceram. Int.* **2018,** *44* (5), 5017–5024.
7. Hsu, C. L.; Chen, K. C. Improving Piezoelectric Nanogenerator Comprises ZnO Nanowires by Bending the Flexible PET Substrate at Low Vibration Frequency. *J. Phys. Chem. C.* **2012,** *116* (16), 9351–9355.
8. Lončar, B.; Vujisić, M.; Stanković, K.; Osmokrović, P. Stability of Metal-Oxide Varistor Characteristics in Exploitation Conditions. *Acta Phys. Polon. A.* **2009,** *116* (6), 1081–1084.
9. Goel, S.; Kumar, B. A Review on Piezo-/Ferro-Electric Properties of Morphologically Diverse ZnO Nanostructures. *J. Alloys Comp.* **2020,** *816*, 152491.
10. D. G. Tong, P. Wu, P. K. Su, D. Q. Wang, H. Y. Tian, Preparation of Zinc Oxide Nanospheres by Solution Plasma Process and Their Optical Property, Photocatalytic and Antibacterial Activities. *Mater. Lett.* **2012,** *70*, 94–97.
11. Boruah, B. D.; Misra, A. Energy-Efficient Hydrogenated Zinc Oxide Nanoflakes for High-Performance Self-Powered Ultraviolet Photodetector. *ACS Appl. Mater. Interf.* **2016,** *8*, 18182–18188.
12. Bhati, V. S.; Hojamberdiev, M.; Kumar, M. Enhanced Sensing Performance of ZnO Nanostructures-Based Gas Sensors: A Review. *Energy Rep.* **2020,** *6*, 46–62.
13. Aal, N. A.; Al-Hazmi, F.; Al-Ghamdi, A. A.; El-Tantawy, F.; Yakuphanoglu, F. Novel Rapid Synthesis of Zinc Oxide Nanotubes via Hydrothermal Technique and Antibacterial Properties. *Spectrochim. Acta Part A: Mol. Biomol. Spectrosc.* **2015,** *135*, 871–877.
14. Asthanal, A.; Yap, Y. K.; Yassar, R. S. Field Emission from Zinc Oxide Nanobelts. *J. Nanosci. Nanotechnol.* **2015,** *15*, 2277–2282.
15. Ko, S. H.; Lee, D.; Kang, H. W.; Nam, K. H.; Yeo, J. Y.; Hong, S. J.; Grigoropoulos, C. P.; Sung, H. J. Nanoforest of Hydrothermally Grown Hierarchical ZnO Nanowires for a High Efficiency Dye-Sensitized Solar Cell. *Nano Lett.* **2011,** *11*, 666–671.

16. Okada, T.; Agung, B.; Nakata, Y. ZnO Nano-Rods Synthesized by Nano-Particle-Assisted Pulsed-Laser Deposition. *Appl. Phys. A*. **2004,** *79*, 1417–1419.
17. Hirate, T.; Sasaki, S.; Li, W.; Miyashita, H.; Kimpara, T.; Satoh, T. Effects of Laser Ablated Impurity on Aligned ZnO Nanorods Grown by Chemical Vapor Deposition. *Thin Solid Films* **2005,** *487*, 35–39.
18. Wu, C. L.; Chang, L.; Chen, H. G.; Lin, C. W.; Chang, T. F.; Chao, Y. C.; Yan, J. K. Growth and Characterization of Chemical-Vapor-Deposited Zinc Oxide Nanorods. *Thin Solid Films* **2006,** *498*, 137–141.
19. Rusli, N. I.; Tanikawa, M.; Mahmood, M. R.; Yasui, K.; Hashim, A. M. Growth of Highdensity Zinc Oxide Nanorods on Porous Silicon by Thermal Evaporation. *Materials* **2012,** *5*, 2817–2832.
20. Giri, P. K.; Patel, P. K.; Panchal, C. J.; Bhattacharyya, S.; Kumari, S.; Singh, D. K.; Kheraj, V.A; Shah, N. M.; Vakil, P. D.; Patel, K. J.; Desai, M. S.; Rehani, B.; Rao, V. J.; Desai, R. R.; Lakshminarayana, D.; Patel, P. B. Studies on Zinc Oxide Nanorods Grown by Electron Beam Evaporation Technique. *Syn. React. Inorg. Metal-Org. NanoMetal. Chem*. **2007,** *37*, 437–441.
21. Yang, Z.; Shi, Y. Y.; Sun, X. L.; Cao, H. T.; Lu, H. M.; Liu, X. D. The Competition Growth of ZnO Microrods and Nanorods in Chemical Bath Deposition Process. *Mater. Res. Bull.* **2010,** *45*, 474–480.
22. Xu, F.; Lu, Y.; Xie, Y.; Liu, Y. Seed Layer-Free Electrodeposition and Characterization of Vertically Aligned ZnO Nanorod Array Film. *J. Solid State Electrochem*. **2009,** *14*, 63–70.
23. Mousavi, S. F.; Davar, F.; Loghman-Estarki, M. R. Controllable Synthesis of ZnO Nanoflowers by the Modified Sol–Gel Method. *J. Mater. Sci.: Mater. Electron*. **2016,** *27*, 12985–12995.
24. Kim, K. H.; Utashiro, K.; Abe, Y.; Kawamura, M. Growth of Zinc Oxide Nanorods Using Various Seed Layer Annealing Temperatures and Substrate Materials. *Int. J. Electrochem. Sci.* **2014,** *9*, 2080–2089.
25. Krunks, M.; Katerski, A.; Dedova, T.; Acik, I. O.; Mere, A. Nanostructured Solar Cell Based on Spray Pyrolysis Deposited ZnO Nanorod Array. *Sol. Energy Mater. Sol. Cells*. **2008,** *92*, 1016–1019.
26. Po, R.; Carbonera, C.; Bernardi, A.; Camaioni, N. The Role of Buffer Layers in Polymer Solar Cells. *Energy Environ. Sci.*, **2011,** *4*, 285–310.
27. White, M. S.; Olson, D. C.; Shaheen, S. E.; Kopidakis, N.; Ginley, D. S. Inverted Bulk-Heterojunction Organic Photovoltaic Device Using a Solution-Derived ZnO Underlayer. *Appl. Phys. Lett.* **2006,** *89*, Article 143517.
28. Tsay, C. Y.; Fan, K. S.; Wang, Y. W.; Chang, C. J.; Tseng, Y. K.; Lin, C. K. Transparent Semiconductor Zinc Oxide Thin Films Deposited on Glass Substrates by Sol-Gel Process. *Ceram. Int.* **2010,** *36*, 1791–1795.
29. Zhang, Y.; Bai, K.; Fu, Z.; Zhang, C.; Zhou, H.; Wang, L.; Zhu, S.; Guan, S.; Li, D.; Hu, J. Composite Coating Prepared by Micro-Arc Oxidation Followed by Sol–Gel Process and in Vitro Degradation Properties. *Appl. Surf. Sci.* **2012,** *258*, 2939–2943.
30. Rani, S.; Suri, P.; Shishodia, P. K.; Mehra, R. M. Synthesis of Nanocrystalline ZnO Powder via Sol–Gel Route for Dye-Sensitized Solar Cells. Sol. *Energy Mater. Sol. Cells*. **2008,** *92*, 1639–1645.

31. Spanhel, L.; Anderson, M. A. Semiconductor Clusters in the Sol–Gel Process: Quantized Aggregation, Gelation, and Crystal Growth in Concentrated ZnO Colloids. *J. Am. Chem. Soc.* **1991,** *113*, 2826–2833.
32. Owens, G. J.; Singh, R. K.; Foroutan, F.; Alqaysi, M.; Han, C. M.; Mahapatra, C.; Kim, H.-W.; Knowles, J. C. Sol–Gel Based Materials for Biomedical Applications. *Progress Mater. Sci.* **2016,** *77*, 1–79.
33. Graham, T. J. On the Properties of Silicic Acid and Other Analogous Substances. *Chem. Soc.* **1864,** *17*, 318–327.
34. Niederberger, M.; Pinna, N. *Metal Oxide Nanoparticles in Organic Solvents Synthesis, Formation, Assembly and Application*; Springer: London, 2009.
35. Hay, J. N.; Raval, H. M. Synthesis of Organic-Inorganic Hybrids via the Non-Hydrolytic Sol-Gel Process. *Chem. Mater.* **2001,** *13*, 3396–3403.
36. Scholz, G.; Kemnitz, E. Sol-Gel Synthesis of Metal Fluorides: Reactivity and Mechanism. In *Modern Synthesis Processes and Reactivity of Fluorinated Compounds*; Groult, H., Leroux, F., Tressaud, A., Eds.; Elsevier Inc.: Amsterdam, The Netherlands, 2017; pp 609–650.
37. Rüdiger, S.; Gross, U.; Kemnitz, E. Non-Aqueous Sol–Gel Synthesis of Nano-Structured Metal Fluorides. *J. Fluorine Chem.* **2007,** *128*, 353–368.
38. Kemnitz, E.; Noack, J. The Non-Aqueous Fluorolytic Sol–Gel Synthesis of Nanoscaled Metal Fluorides. *Dalton Trans.* **2015,** *44*, 19411.
39. Krahl, T.; Broßke, D.; Scheurell, K.; Lintner, B.; Kemnitz, E. Novel Aspects in the Chemistry of the Non-Aqueous Fluorolytic Sol–Gel Synthesis of Nanoscaled Homodisperse MgF_2 Sols for Antireflective Coatings. *J. Mater. Chem. C* **2016,** *4*, 1454–1466.
40. Rondinini, S.; Ardizzone, S.; Cappelletti, G.; Minguzzi, A.; Vertova, A. Materials Sol–Gel Synthesis. *Encyclopedia Electrochem. Power Sour.* **2009,** 613–624.
41. Debecker, D. P.; Mutin, P. H. Non-Hydrolytic Sol-Gel Routes to Heterogeneous Catalysts. *Chem. Soc. Rev.* **2012,** *41*, 3624–3650.
42. Debecker, D. P.; Hulea, V.; Mutin, P. H. Mesoporous Mixed Oxide Catalysts via Non-Hydrolytic Sol–Gel: A Review. *Appl. Catal. A.* **2013,** *451*, 192–206.
43. Mutin, P. H.; Vioux, A. Nonhydrolytic Processing of Oxide-Based Materials: Simple Routes to Control Homogeneity, Morphology, and Nanostructure. *Chem. Mater.* **2009,** *21*, 582–596.
44. Pinna, N.; Niederberger, M. Surfactant-Free Nonaqueous Synthesis of Metal Oxide Nanostructures. *Angew. Chem. Int. Ed. Engl.* **2008,** *47*, 5292–5304.
45. Bilecka, I.; Niederberger, M. New Developments in the Nonaqueous and/or Non-Hydrolytic Sol–Gel Synthesis of Inorganic Nanoparticles. *Electrochim. Acta.* **2010,** *55*, 7717–7725.
46. Nagarajan, S.; Mohana, M.; Sudhagar, P.; Raman, V.; Nishimura, T.; Kim, S.; Kang, Y. S.; Rajendran, N. Nanocomposite Coatings on Biomedical Grade Stainless Steel for Improved Corrosion Resistance and Biocompatibility. *ACS Appl. Mater. Interf.* **2012,** *4*, 5134–5141.
47. Sudha, M.; Senthilkumar, S.; Hariharan, R.; Suganthi, A.; Rajarajan, M. Synthesis, Characterization and Study of Photocatalytic Activity of Surface Modified ZnO Nanoparticles by PEG Capping. *J. Sol Gel Sci. Technol.* **2012,** *65*, 301–310.

48. Morselli, D.; Niederberger, M.; Bilecka, I.; Bondioli, F. Double Role of Polyethylene Glycol in the Microwaves-Assisted Non-Hydrolytic Synthesis of Nanometric TiO_2: Oxygen Source and Stabilizing Agent. *J. Nanopart. Res.* **2014,** *16*, 1–11.
49. Boudjouan, F.; Chelouche, A.; Touam, T.; Djouadi, D.; Khodja, S.; Tazerout, M.; Ouerdane, Y.; Hadjoub, Z. Effects of Stabilizer Ratio on Photoluminescence Properties of Sol–Gel ZnO Nano-Structured Thin Films. *J. Lumin.* **2015,** *158*, 32–37.
50. Sagar, P.; Shishodia, P. K.; Mehra, R. M. Influence of pH Value on the Quality of Sol–Gel Derived ZnO Films. *Appl. Surf. Sci.* **2007,** *253*, 5419–5424.
51. Livage, J.; Ganguli, D. Sol-Gel Electrochromic Coatings and Devices: A Review. *Sol. Energy Mater. Sol. Cells.* **2001,** *68*, 365–381.
52. Yoon, S. H.; Liu, D.; Shen, D.; Park, M.; Kim, D. J. Effect of Chelating Agents on the Preferred Orientation of ZnO Films by Sol–Gel Process. *J. Mater. Sci.* **2008,** *43*, 6177–6181.
53. Marin, O.; Tirado, M.; Budini, N.; Mosquera, E.; Figueroa, C.; Comedi, D. Photoluminescence from C-Axis Oriented ZnO Films Synthesized by Sol-Gel with Diethanolamine as Chelating Agent. *Mater. Sci. Semicond. Proces.* **2016,** *56*, 59–65.
54. Singh, N.; Pandey, P.; Haque, F. Z. Effect of Heat and Time-Period on the Growth of ZnO Nanorods by Sol–Gel Technique. *Optik* **2012,** *123* (15), 1340–1342.
55. Parra, M. R.; Haque, F. Z. Optical Study and Ruthenizer (II) N3 Dye-Sensitized Solar Cell Application of ZnO Nanorod-Arrays Synthesized by Combine Two-Step Process. *Opt. Spectrom.* **2015,** *119*, 672.
56. Parra, M. R.; Haque, F. Z. Poly (Ethylene Glycol) (PEG)-Assisted Shape-Controlled Synthesis of One-Dimensional ZnO Nanorods. *Optik* **2015,** *126*, 1562–1566.
57. Siddiqui, H.; Parra, M. R.; Qureshi, M. S.; Malik, M. M.; Haque, F. Z. Studies of Structural, Optical, and Electrical Properties Associated with Defects in Sodium-Doped Copper Oxide (CuO/Na) Nanostructures. *J. Mater. Sci.* **2018,** *53*, 8826–8843.
58. Siddiqui, H.; Parra, M. R.; Malik, M. M.; Haque, F. Z. Structural and Optical Properties of Li Substituted CuO Nanoparticles. *Opt. Quant. Electron.* **2018,** *50*, 260.
59. Siddiqui, H.; Parra, M. R.; Pandey, P.; Qureshi, M. S.; Haque, F. Z. Utility of Copper Oxide Nanoparticles (CuO-NPs) as Efficient Electron Donor Material in Bulk-Heterojunction Solar Cells with Enhanced Power Conversion Efficiency. *J. Sci. Adv. Mater. Devices* **2020,** *5*, 104–110.
60. Siddiqui, H.; Parra, M. R.; Pandey, P.; Qureshi, M. S.; Haque, F. Z. Performance Evaluation of Optimized Leaf-Shaped Two-Dimension (2D) Potassium Doped CuO Nanostructures with Enhanced Structural, Optical and Electronic Properties. *Ceram. Int.* **2020,** *46*, 20404–20414.
61. Parra, M. R.; Pandey, P.; Siddiqui, H.; Sudhakar, V.; Krishnamoorthy, K.; Haque, F. Z. Evolution of ZnO Nanostructures As Hexagonal Disk: Implementation as Photoanode Material and Efficiency Enhancement in Al: ZnO Based Dye Sensitized Solar Cells. *Appl. Surf. Sci.* **2019,** *470*, 1130–1138.
62. Siddiqui, H.; Qureshi, M. S.; Haque, F. Z. Hexamine (HMT) Assisted Wet Chemically Synthesized CuO Nanostructures with Controlled Morphology and Adjustable Optical Behavior. *Opt. Quant. Electron.* **2016,** *48*, 349.
63. Siddiqui, H.; Qureshi, M. S.; Haque, F. Z. One-Step, Template-Free Hydrothermal Synthesis of CuO Tetrapods. *Optik* **2014,** *125*, 4663–4667.

64. Siddiqui, H.; Qureshi, M. S.; Haque, F. Z. Surfactant Assisted Wet Chemical Synthesis of Copper Oxide (CuO)Nanostructures and Their Spectroscopic Analysis. *Optik* **2016,** *127*, 2740–2747.
65. Shaban, Z.; Ara, M. H. M.; Falahatdoost, S.; Ghazyani, N. Optimization of ZnO Thin Film through Spray Pyrolysis Technique and Its Application as a Blocking Layer to Improving Dye Sensitized Solar Cell Efficiency. *Curr. Appl. Phys.* **2016,** *16*, 131–134.
66. Parra, M. R.; Haque, F. Z. Structural and Optical Properties of Poly-Vinylpyrrolidone Modified ZnO Nanorods Synthesized through Simple Hydrothermal Process. *Optik* **2014,** *125*, 4629–4632.
67. Siddiqui, H.; Qureshi, M. S.; Haque, F. Z. Structural and Optical Properties of CuONanocubes Prepared Through Simple Hydrothermal Route. *Int. J. Sci. Eng. Res.* **2014,** *5* (3), 173–177.
68. Parra, M. R.; Haque, F. Z. Optical Investigation of Various Morphologies of ZnO Nanostructures Prepared by PVP-Assisted Wet Chemical Method. *Opt. Spectrosc.* **2015,** *118*, 765.
69. Haque, F. Z.; Parra, M. R.; Siddiqui, H.; Singh, N.; Pandey, P.; Mishra, K. M. PVP Assisted Shape-Controlled Synthesis of Self-Assembled 1D ZnO and 3D CuO Nanostructures. *Opt. Spectrosc.* **2016,** *120* (3), 408–414.
70. Shrivastava, M.; Kumari, R.; Parra, M. R.; Pandey, P.; Siddiqui, H.; Haque, F. Z. Electrochemical Synthesis of MoS_2 Quantum Dots Embedded Nanostructured Porous Silicon with Enhanced Electroluminescence Property. *Opt. Materi.* **2017,** *73*, 763–771.
71. Siddiqui, H.; Parra, M. R.; Haque, F. Z. Optimization of Process Parameters and Its Effect on Structure and Morphology of CuO Nanoparticle Synthesized via the Sol−Gel Technique. *J. Sol-Gel Sci. Technol.* **2018,** *87*, 125–135.
72. Parra, M. R.; Pandey, P.; Siddiqui, H.; Qadri, S. B.; Haque, F. Z. New-Insight Into the Physical Properties of Zn1xBxO Two-Dimensional Hexagonal Nanodisks: An Efficient Material for Dye Sensitized Solar Cells. *Mater. Lett.* **2019,** *238*, 194–197.
73. Pandey, P.; Parra, M. R.; Haque, F. Z.; Kurchania, R. Effects of Annealing Temperature Optimization on the Efficiency of ZnO Nanoparticles Photoanode Based Dye Sensitized Solar Cells. *J Mater Sci: Mater Electron.* **2017,** *28*, 1537–1545.
74. Pandey, P.; Singh, N.; Haque, F. Z. Enhancement in Structural and Optical Properties of Boron Doped ZnO Nanostructures Synthesized by Simple Aqueous Solution Growth Technique. *J. Adv. Phy.* **2017,** *6* (4), 358–366.
75. Pandey, P.; Kurchania, R.; Haque, F. Z. Structural, Diffused Reflectance and Photoluminescence Study of Cerium Doped ZnO Nanoparticles Synthesized through Simple Sol-Gel Method. *Optik* **2015,** *126* (21), 3310–3315.
76. Xia, J.; Masaki, N.; Jiang, K.; Yanagida, S. Sputtered Nb_2O_5 as a Novel Blocking Layer at Conducting Glass/TiO_2 Interfaces in Dye-Sensitized Ionic Liquid Solar Cells. *J. Phys. Chem. C.* **2007,** *111* (22), 8092–8097.
77. Znaidi, L. Sol–Gel-Deposited ZnO Thin Films: A Review. *Mater. Sci. Eng. B.* **2010,** *174*, 18–30.
78. Livage, J.; Ganguli, D. Sol–Gel Electrochromic Coatings and Devices: A Review. *Sol. Energy Mater. Sol. Cells.* **2001,** *68*, 365–381.
79. Van de Lagemaat, J.; Frank, A. J. Nonthermalized Electron Transport in Dye-Sensitized Nanocrystalline TiO_2 Films: Transient Photocurrent and Random-Walk Modeling Studies. *J Phys Chem B.* **2001,** *105*, 11194–11205.

80. Law, M.; Greene, L. E.; Johnson, J. C.; Saykally, R.; Yang. P. D. Nanowire Dye-Sensitized Solar Cells. *Nat Mater.* **2005,** *4*, 455–459.
81. Martinson, A. B. F.; McGarrah, J. E.; Parpia, M. O. K.; Hupp, J. T. Dynamics of Charge Transport and Recombination in ZnO Nanorod Array Dye-Sensitized Solar Cells. *Phys. Chem. Chem. Phys.* **2006,** *8*, 4655–4659.
82. Smiglak, M.; Pringle, J.; Lu, X.; Han, L.; Zhang, S.; Gao, H.; MacFarlane, D.; Rogers, R. Ionic Liquids for Energy, Materials, and Medicine. *Chem. Commun.* **2014,** *50*, 9228–9250.
83. Mane, R. S.; Lee, W. J.; Pathan, H. M.; Han, S.-H. Nanocrystalline TiO_2/ZnO Thin Films: Fabrication and Application to Dye-Sensitized Solar Cells. *J. Phys. Chem. B.* **2005,** *109* (51), 24254–24259.
84. Lepoutre, S.; Julián-López, B.; Sanchez, C.; Amenitsch, H.; Lindend, M.; Grosso, D.; Nanocasted Mesoporous Nanocrystalline ZnO Thin Films. *J. Mater. Chem.* **2010,** *20*, 537–542.
85. Ariano, P.; Budnyk, O.; Dalmazzo, S.; Lovisolo, D.; Manfredotti, Ch.; Rivolo, P.; Vittone, E. On Diamond Surface Properties and Interactions with neurons. *Eur. Phys. J. E.* **2009,** *30*, 149–156.
86. Raoufi, D.; Kiasatpour, A.; Fallah, H. R.; Rozatian, A. S. H. Surface Characterization and Microstructure of ITO Thin Films at Different Annealing Temperatures. *Appl. Surf. Sci.* **2007,** *253*, 9085–9090.
87. Schmidt-Mende, L.; MacManus-Driscoll, J. L. ZnO–Nanostructures, Defects, and Devices. *Mater. Today* **2007,** *10* (5), 40–48.
88. Amala Rani, A.; Ernest, S. Characterization of Spray-Deposited ZnO Thin Films for Dye-Sensitized Solar Cell Application. *Appl. Phys. A.* **2016,** *122*, 647.
89. Cameron, P. J.; Peter, L. M. Characterization of Titanium Dioxide Blocking Layers in Dye-Sensitized Nanocrystalline Solar Cells. *J. Phys. Chem. B.* **2003,** *107*, 14394–14400.
90. Klein, A.; Körber, C.; Wachau, A.; Säuberlich, F.; Gassenbauer, Y.; Harvey, S. P.; Proffit, D. E.; Mason, T. O. Transparent Conducting Oxides for Photovoltaics: Manipulation of Fermi Level, Work Function and Energy Band Alignment. *Materials.* **2010,** *3*, 4892–4914.
91. Tsai, C. L.; Lin, Y. J.; Wu, P. H.; Chen, S. Y.; Liu, D. S.; Hong, J. H.; Liu, C. J.; Shih, Y. T.; Cheng, J. M.; Chang, H. C. Induced Increase in Surface Work Function and Surface Energy of Indium Tin Oxide-Doped ZnO Films by (NH4) 2Sx Treatment. *J. Appl. Phys.* **2007,** *101*, 113713.
92. Sharma, S.; Periasamy, C.; Chakrabarti, P. Thickness Dependent Study of RF Sputtered ZnO Thin Films for Optoelectronic Device Applications. *Electron. Mater. Lett.* **2015,** *11* (6), 1093–1101.
93. Daniel, G. P.; Justinvictor, V. B. Effect of Annealing Temperature on the Structural and Optical Properties of ZnO Thin Films Prepared by RF Magnetron Sputtering. *Phys. B.* **2010,** *405* (7), 1782–1786.
94. Al-Hardan, N. H.; Abdullah, M. J. The Effect of Oxygen Ratio on the Crystallography and Optical Emission Properties of Reactive RF Sputtered ZnO Films. *Phys. B.* **2010,** *405* (4), 1081–1085.
95. Pandey, P.; Kurchania, R.; Haque, F. Z. Controlled Hydrothermal Synthesis, Structural and Optical Analysis of Nanometer-Sized ZnO Spheres. *Optik* **2015,** *126* (2), 301–303.

96. Pandey, P.; Kurchania, R.; Haque, F. Z. Optical Studies of Europium Doped ZnO Nanoparticles Prepared by Sol-Gel Technique. *J. Adv. Phys.* **2014,** *3,* 104–110.
97. Parra, M. R.; Haque, F. Z. Aqueous Chemical Route Synthesis and the Effect of Calcination Temperature on the Structural and Optical Properties of ZnO Nanoparticles. *J. Mater. Res. Technol.* **2014,** *3* (4), 363–369.
98. Haque, F. Z.; Singh, N.; Pandey, P.; Parra, M. R. Study of Zinc Oxide Nano/Micro Rods Grown on ITO and Glass Substrates. *Optik* **2013,** *124,* 4167–4171.
99. Singh, N.; Haque, F. Z. Ionic Liquid Controlled Growth of Zinc Oxide Nano Particles and Their Fluorescence Study in the Presence of NH_3 Gas. *Mater. Sci. Res. India* **2014,** *11* (1), 27–34.
100. Parra, M. R.; Syed, F.; Husain, M.; Haque, F. Z. Structural and Optical Characterization of ZnO Nanoparticles Synthesized via Low Temperature Precipitation Method, In Jain, V., Verma, A., Eds.; *Physics of Semiconductor Devices. Environmental Science and Engineering*; Springer: Cham, 2014; pp 597–598.
101. Johnson, J. C.; Yan, H.; Yang, P.; Saykally, R. J. Optical Cavity Effects in ZnO Nanowire Lasers and Waveguides. *J. Phys. Chem., B.* **2003,** *107,* 8816.
102. Pandey, P.; Singh, N.; Haque, F. Z. Development and Optical Study of Hexagonal Multi-Linked ZnO Micro-Rods Grown Using Hexamine as Capping Agent. *Optik* **2013,** *124* (12), 1188–1191.

CHAPTER 7

Effect of Rare Earth Y^{3+} Doping on the Formation of Anatase Phase of TiO_2 Nanoparticles

JYOTI BAMNE[1], NITU SINGH[2], ARVIND MITTAL[3], KAJOL TAIWADE[1,2], P. K. SHARMA[1], and FOZIA Z. HAQUE[2,*]

[1]*Department of Physics, Barkatullah University, Bhopal, Madhya Pradesh, India*

[2]*Optical Nanomaterials Laboratory, Department of Physics, Maulana Azad National Institute of Technology, Bhopal, Madhya Pradesh, India*

[3]*Energy Centre, Maulana Azad National Institute of Technology, Bhopal, Madhya Pradesh, India*

[*]*Corresponding author. E-mail: foziazia@rediffmail.com*

ABSTRACT

This chapter deals with the preparation of pure TiO_2 and rare-earth Y^{3+}-doped TiO_2 nanoparticles by modified sol–gel method using mixed solvent as ethanol and methanol, calcinated at different temperatures. In the present work, the nanostructures of TiO_2 are synthesized by the sol–gel method. The nanostructure, morphology, and optical properties of prepared nanoparticles were investigated by SEM, XRD, and BET. SEM images show well- defined boundaries with spherical shape and less aggregation of particles calcination at 400°C and 500°C. Pure TiO_2 and Y^{3+}-doped TiO_2 nanoparticles have been investigated through UV–visible,

Advanced Microscopy: A Strong Analytical Tool in Materials Science. Merin Sara Thomas, Jozef T. Haponiuk, Sabu Thomas, & Anne George (Eds.)

photoluminescence (PL), and FTIR spectroscopic methods that confirms their applicability as adsorbents and photocatalyst. X-ray diffraction patters of 0.2 mol% Y^{3+}-doped TiO_2 sample indicated that the samples were crystalline with broad peaks of anatase phase calcinated at 400°C and 500°C. The anatase particles had a crystalline size of about 8.46 nm at 400°C and 12.51 nm at 500°C for 0.2 mol% Y^{3+}-doped TiO_2 nanoparticles. About 0.2 mol% Y^{3+}-doped TiO_2 shows higher surface area of 257.29 m^2/g and pore size 1.678 calcinated at 500°C. At higher calcination temperatures, Y^{3+}-doped TiO_2 shifts its optical absorption from UV to nearby a visible region. The emission spectra of Y^{3+}-doped TiO_2 nanoparticles of samples calcinated at 400°C and 500°C have been split into many sharp peaks confirming well-defined lattices site in particles. The sharp emission peak and lower PL intensity is due to reduced recombination rate which envisioned good photocatalytic activity. Y^{3+}-doped TiO_2 nanoparticles at 400°C show good photocatalytic activity as compared to yttrium-doped element at 500°C because of lower PL intensity. The present results will help to understand the effect of yttrium element in the growth of TiO_2 nanoparticles, its transformation into anatase, and their possible application as antibacterial or antiviral catalysts.

7.1 INTRODUCTION

In the recent years, titanium dioxide (TiO_2) nanomaterials have received great attention as a promising semiconductor photocatalyst. It is used in a wide range of applications such as paints, photocatalyst, separations, sensor devices,[1,2] dye-sensitized solar cells,[3] and memory devices.[4–7] TiO_2 materials with low cost, strong oxidizing power, less toxicity, and biodegradable material,[1,8] have been extensively used in environmental applications. TiO_2 has been considered as a promising photocatalyst for air purification and water pollution control due to its potential applications in the complete mineralization of organic pollutant in air and wastewater. Many studies focused on particularly in nano-anatase TiO_2 have been extensively investigated and utilized in many technological applications due to its excellent physicochemical properties. Because of its wide band gap 3.2 eV for anatase, there is still a problem that TiO_2 can only be activated by UV light and only utilize 3–5% of the solar spectrum that can reach the Earth. The modification of doping is mainly performed from two aspects, one is extending the light adsorption edge of TiO_2 toward the visible light region and the other is preventing the recombination rate of photogenerated electron and hole pairs.[9] Recently, doping of rare-earth metals has great potential because of 4f electron configuration which can interact with its f-orbitals. However, rare-earth metals are extensively used in many applications such as fluorescent devices, computer, catalytic conversion, and good photocatalyst activity. Rare earths such as europium, lanthanum, neodymium, erbium, yttrium, ytterbium, samarium, and thulium have been dopant for TiO_2 nanoparticles. In previous literature, rare-earth metals Y^{3+}-doped TiO_2 nanoparticles, due to its outstanding properties, have been used in many applications prepared by the sol–gel method using titanium tetra isopropoxide as a precursor.

Y^{3+} doping could effectively reduce the crystalline size, obstruct the anatase to rutile phase transformation, and can provide strong adsorption capacity, and greatly improve the separation rate of photogenerated charge carriers. Thus, doping rare-earth ions with 4f electron configuration can significantly enhance the photocatalytic activity of TiO_2 nanoparticles. Rare-earth Y^{3+} doping could stabilize the anatase phase of TiO_2 reduce the crystallite size and may be favorable for the separation of photogenerated electron–hole pairs in photocatalytic application. Smaller crystallite size and larger surface area were beneficial to absorbed more organic molecules efficiently onto the surface of catalyst, resulting in the enhancement of photocatalytic activity.[10] It is of great interest finding ways to extend the wavelength range absorption of TiO_2 into the visible region or solar light where less expensive light sources are available. Several approaches have been developed to extend the absorption band-edge of TiO_2 from UV to the visible light region, namely increasing its surface-to-volume ratio, sensitization using dye molecules.[11]

The primary purpose of the TiO_2 doping is as follows: (1) to retard the fast charge recombination and (2) to enable visible light absorption by creating defect states in the band gap. In the first case, the conduction band electrons or the valence band holes are trapped in the defect sites, inhibiting the recombination and enhancing the interfacial charge transfer. Moreover, smaller TiO_2 nanoparticles are also beneficial for more effective separation of photogenerated carriers (electron–hole pair) and greater photocurrent, resulting in higher photocatalytic and photoelectrical chemical conversion. Nevertheless, TiO_2 nanoparticles are insufficient for industrial purposes, mainly limited by the low specific surface area and detrimental electron–hole recombination within nanoseconds.[11] Various techniques are available for the preparation of TiO_2 nanoparticles, such as thin film, spin coating, dip coating, two step wet chemical, precipitation, thermal (hydrothermal and solvothermal), chemical solvent and chemical vapor decomposition (CSV & CVD), ultrasonic irradiation, extremely low temperature and aerogel and xerogel, other method different from sol–gel, modified sol–gel, two route sol–gel.[12] However, the sol–gel method has been found to be a versatile way of preparation of entirely new generations of advanced materials for many applications.[13,14] The sol–gel method is a very cheap, effective and easy technique. Additionally, pure anatase phase shows superior photocatalytic activity and photovoltaic efficiency than the rutile phase.[15]

In this chapter, we investigated the phase control of TiO_2 nanoparticles by Y^{3+} doping by the sol–gel method, the effect of Y^{3+} doping on grain size, phase transformation (anatase to rutile), absorption band gap, and photoluminescence (PL) was investigated in detail. This study deals with the structural and morphological properties of pure and yttrium-doped titanium oxide nanoparticles prepared by the sol–gel method.

7.2 EXPERIMENTAL

7.2.1 CHEMICALS USED

All the materials used in this study were analytical reagent grade purity. In this synthesis procedure, we use titanium tetra isopropoxide (97.9%, Sigma Aldrich) as a precursor material of TiO_2 is taken as solvent absolute ethanol (Et-OH 99.9%, CDH grade), methanol (purity 99.9%, Merck), deionized water, and using concentrated nitric acid (HNO_3 99.9%, Merck) as a catalyst. The synthesis was carried out in the conditions of continuous stirring at the room temperature. All the solutions were prepared using de-ionized water (purity 99.9%, CDH grade). The chemicals were used without any further purification.

7.2.2 SYNTHESIS OF TIO_2 NANOPARTICLES

TiO_2 nanoparticles have been prepared through nonaqueous sol–gel technique using ethanol and methanol as a mix solvent, rare-earth element (yttrium nitrate hexahydrate) used as a dopant material and titanium tetra-isopropoxide precursor at a room temperature. A flow diagram is shown in Figure 7.1.

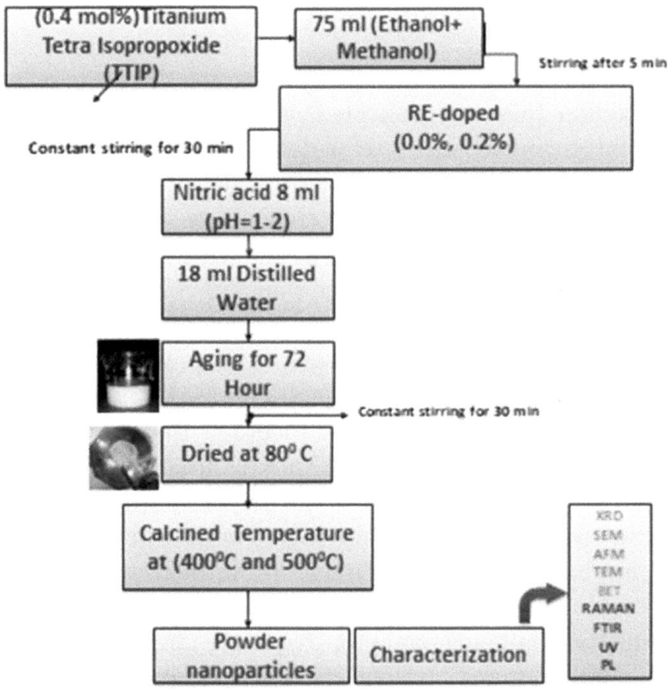

FIGURE 7.1 Flowchart for the preparation of pure and Y^{3+}-doped TiO_2 nanoparticles calcinated at 400°C and 500°C.

TiO_2 nanoparticles were prepared by via nonaqueous sol–gel process. About 0.042 mol of titanium tetra-isopropoxide mixture was prepared by dissolving it in 75 mL ethanol and methanol solvent to obtain homogeneous transparent solution stirrer at room temperature with 400 rpm. We added dopant element (yttrium nitrate hexahydrate) with 0.2 mol% concentration, after 1 h we added nitric acid (HNO_3) drop by drop with a syringe to change the pH to 1–2. We obtained yellowish transparent homogenous stable solution and also added distilled water (18 mL). The prepared sol was kept for half an hour for a gel preparation at room temperature. The gel thus obtained was dried at 60°C in an oven for 72 h. Powder sample of TiO_2 was calcined at 400°C and 500°C calcination was carried out in a programmable muffle furnace; the heating rate was increasing at 140°C/h. The temperature was increased to a certain temperature at 400°C and 500°C held for 3 h to completely remove water and solvent and cooled down naturally.[11]

7.2.3 CHARACTERIZATION TOOLS

The prepared TiO_2 nanoparticles were characterized by using powder X-ray diffraction that was collected using Bruker D8 Advance X-ray diffractometer with nickel-filtered Cu-Ka radiation (k= 1.5406 Å) operating at 40 kV and 40 mA and a position sensitive detector (Lyn xEye) based on Bruker AXS compound silicon strip technology was used. In each case, scanning has been performed from 20° to 80°, 2θ range. The crystalline phase was compared with those of the standard compound reported in the JCPDS (Joint Committee on Powder Diffraction Standards) data files. Surface morphology and elemental compositions were investigated using scanning electron microscope (SEM) (JEOL-JSM-6390), operating at 10 kV. BET surface analysis data were recorded from Brunauer, Emmett, and Teller (BET) [Autosorb (iQ)]. The structural information about the various vibrational modes present in the prepared samples was examined by recording their IR spectra employing a Perkin Elmer Model Vertex-70 FTIR spectrophotometer. Fourier transform infrared spectroscopy (FTIR) spectra were acquired scanning the powder samples embedded in KBr matrix in the transmittance mode for frequency range of 400—4000 cm^{-1}. Structural properties and defects chemistry of prepared samples were investigated by Raman spectra performed in backscattering geometry using micro-Raman spectrometer (Renishaw, Gloucestershire, GL 12 7 DW, UK). Optical properties were studies via UV–vis absorption spectra recorded with a Carry 5000 UV–vis–NIR spectrophotometer. The UV–vis absorption spectra technique can be used to determine the band gap of the semiconductor materials. Optical properties were also studied through PL spectra at room temperature and the data was recorded using fluorescence spectrometer (F-7000 Hitachi).[11,16]

7.3 RESULTS AND DISCUSSIONS

7.3.1 X-RAY DIFFRACTION ANALYSIS (XRD)

The phase identification with estimation of anatase and rutile ratio and crystal structure of pure TiO_2 and Y^{3+}-doped TiO_2 nanoparticles with concentration 0.2 mol%, calcination at 400°C and 500°C temperature modified by nonaqueous sol–gel technique were studied by powder XRD pattern as shown in Figure 7.2a–d. In Figure 7.2a,b, the pure TiO_2 and Y^{3+}-doped

TiO_2 nanoparticles calcinated at 400°C show pure active anatase (101) peak at 2θ = 25.30° and complete anatase peak at 2θ = 25.28° for yttrium-doped element with 0.2 mol% concentration. When we increase calcination temperature at 500°C it shows mixed phase, anatase (101) peak at 2θ = 25.28° and the rutile (110) peak at 27.42° has been analyzed by #JCPDS card for pure TiO_2 and Y^{3+}-doped TiO_2 nanoparticles. The percentage of anatase (35.04%) and rutile (64.6%) is obtained for pure TiO_2 calcinated at 500°C. The peak of mixed of anatase and rutile peak at 25.28° (101), 27.42° (110), 36.05° (101), 37.77° (004), 41.23° (111), 48.06° (200), 54.32° (211), and 56.62° (220). The peaks of anatase Y^{3+}-doped TiO_2 appeared at 25.28° (101), 37.82° (004), 48.01° (200), 53.92° (105), and 55.02° (211) appeared in Figure 7.2c,d. The powder samples showed the crystalline pattern and the observed d_{101} lines matches very well with reported peaks for the anatase phase. Within the anatase tetragonal TiO_2 crystal structure and in good agreement with the standard #JCPDS card No. 211272. Y^{3+}-doped TiO_2 result show anatase phase (101) with compromised crystallinity of the material and decreased crystallite size and increased surface area.

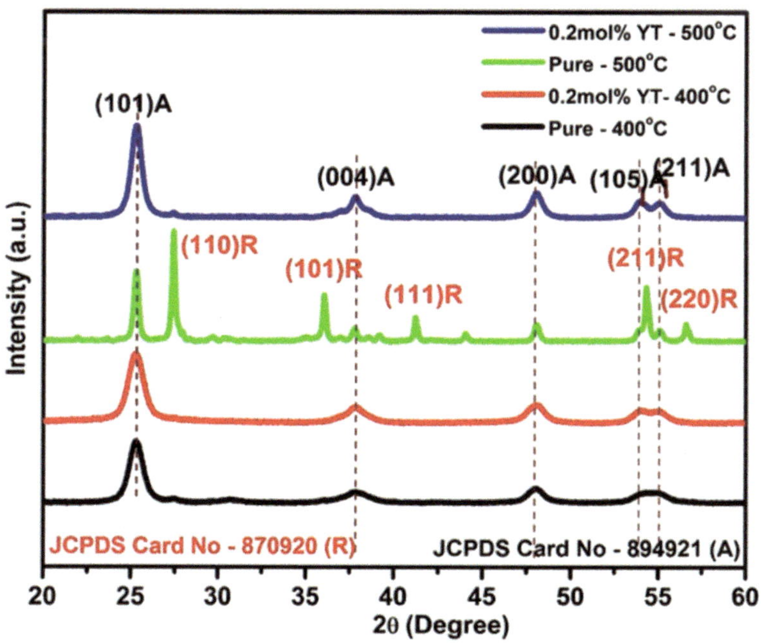

FIGURE 7.2 XRD pattern of TiO_2 nanoparticles samples (a) TiO_2—400°C, (b) 0.2mol% YT—400°C, (c) TiO_2—500°C (d) 0.2mol% YT—500°C nanoparticles.

The crystallite size can be determined by Scherrer's formula[17,18]

$$D = \frac{0.9\lambda}{\beta \cos\theta} \tag{7.1}$$

where D is the crystalline size, λ is the X-ray wavelength, k is usually taken as 0.9, β is the full-width at half maximum (FWHM) of the diffraction line, after subtraction of equipment broadening, and θ is the Bragg's angle.[19] The crystalline sizes of pure TiO_2, Y^{3+}-doped TiO_2 samples synthesized using nonaqueous sol–gel technique are found to be pure anatase phase for pure and doped TiO_2 at 400°C temperature. We observed crystalline size is 9.67 nm for pure TiO_2 sample and crystalline size is 8.46 nm for 0.2 mol% doped sample. Similarly, the pure TiO_2 nanoparticles sample calcinated at 500°C show mixed phase of anatase and rutile phase with crystalline size 29.10 nm (anatase) and 38.80 nm (rutile) and also we observed crystalline size of Y^{3+}-doped sample is found to be 12.51 nm anatase with high crystallinity material calcined at same temperature and the strain induced in powder due to crystal imperfection and distortion was calculated using the formula (7.4)[19] and all results are shown in Table 7.1.

The percentage of rutile in the sample can be estimated from the respective integrated XRD peak intensity using eqs 7.1 and 7.2[1,16]

$$\text{anatase}(\%) = \left[\frac{0.79 I_R}{I_R + 0.79 I_A}\right] \times 100 \tag{7.2}$$

$$\text{Rutile}(\%) = \left\{\frac{1}{[(I_R + \frac{0.79 I_A}{I_R}]}\right\} \times 100 \tag{7.3}$$

where I_A and I_R are the X-ray peak intensities of the anatase [101] and rutile [110], respectively.[20]

$$\varepsilon = \beta_{hkl} / 4\tan\theta \tag{7.4}$$

7.3.2 SCANNING ELECTRON MICROSCOPY (SEM) ANALYSIS

SEM is used to study the morphologies of the synthesized samples of pure TiO_2 and Y^{3+}-doped TiO_2 nanoparticles prepared using different solvents and all samples were calcined at 400°C and 500°C for 3 h.

TABLE 7.1 Summary of the Properties of Pure TiO_2 and Doped TiO_2 Nanoparticles Samples Calcinated at 400°C and 500°C.

Sample 500°C	Crystalline phase	Angle θ (degree)	*hkl* value	Crystalline size (nm)	*d*-spacing (Å)	Micro-strain ($lines^{-2}/m^4$)	Band gap (eV)	PL peak Intensity (nm)
TiO_2—400°C	Anatase (100%)	25.30	(101)	9.67	3.527	0.0164	3.54	426
YT—400°C	Anatase (100%)	25.28	(101)	8.46	3.530	0.0187	3.12	424
TiO_2—500°C	Anatase (35.06%)	25.28	(101)	29.10	3.530	0.0053	3.33	424
	Rutile (65.94%)	27.42	(110)	38.80	3.259	0.0037		
YT—500°C	Anatase (100%)	25.28	(101)	12.51	3.522	0.0126	2.91	424

Figure 7.3a shows the SEM images of pure TiO_2 and doped TiO_2 observed spherical in nature and well-defined boundary and less aggregation of the particles. When we used rare-earth-doped element show smaller particle size with high crystallinity which is confirmed by XRD result and increases surface area confirmed with the BET result. These results are good in agreement with photodegradation activity. At higher calcination temperature at 500°C show spherical in shape with agglomeration of the particles, this may cause the surface to be nonuniform.

As a result, the particle size of this film increased and the surface area obviously decreased as shown in Figure 7.3c. When we used Y^{3+}-doped TiO_2 nanoparticles we observed that they were in spherical shape with no agglomeration. SEM images show that the particle size of the TiO_2 nanoparticle samples is in the range of nanometer and nanoparticles are homogeneous (Fig. 7.3d). It is because of two reasons, (1) in the sol–gel process, the precursor of titanate reacted with NO^{3-} and obtained $TiO(NO_3)_2$ when doping source $Y\ (NO_3)_3$ was added in. Meanwhile, NO^{3-} can also acts as the action of a chelating agent,[21] both of which can slow down the condensation reaction rate; thereby controlling the nucleation and growth process. (2) Yttrium elements can strongly inhibit the conversion of TiO_2 crystal phase; thus improve the temperatures of phase transformation of TiO_2 from anatase to rutile,[34] and hindered the TiO_2 crystalline growth during calcination.

7.3.3 BET SURFACE AREA

The specific surface area and porosity of the mesoporous anatase TiO_2 nanoparticles were evaluated by using the nitrogen adsorption and desorption isotherms after calcination as shown in Figure 7.4a,d. The isotherm of sample reveals the stepwise adsorption and desorption, indicating the presence of mesoporous material having a three-dimensional (3D) intersection according to the IUPAC classification.

A hysteresis loop with a stepwise adsorption and desorption branch is observed at a wide range of pressure (P/P_0). BET surface area analysis was used to establish the surface area, total pore volume, and average pore diameter of the material. The BET surface area of synthesized pure TiO_2 nanoparticles calcinated at 400°C was found to be 306.73 m^2/g, analyses of pore volume and a pore diameter were found to be 0.836 cc/g and 1.721 nm respectively. Similarly, same preparation condition of pure TiO_2 and doped TiO_2 at 500°C were found to be 149.9 and 257.29 m^2/g, analyses of pore volume 0.0026 and 0.0506 cc/g and pore diameter was found to be 1.891 and 1.678 nm, respectively, as shown in Figure 7.4(a,c,d).

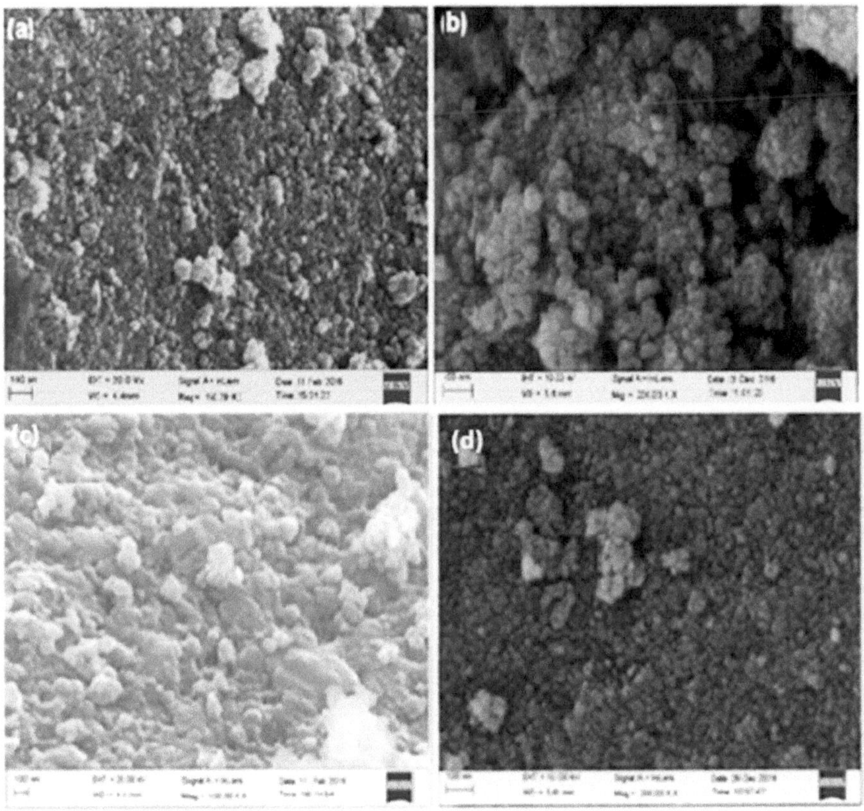

FIGURE 7.3 SEM pattern of TiO_2 nanoparticles samples (a) TiO_2—400°C, (b) 0.2mol% YT—400°C, (c) TiO_2—500°C (d) 0.2mol% YT—500°C nanoparticles.

7.3.4 FOURIER TRANSFORM INFRARED (FTIR) ANALYSIS

FTIR is performed on the sample to identify materials composition and functional group present in the sample. It is fingerprint of all types of organic compounds and some inorganic compounds. FTIR spectra were measured in the spectral range from 400 to 4000 cm^{-1} at room temperature. The FTIR spectrums of the pure TiO_2 and doped TiO_2 sample calcinated at 400°C and 500°C are shown in Figure 7.5a–d. From Figure 7.5a,b, the peak observed at 3200 and 3300 cm^{-1} is characteristic of the –OH (hydroxyl) stretching vibration of absorbed water on the surface. The band at 1630 cm^{-1} is associate with H–O–H bending vibration mode because of hydroxyl group present on the surface of titania and the peak at 618 cm^{-1}, 638 cm^{-1}, strong absorption peak 716 cm^{-1} and additional peak around at 975–993 cm^{-1} at 400°C.

TABLE 7.2 BET Surface Area Analysis Results for Synthesized (a) Pure TiO_2, (b) 0.2 mol% Y^{3+}-doped TiO_2 samples Calcinated at 400°C and 500°C.

Sr. No.	Sample name	BET surface area (in m^2/g)	Pore-size distribution (in nm)	Pore volume (in cc/g)
1	Pure-TiO_2 (400°C)	306.7	1.721	0.836
1	Pure-TiO_2 (500°C)	149.9	1.891	0.0026
2	0.2 mol% YT—500°C	257.29	1.678	0.050

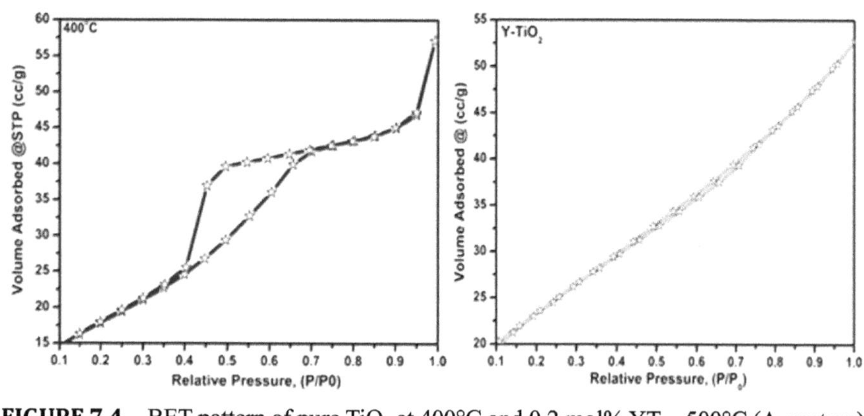

FIGURE 7.4 BET pattern of pure TiO_2 at 400°C and 0.2 mol% YT—500°C (A-anatase).

In Figure 7.5c,d, the absorption spectrum of pure and doped TiO_2 nanoparticles was calcinated at 500°C. The peak at 3304–3300 cm^{-1} characterization of the –OH (hydroxyl) stretching vibration of absorbed water on the surface. The band at 1660 cm^{-1} is assigned to the H–O–H bending vibration mode due to adsorbed water and hydroxyl groups on metal surface. These two absorption bands are attributed to the characteristics vibration of O–H bond in water molecules,[22] and indicating that the drying process at 60°C does not completely draped the water molecules from the pores of titania network. Peak at 1540 cm^{-1} is due to the H–O–H bending of physically absorbed water, indicating the existence of water absorbed on the surface of nanocrystalline titania powders.[23,24] In the spectrum of TiO_2, the peak at 630–411 cm^{-1} shows stretching vibration of Ti–O^{II}, Ti–O^{IV}. The sample exhibits an additional weak band around 993 cm^{-1}. A very strong absorption at 778 cm^{-1} has been observed, which is the characteristic peak of anatase titania due to Ti–O bond.[14] A small hump can be identified around 500 cm^{-1} as the Ti–O–Ti stretching vibration in TiO_2 network. These bonds are important due to their relation to the photocatalytic activity.[25]

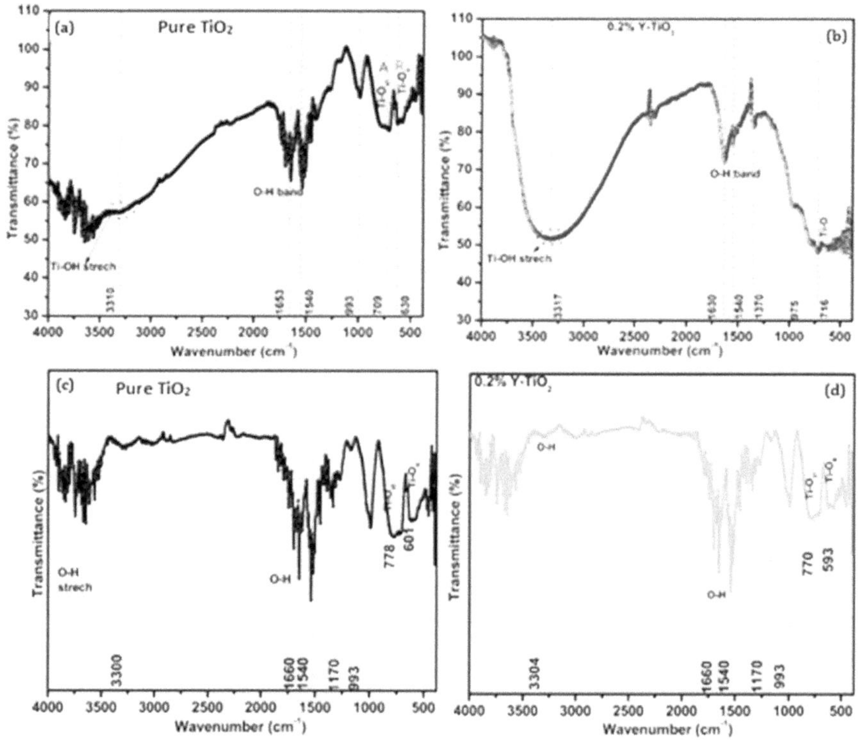

FIGURE 7.5 FTIR spectra graph of (a) TiO_2—400°C, (b) 0.2mol% YT—400°C, (c) TiO_2—500°C (d) 0.2mol% YT—500°C nanoparticles.

7.3.5 MICRO-RAMAN SPECTROSCOPIC ANALYSIS

Raman spectroscopy is a sensitive, fast, and nondestructive method. It is one of the most effective tools for the study of crystallinity and defects structure associated with the materials. The change of Raman spectra is related to nonstoichiometry and photon confinement effect in nanostructured materials. Oxygen deficiency within the material, however, strongly affects the Raman spectrum by producing frequency shift and broadening of some spectral peaks. The crystallite size in the nanoscale range may affect the frequency shift and broadening of Raman peaks due to the phonon confinement. The Raman signal of TiO_2 is very sensitive to the vibrational mode of oxygen ions of the Ti–O bond. The strength of the Raman signal depends upon the polarization of oxygen ions surrounding Ti^{3+} in the basic TiO_6 octahedral. The presence of oxygen vacancies affect the vibrational of Ti–O bond and the direct consequence of these vacancies can be revealed in the intensity, position, and width of the Raman signal.

The pure TiO_2 and Y^{3+}-doped TiO_2 nanoparticles with 0.2 mol% concentration were characterization by micro-Raman spectroscopy as shown in Figure 7.6a–d and it gave phase complimentary analysis. The group theoretical analysis gives that TiO_2 anatase phase contains six Raman active modes (A_{1g} + $2B_{1g}$ + $3E_g$) three infrared (IR) active modes (A_{1g} and $2E_g$), and a silent B_{1g} mode. We are emphasizing only Raman active modes. The Raman spectra of single crystal anatase phase were investigated by Huarong Zhang who concluded that six allowed modes appear at 141 cm^{-1} (E_g) 197 cm^{-1} (E_g) 399 cm^{-1} (B_{1g}), 513 cm^{-1} (A_{1g}) 519 cm^{-1} (B_{1g}), and 639 cm^{-1} (E_g).[26]

In Figure 7.6a,d, it is observed that nanoparticles have three E_g modes, appeared at 141, 196, and 639 cm^{-1}, whereas one B_{1g} mode has appeared at 394 cm^{-1} and another mode B_{1g} has overlapped with A_{1g} at 515 cm^{-1} as they are very close to each other. We observed mixed phase of anatase and rutile at 500°C for pure TiO_2 sample. In the present Raman spectra, the absence of Raman active modes at 612, 443 (intense line), and 235 cm^{-1} which corresponding to the rutile phase of titania, reconfirms the anatase phase purity of synthesized for pure TiO_2 nanoparticle at 400°C and for Y^{3+}-doped TiO_2 nanoparticle at 500°C. In the case of 0.2 mol% Y^{3+}-doped TiO_2 nanoparticles calcinated at 400°C and 500°C, observed active three modes, appeared at 141 cm^{-1} (E_g), 396 cm^{-1} (B_{1g}), and 638 cm^{-1} (E_g) which confirmed anatase phase.[27,28] However, the most intense E_g, this is also consistent with XRD result.

7.3.6 UV-VISIBLE ANALYSIS

To examine the optical properties, energy band gap, the synthesized pure, and Y^{3+}-doped TiO_2 nanoparticles with different calcinated temperature were characterized by UV–visible spectroscopy at room temperature and observed all results shown in Figure 7.7. The optical properties of semiconductor are generally determined by its electronic structure, so the essence of light absorption will be found by investigating the relationship between electronic structure and optical structure of titania. We have explored the optical properties and analyzed the energy band gap of pure and surfactant-assistant TiO_2 and yttrium-doped TiO_2 nanoparticles calcinated at 400°C and 500°C using UV–visible absorption spectroscopy shown in Figure 7.7a–d. The absorption can vary depending in material prepared, etc. The band gap energy (E_g) have calculated direct band gap (E_g), with the help of Tauc's relation[29]

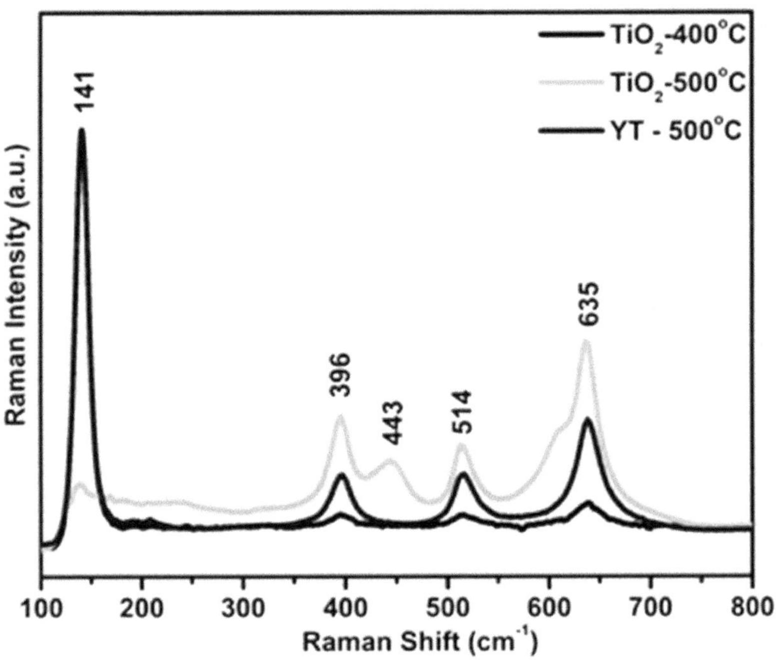

FIGURE 7.6 Raman spectra of (a) TiO_2—400°C, (c) TiO_2—500°C (d) 0.2mol% YT—500°C nanoparticles.

$$\alpha h\nu = A(h\nu - E_g)^n \quad (7.5)$$

where α is the absorption coefficient, A is a constant, and n = ½ for direct band gap n = 2 for indirect allowed transition of the semiconductor. An extrapolation of the liner region to $\alpha = 0$ of a plot of $(\alpha h\nu)^2$ vs $h\nu$ gives the value of the optical band gap E_g.

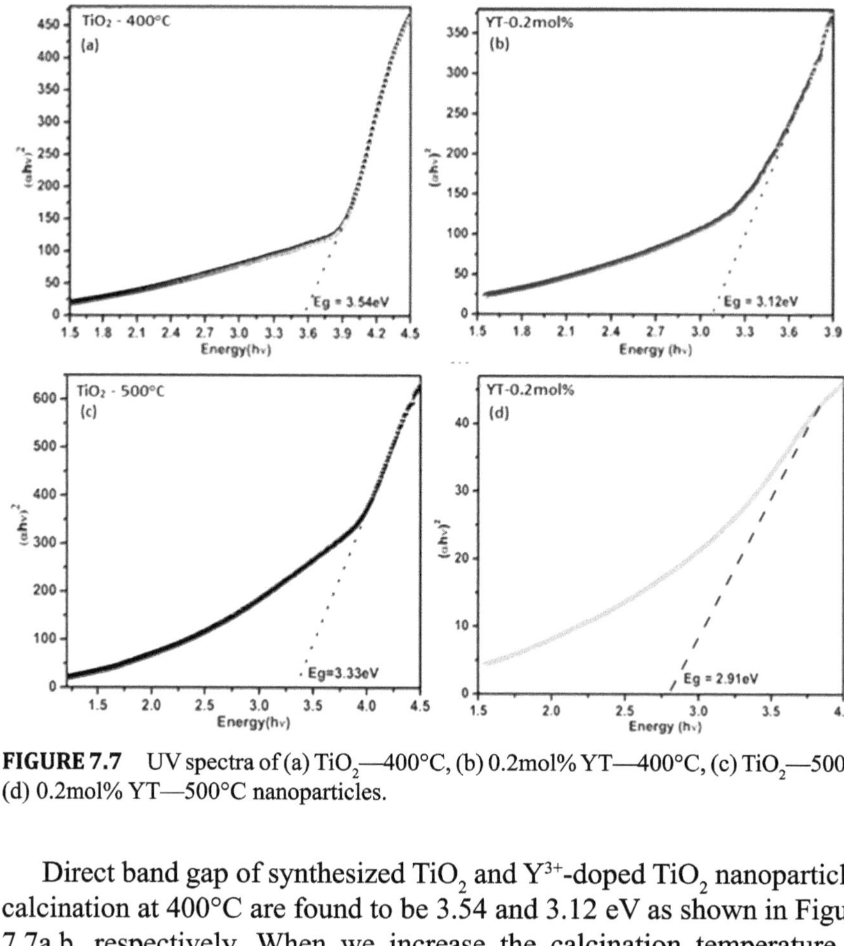

FIGURE 7.7 UV spectra of (a) TiO_2—400°C, (b) 0.2mol% YT—400°C, (c) TiO_2—500°C (d) 0.2mol% YT—500°C nanoparticles.

Direct band gap of synthesized TiO_2 and Y^{3+}-doped TiO_2 nanoparticles calcination at 400°C are found to be 3.54 and 3.12 eV as shown in Figure 7.7a,b, respectively. When we increase the calcination temperature at 400°C of pure and doped TiO_2 nanoparticles, the energy band gap of 3.33 and 2.91 eV, respectively, are observed as shown in Figure 7.7c,d. The literature reports the value of 3.0 eV for pure rutile phase titania and absorption decreased in band gap with increasing Y^{3+} doping concentration.[11]

The kind of red shift in absorption spectra observed in literature with the decrease in particle size has been reported quantum size effect for semiconductor for the titania particle size below 10 nm. It may also be due to strong Coulomb interaction effect on the optical process involving the doped due to limited solubility in titania host matrix. The absorption spectra of pure and doped TiO_2 nanoparticles acquire the ultraviolet range of photon energy. In Y^{3+}-doped TiO_2, the absorption spectra are observed in the UV region and energy band gap found to be 2.91 eV. The ionic radius of Y^{3+} is 0.088 nm, which is much larger than that of Ti^{4+} (0.068 nm) thus 100% anatase phase at higher temperature treatment.

7.3.7 PHOTOLUMINESCENCE SPECTROSCOPY

The observed optical properties of the synthesized pure and doped TiO_2 nanoparticles were investigated by room temperature PL in the range of 350–700 nm and all results are shown in Figure 7.8. The emission spectrum (424–493 nm) excited by 350 nm wavelength gives half a dozen peaks at various wavelengths. During the PL process, both radiative and nonradiative combinations happen. PL spectroscopy is an effective method to study the electronic structure and optical properties of nano-semiconductor materials, it is possible to obtain the information of optical carriers' migration, capture and composite, and surface oxygen defects. The PL technique can be used to determine impurity levels, defects, and recombination process in the semiconductor materials.[30] Figure 7.8a–d, shows the PL spectra graph of pure TiO_2 and 0.2 mol% Y^{3+}-doped TiO_2 nanoparticles and all sample calcined at 400°C and 500°C temperature showing the PL spectrum intensity has sharp peaks due to well-defined lattices site. All graphs of TiO_2 samples calcined at 400°C and 500°C, which exhibit a broad luminescence band due to the agglomerated lattice sites. The result indicated that the PL intensity has significantly increased when sample is prepared using Y^{3+} is doped. When TiO_2 nanoparticle is excited at 350 nm, a strong and broad emission peaks are observed at 424 nm. Other emissions around 452, 469, 483, and 493 nm are having less intensity as compared to the peak 424 nm. All these peaks belong to the free exciton luminescence. The emission at 483 nm is due to O^{2-} vacancies and appeared due to bound exciton luminescence. In this case, the violet emission is attributed because of the recombination of an electron and hole in the valance band. The PL peaks exhibited consistency in samples with Y^{3+} doping without forming any new PL feature peaks, confirming no further splitting in energy level. But the height of all individual peaks is significantly increased for both samples in comparison to pure sample showing the increase of recombination resulting from increased surface defects. When Ti^{4+} is replaced with Y^{3+}, an additional electron is injected from O_2.

Whereas the PL intensity of the nanoparticles synthesized using yttrium doped at lower temperature same concentration is slightly lower than those of the Y^{3+}-doped probably due to the existence of some defects in the crystals.[31–34]

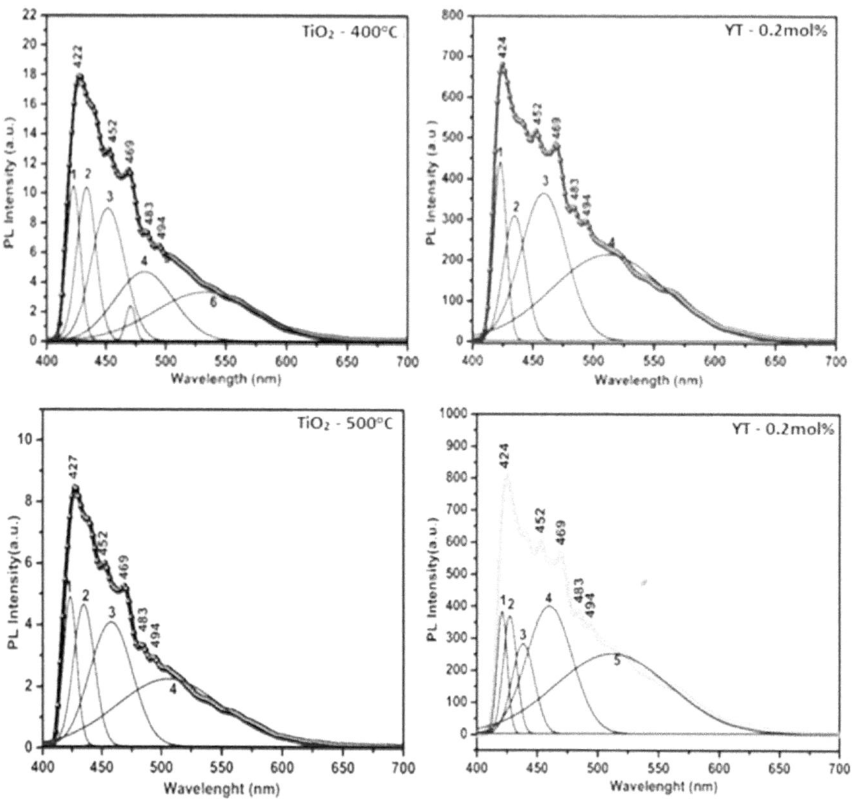

FIGURE 7.8 PL spectroscopy emission spectra of (a) TiO_2—400°C, (b) 0.2mol% YT—400°C, (c) TiO_2—500°C (d) 0.2mol% YT—500°C nanoparticles.

7.4 CONCLUSION

TiO_2 nanoparticles have been successfully synthesized by nonaqueous sol–gel technique at room temperature using (1) pure) and (2) rare-earth ion Y^{3+}doping followed by calcination at 400°C and 500°C temperature.

Y^{3+}-doped TiO_2 nanoparticles show less agglomeration of particles, reduced particle size, geometrical shape, increased surface area, and also control the phase transformation which is justified by X-ray diffraction pattern and SEM morphology. The Y^{3+}-doped TiO_2 samples give pure anatase phase and improve material crystallinity. While the pure and doped TiO_2 nanoparticles, calcined at 500°C, show mixed phase of anatase (35.06%) and rutile (65.94%) for pure TiO_2 and 0.2 mol% doped TiO_2 show 100% pure anatase phase. SEM images show that the particles of pure and doped samples calcined at 400°C and 500°C temperatures are nearly spherical in shape. All the TiO_2 samples show absorption in the ultraviolet region. Photoluminescence intensity of TiO_2 nanoparticle shows sharp peaks with reduced recombination rate, which envisioned good photocatalytic activity. Y^{3+}-doped TiO_2 nanoparticles at 400°C show good photocatalytic activity as compared to yttrium-doped element at 500°C because of lower PL intensity.

ACKNOWLEDGMENTS

Authors are pleased to acknowledge the Director, MANIT, Bhopal, for providing necessary facilities for this work, and the Director, UGC-DAE-Consortium for Scientific Research, Indore Centre (M.P.), India for XRD, FTIR, and Micro-Raman characterization facility.

COMPETING INTERESTS

The authors declare that they have no competing interests and are alone responsible for the content and the writing of this chapter.

KEYWORDS

- **TiO_2 nanoparticles**
- **Y^{3+} doping element**
- **anatase and rutile phase of TiO_2**
- **sol-gel technique**

REFERENCES

1. Duong, T.; Phan, N.; Dinh Pham, H. Surfactant Removal from Mesoporous TiO_2 Nanocrystals by Supercritical CO_2 Fluid Extraction. *J. Indust. Eng. Chem.* **2010,** *16*, 823–828.
2. Haque, F. Z.; Nandanwar, R.; Singh, P. Evaluating Photodegradation Properties of Anatase and Rutile TiO_2 Nanoparticles for Organic Compounds. *Optik* **2017,** *128*, 191–200.
3. Nandanwar, R.; Singh, P.; Haque, F. Z. Synthesis and Characterization of SiO_2 Nanoparticles by Sol-Gel Process and Its Degradation of Methylene Blue. *Am. Chem. Sci. J.* **2015,** *1* (5), 1–10.
4. Pichat, P.; Disdier, J.; Hoang-Van, C.; Mas, D.; Goutailler, G.; Gaysse, C. Purification/Deodorization of Indoor Air and Gaseous Effluents by TiO_2 Photocatalysis. *Catalys. Today* **2000,** *63* (2–4), 363–369.
5. Hisanaga, T.; Tanaka, K. Photocatalytic Degradation of Benzene on zeolite-Incorporated TiO_2 Film. *J. Hazard. Mater. B.* **2002,** *93* (3), 331–337.
6. Demeestere, K.; Dewulf, J.; Langenhove, H. V.; Sercu, B. Gas–Solid Adsorption of Selected Volatile Organic Compounds on Titanium Dioxide Degussa P25. *Chem. Eng. Sci.* **2003,** *58*, 2255–2267.
7. Paul, T.; Miller, P. L.; Strathmann, T. J. Visible-Light-Mediated TiO_2 Photocatalysis of Fluoroquinolone Antibacterial Agents. *Environ. Sci. Technol.* **2007,** *41* (13), 4720–4727.
8. Visa, M.; Carcel, R. A.; Andronic, L.; Duta, A. Advanced Treatment of Wastewater with Methyl Orange and Heavy Metals on TiO_2, Fly Ash and Their Mixtures. *Catal. Today*. **2009,** *144*, 137–142.
9. Xu, A W.; Gao Y, Liu H Q. The Preparation, Characterization, and Their Photocatalytic Activities of Rare-Earth-Doped TiO_2 Nanoparticles. *J. Catal.* **2002,** *207* (2), 151–157.
10. Xinshu, N.; Sujuan, L.; Huihui, C.; Jianguo, Z. Preparation, Characterization of Y^{3+}-Doped TiO_2 Nanoparticles and Their Photocatalytic Activities for Methyl Orange Degradation. *J. Rare Earth* **2011,** *29* (3), 225.
11. Bamne, J.; Sharma, P. K.; Haque, F. Z. Effect of Solvent Mixing and Calcination Temperature on the Growth of TiO_2 Nanoparticle Prepared via Sol–Gel Method. *J. Mater. Focus* **2018,** *7*, 1–10.
12. Park, H.; Park, Y.; Kim, W.; Choi, W. Surface Modification of TiO_2 Photocatalyst for Environmental Applications. *J. Photochem. Photobiol. C: Photochem. Rev.* **2012.**
13. Akpan, U.G; Hameed, B. H. A Review: The Advancements in Sol–Gel Method of Doped-TiO_2 Photocatalyst. *J. Appl. Cataly. A: Gen.* **2010,** *375*, 1–11.
14. Nandanwar, R.; Singh, P.; Haque, F. Z. Effect of Solvents on the Morphological and Optical Properties of SiO_2 Film Prepared by Sol-Gel Method. *Int. J. Chem. Phys. Sci.* **2014,** *3* (3), 40–45.
15. Nandanwar, R.; Singh, P.; Haque, F. Z. Preparation of TiO_2/SiO_2 Nanocomposite with Non-Ionic Surfactants via Sol-Gel Process and Their Photocatalytic Activity. *Orient. J. Chem.* **2014,** *30* (4), 1–8.

16. Bamne, J.; Taiwade, K.; Sharma, P. K.; Haque, F. Z. Effect of Calcination Temperature on the Growth of TiO_2 Nanoparticle Prepared via Sol-Gel Method Using Triton X-100 as Surfactant. *AIP Conf. Proc.*, **2018,** *2039* (1), 020076.
17. Joseph, K.; Raj, A.; Visvanathan, B. Temp Effect of Surface Area, Pore Volume and Particle Size of P25 Titania on the Phase Transformation of Anatase to Rutile. *Indian J. Chem.* **2009,** *48A*, 1378–1382.
18. Siddiqui, H.; Parra, M. R.; Qureshi, M. S.; Malik, M. M.; Haque, F. Z. Studies of Structural, Optical, and Electrical Properties Associated with Defects in Sodium-Doped Copper Oxide (CuO/Na) Nanostructures. *J. Mater. Sci.* **2018,** *53*, 8826–8843.
19. Parra, M. R.; Haque, F. Z. Poly (Ethylene Glycol) (PEG)-Assisted Shape-Controlled Synthesis of One-Dimensional ZnO Nanorods, *Optik* **2015,** *126*, 1562–1566.
20. Singh, N.; Haque, F. Z. Synthesis of Zinc Oxide Nanoparticles with Different pH by Aqueous Solution Growth Technique. *Optik* **2016,** *127*, 174–177.
21. Crisan, M.; Malina, R.; Dragan, N.; Crisan, D. Sol–Gel Iron-Doped TiO_2 Nanopowders with Photocatalytic Activity. *Appl. Catal. A-Gen.* **2015,** *504*, 130–133.
22. Haque, F. Z.; Nandanwar, R.; Singh, P.; Evaluating Photodegradation Properties of Anatase and Rutile TiO_2 Nanoparticles for Organic Compounds. *Optik* **2017,** *128*, 191–200.
23. Ruifen, W.; Fuming, W.; Shengli, A. N.; Jnling, S. Y/Eu Co-Doped TiO_2: Synthesis and Photocatalytic Activities under UV-Light. *J. Rare Earths* **2015,** *33*, 154.
24. Nithya, N.; Bhoopathi, G.; Magesh, G.; Balasundaram, O. N. Synthesis and Characterization of Yttrium Doped Titania Nanoparticles for Gas Sensing Activity. *Mater. Sci Semicond. Process.* **2019,** *99*, 14–22.
25. Tang, J.; Chen, X.; Liv, Yu.; Gong, W.; Peng, Z.; Cai, T. Europium-Doped Mesoporous Anatase with Enhanced Photocatalytic Activity toward Elimination of Gaseous Methanol. *J. Phys. Chem. Solids* **2012,** *73*, 198–203.
26. Haque, F. Z.; Nandanwar, R.; Singh, P.; Dharavath, K.; Syed, F. F. Effect of Different Acids and Solvents on Optical Properties of SiO_2 Nanoparticles Prepared by the Sol-Gel Process. *Silicon* **2016,** *10*, 413.
27. Zhang, H. Tan, K.; Zheng, H.; Gu, Y.; Zhang, W. F. Preparation, Characterization and Photocatalytic Activity of TiO_2 Co-Doped with Yttrium and Nitrogen. *Mater. Chem. Phys.* **2011,** *125*, 156–160.
28. Kallel. W.; Bouattour, S. Y-Dy Doped and Co-Doped TiO_2. Enhancement of Photocatalytic Activity under Visible Light Irradiation. *Physicochem. Probl. Miner. Process* **2017,** *53* (1), 427–442.
29. Khan, M.; Cao, W. Preparation of Y-doped TiO_2 by Hydrothermal Method and Investigation of Its Visible Light Photocatalytic Activity by the Degradation of Methylene Blue. *J. Mol. Catal. A Chem.* **2013,** *376*, 71–77.
30. Siva Kumar, K.; Song, C-G.; Myeon Bak, G.; Gaeun, H.; Seong, M-J.; Yoon, J-W. Phase Control of Yttrium (Y)-Doped TiO_2 Nanofibers and Intensive Visible Photoluminescence. *J. Alloys Compounds.* **2014,** *617*, 683–687.
31. Singh, N.; Umar, A.; Singh, N.; Fouad, H.; Alothman, O. Y.; Haque, F. Z. Highly Sensitive Optical Ammonia Gas Sensor Based on Sn Doped V_2O_5 Nanoparticles. *Mater. Res. Bull.* **2018,** *108*, 266–274.

32. Gupta, S. M.; Tripathi, M. A review of TiO_2 nanoparticles. *J. Chinese Sci. Bull.* **2011,** *56* (16), 1639–1657.
33. Singh, N.; Pandey, V.; Singh, N.; Malik, M. M.; Haque, F. Z. Application of TiO_2/SnO_2 Nanoparticles in Photoluminescence Based Fast Ammonia Gas Sensing. *J. Opt.* **2017,** *46* (3), 199–203.
34. Kumar, K. S.; Song, C. G.; Bak, G. M.; Heo, G.; Seong, M. J.; Yoon, J. W. Phase Control of Yttrium (Y)-Doped TiO_2 Nanofibers and Intensive Visible Photoluminescence. *J Alloy Compd.* **2014,** *617*, 683–687.

CHAPTER 8

Gas Sensing through Photluminescence Method Using Cr_2O_3 Nanostructures for NH_3

NITU SINGH[1], NEHA SINGH[1], JYOTI BAMNE[1,3], K. M. MISHRA[2], VIVEK CHANDEL[1], and FOZIA Z. HAQUE[1,*]

[1]*Optical Nanomaterials Lab, Department of Physics, Maulana Azad National Institute of Technology, Bhopal, Madhya Pradesh, India*

[2]*Department of Physics, D.D.U. Gorakhpur University, Gorakhpur, Uttar Pradesh, India*

[3]*Department of Physics, Barkatullah University, Bhopal, Madhya Pradesh, India*

[*]*Corresponding author. E-mail: foziazia@rediffmail.com*

ABSTRACT

In this research article, enhanced optical sensing response for ammonia gas by pure and Li-doped Cr_2O_3 nanoparticles (NPs) is reported. Doped and pure chromium oxide NPs were synthesized using the co-precipitation method. The physical characterization of synthesized NPs was carried out using X-ray diffraction (XRD), atomic force microscopy (AFM), Transmission electron microscopy (TEM), and photoluminescence (PL) spectroscopy. The XRD patterns of so produced NPs confirm the formation of a single phase with rhombohedral structure. The average crystallite size calculated by Debye–Scherer equation for these NPs was found to be

Advanced Microscopy: A Strong Analytical Tool in Materials Science. Merin Sara Thomas, Jozef T. Haponiuk, Sabu Thomas, & Anne George (Eds.)

around 34.97–35.68 nm. AFM (grain size) and TEM results are consistent with the XRD results. The morphology of the obtained NPs shows the irregular shape and agglomerated particles. Thin films of synthesized powders were developed onto quartz substrate and examined for ammonia optical sensing at room temperature. The PL spectroscopy results showed that the pure and doped Cr_2O_3 NPs depicted strong emission ranging violet to green region. The highest value of sensitivity for 45 ppm ammonia is 89.53% observed for 2 wt.% Li-doped chromium oxide NPs.

8.1 INTRODUCTION

Optical sensors based on metal oxides have received a great attention in the present time they can monitor low concentrations of certain gases by a change of peak intensity of a metal oxide thin layer. Over the last few years, substantial attention has been focused on the use of semiconductor metal oxides (SMOs) for optical gas sensing application. Chromium oxide (Cr_2O_3), a wide bandgap p-type semiconductor (Eg = 3.5 eV), has many applications including gas sensors, protective coatings, cathodes for lithium cells, etc. both in doped and pure forms.[1–4] There is also exceeded interest in the synthesis of wide bandgap semiconductors, as they have many applications in optical and electronic devices.[5]

Chromium oxide has extensive applications such as catalyst, dye and pigment, storage, wear resistance materials, advanced colorants, digital recording system, solar energy application, coating materials for thermal protection, chemo-resistive gas sensor, humidity sensor, and electrochromic materials.[5–10] Many synthesis procedures have been described in the literature for the synthesis of Cr_2O_3 nanoparticles (NPs) as hydrothermal reduction,[11] sol–gel,[12] solution-combustion method,[13] participation-gelation,[14] urea-assisted homogeneous precipitation,[15] gas condensation,[16] and microwave plasma.[17] Cr_2O_3 is a promising wide gap p-type semiconductor owing to its corundum structure and good catalytic activity toward various gases.[17–19] Precipitation is a significant method for optical gas sensing, as this method is environmentally friendly and less expensive than other methods.

The SMO optical gas sensing devices have several unique advantage such as a chip, convenient fabrication, small size, and extremely low detection limits (<ppm levels). In this chapter, pure and Li-doped Cr_2O_3 NPs are prepared successfully by a simple co-precipitation method and then characterized for further studies. The detailed sensing studies revealed

that the optical sensitivity at 2.0 wt.% Li-doped Cr_2O_3 NPs is 89.53% for 45 ppm concentration of NH_3 when carried at room temperature. Also, an increase in optical sensitivity is observed with increasing doping concentration.

In principle, gas detection with SMO-based gas sensors is quite simple. There is adsorption of oxygen by trapping electrons residing at the surface of the metal oxides, as a result of which the resistance of the sensor increases (for n-type layer), or decreases (for p-type). Many n-type or p-type metal oxides are found appropriate for sensing combustible, both gases (reducing or oxidizing) by conductive measurements. The most important part of the sensor assures main four parameters selectivity, speed, sensitivity, and stability. The activity of the sensor is determined by measuring the changes in the physical condition (resistivity/conductivity) or chemical constitutes occurring as a result of interaction between target gas adsorbed and surface atoms (O^-, O^{2-}, H^+, and OH^-) at a specific at room temperature.[20–22]

There are two main types of SMO-based sensors, including n-type in which majority carriers are electrons and p-type in which majority carriers are holes.[21–23] The chemo-resistance is the working principle of metal oxide gas sensor. A change in electrical conductivity or resistivity of thin films occurs when a gas molecule interacts with metal oxides.[24] The width of the space charge region formed on the crystallites changes due to the transfer of electrons during the adsorption and desorption of gas molecules resulting in chemo-resistivity of the metal film. The wider the space charge region, larger is the potential barrier for the conduction process between the grains. This changes the position of the Fermi level and subsequent band bending. Because of advantages such as low fabrication expense, easy preparation, compact size, and simple measuring electronics, metal oxide layers are widely used in portable gas detection systems. However, the morphology and composition of sensing materials have a significant impact on the performance of such sensors. Due to this, gas sensors based on bulk materials or dense films cannot achieve high sensitivity. Summarizing, the response of a metal oxide gas sensor to a given gas is determined by the speed of the chemical reaction on the sensor's surface as well as the speed of gas molecules diffusion to the surface. The sensor's response is limited by the speed of the chemical reaction at low temperatures while at high temperatures it is restricted by the speed of diffusion of gas molecules. Therefore, at a certain intermediate temperature, the speed of the two processes becomes equal and at that point, the sensor's response reaches the maximum.[26]

The following oxides appearance a gas response in their conductivity: Cr_2O_3, In_2O_3, WO_3, TiO_2, V_2O_3, Fe_2O_3, Mn_2O_3, Co_3O_4, NiO, CuO, MoO_3, Nd_2O_3, Ta_2O_5, La_2O_3, CeO_2, GeO_2, and Nb_2O_5.[28] Metal oxides selected for gas sensors are based on their electronic structure. Metal oxides are divided into the following categories[27] as tabulated in Table 8.1.

1. Transition-metal oxides (Fe_2O_3, NiO, Cr_2O_3, etc.).
2. Non-transition-metal oxides,
   a) pre-transition metal oxides and
   b) post-transition-metal oxides (ZnO, SnO_2, In_2O_3, etc.).[28]

Metal oxide gas sensors use an appropriate material, either in bulk form or thin film, as a gas sensing element. The working principle for gas detection of these sensors based on changes in-

i) **Work function:** Work function is a simple and fast technique to monitor the variation of surface potential during adsorption and desorption of gasses from the environment.
ii) **Resistance:** Gas sensor's response is measured with an automatic logging meter, which records the values of the resistance of thin film with time.
iii) Mass of the sensing element due to adsorption of gas.

According to the definition of a gas sensor (chemo-resistive), chemical sensor is a device that transforms the change in chemical structure of adsorbent due to the adsorption of a target gas into an electrical signal. Classification of chemical sensor is given below.

Important of nanostructured films: In nano-sized (nanoribbon, nanorods, nanowire) due to higher surface to volume ratio, higher efficiency is obtained than traditional bulk material based devices.

A—Physical Sensor: Physical sensors measure parameters such as temperature that regulate the transfer of flux of mass or energy. They are commonly used in the study of lakes and other freshwater resources.

B—Chemical sensor: A chemical signal is observed upon the interacting of the analyte to the recognition element. A transducer then results this signal into a quantifiable output signal.

C—Biochemical sensor: A biochemical sensor converts a chemical (or biological) quantity into an electrical signal. Basic components of the sensor include analyte molecule, chemically sensitive layer, and transducer.

The resultant change in physico-chemical properties is measured to detect the presence and percentage of a particular gas. Conventional sensors use bulk or thick films as a gas sensing active material. However, nanostructured thin films offer better control on gas sensing properties of a material due to large surface to volume ratio. According to the material used, there are three types of gas sensors based on thin films:

a. Metal oxides thin-film sensors (SnO_2, ZnO, Ga_2O_3, etc.).
b. Catalytic metal thin-film sensors (Pd and Pt).
c. Special class of organic material thin-film sensor (phthalozyanine, poly-pyrol).

Catalytic sensors impose selectivity, compatibility, and short lifetime issue of organic materials films makes metal oxide thin-film sensors being used for the commercial purpose.[29]

8.1.1 GAS SENSING MECHANISM IN METAL OXIDES

The gas detection in metal oxide sensors is based on the variations of the depletion layer at the grain boundaries in the presence of reducing or oxidizing gases. This leads to a change in the energy barrier for free charge carrier to flow. This leads to a change in the conductivity of the sensing material. The active sensing layer of sintered and thick-film type gas sensors contains numerous interconnected metal oxide grains. As shown in Figure 8.1, the adsorption of O_2 removes the charge carriers (electrons) from the surface region of the grains in forms oxygen ion species and its depending on the temperature, and forms a depletion layer around the grain boundaries. The depth of the depletion layer is determined by the oxygen partial pressure and the surface characteristics of metal oxide materials.

Figure 8.1 shows the schematic diagram of band bending after chemisorptions of charged species (oxygen are three species O_2^-, O^- and O^{2-}) when O_2 molecules are adsorbed on the surface of metal oxides. They will extract electrons from the conduction band (E_C) and trap them in the form of ions at the material's surface, thus electron transferred from conduction band to valance band region. O^- is believed to be a key species at the operating temperature of 300°C and 450°C.[30,31] When gas molecules interact with the surface complexes of the active material of the sensors, they can induce a charge in several physical properties of the material, such as the charge-carriers' density, including variation in the electrical and optical emission. When the sensor is exposed to combustible gases, the surface oxygen species react to form combustion products. This reaction lowers the oxygen species coverage, returns the free electron charge carriers to the bulk of the oxide material and increases electrical conductivity. This surface reaction modified conductance serves as a gas response signal of the sensor. In Figure 8.1, E_C denotes the conduction band, E_V denotes valence band, E_F denotes the Fermi level, and air is the thickness of the space-charge layer and eV surface is the Fermi potential barrier. The conducting electrons are seen as e^- and + as the donors.

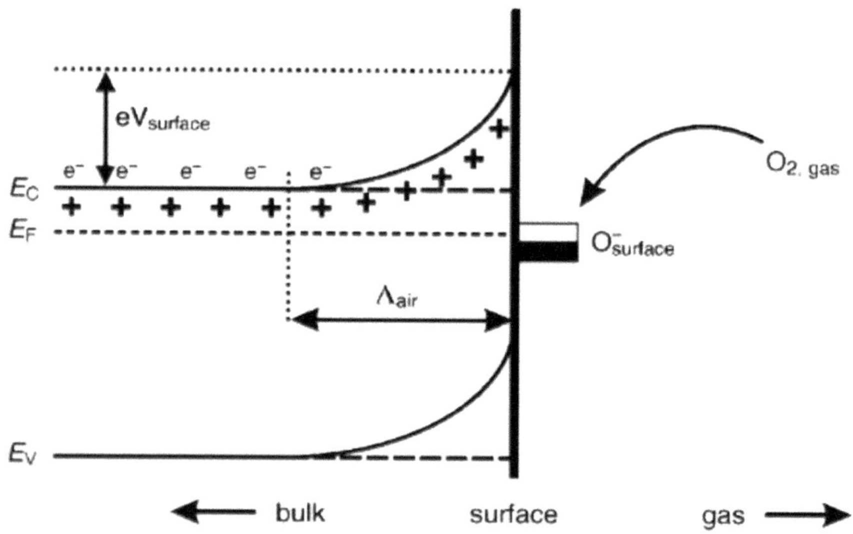

FIGURE 8.1 A schematic diagram of band bending after chemisorptions of charged species.

8.2 GAS SENSING METHODS

There are various methods for gas sensing; we will discuss only two methods for comparison.

8.2.1 PHOTOLUMINESCENCE OPTICAL GAS SENSING

Remote and noninvasive measurements, as well as operating in dangerous conditions, are all possible with optical sensing techniques. Over the last few decades, scientists have been drawn to this technology because of these and other important advantages over other detection technologies, and hence this technology has experienced a high development. Optical sensors can be based on different transduction mechanisms, such as absorption, resonance, or photoluminescence (PL). The latter consists of emission of light by a material as a consequence of its previous absorption at lower wavelengths. Luminescence is known as fluorescence or phosphorescence depending on the lifetime of the emission. The intensity of this emission, as well as its lifetime, can be quenched or enhanced by the variation of different external parameters: temperature, oxygen, or metal ion concentration. The use of external parameters to modulate the intensity has been widely used in the development of luminescence-based sensors. Recently, the optical gas sensing method has shown a significant interest compared to other competitive technologies owing to its safety, stability, sensitivity, and low power consumption. The majority of optical gas sensors produced till date have complex structures or are based on optical fibers that are difficult to integrate with electronic circuits. Simple and inexpensive optical gas sensors, on the other hand, have picked the interest of the gas sensing community because they can gracefully suit a wide range of applications. PL allows for the development of simple, compact, and high-sensitivity sensor methods shown in Figure 8.2. Optical gas sensing is characterized into two categories; the first one is based on optical adsorption when NH_3 is exposed to the light source such as lasers.[32,33] And, the second one is based on a change in absorption/emission spectrum of active material when ammonia reacts with it. The charge transfer from oxygen to chromium ions is responsible for the emission of light.[34–36]

The optical sensing response or sensitivity of the sensing element is presented by sensitivity formula for reducing ammonia gas.[37,38]

$$S\% = \frac{Ig - Ia}{Ig} \times 100 \; I_g > I_a \tag{8.1}$$

where I_g is the intensity of material in the presence of gas and Ia is the intensity of material in presence of air at room temperature.

8.3 EXPERIMENTAL DETAILS

8.3.1 SYNTHESIS OF MATERIALS

All chemicals were of the high purity and analytic reagent (AR) grade. Chromium sulfate $Cr_2(SO_4)_3{\cdot}6H_2O$ (Merck), Lithium oxide Li_2O (Merck),

TABLE 8.1 Comparison of the Sensing Performance of Various Nanoparticles Based NH_3 Optical and Resistive Gas Sensors.

Material	Structure	Sensing method	Temp. (°C)	Concen. (ppm)	Sensitivity	Response time(s)	Recovery time(s)	Ref.
SnO_2	Nanoparticles	PL	RT	25	–	14	11	[35]
TiO_2/SnO_2	Spherical Nanoparticles	PL	RT	20	63	12	50	[39]
SnO_2/SnS_2	Nanotubes	Resistive	RT	100	2.48 (I_g/I_a)	21	110	[40]
MoO_3	Nanoparticle	Resistive	RT	500	69% ($R_{a-}R_g$)/R_a	60	180	[41]
V_2O_5	Nanofiber	Resistive	RT	100	–	100	180	[42]
Co_3O_4/ SnO_2	Nano-spheres	Resistive	200°C C	50	13.6	4	17	[43]
Pt/SnO_2	Thin film	Resistive	230°C	450	25.7	1	–	[44]
SnO_2-SnS_2	Nano-sheets	Resistive	RT	100	2.0	200	200	[45]
In_2O_3/PANI	Nanofibers	Resistive	RT	1000	1.2	500	500	[46]
CeO_2@Pani	Particles	Resistive	RT	2	7.5%	400	600	[47]
Pd/SnO2/RGO	Thin film	Resistive	25°C	100	19.6%	>300	>500	[48]
V_2O_5	Thin film	Resistive	RT	10	60	300	–	[49]
WO3	Nanorods	Resistive	180	100	57	60	–	[50]
TiO_2	Spherical Nanoparticles	PL	RT	25	–	07	05	[51]
SnO_2:Mg	Nanoparticles	PL	RT	80	–	17	49	[38]
SnO_2:Ni	Nanoparticles	PL	RT	10	–	15	30	[38]

RT, room temperature; PL, photoluminescence.

Ammonium hydroxide NH_4OH (CDH), and distilled water were used as precursor materials.

Chromium oxide (Cr_2O_3) NPs were synthesized by the co-precipitation method. A solution was prepared by mixing 0.1 M $Cr_2(SO_4)_3 \cdot 6H_2O$ in 100 mL distilled water followed by sonication for 15 min at room temperature, this solution was then continuously stirred for 30 min. The Lithium dopant was added in (0.5, 1.0, 1.5, and 2.0 wt.%) chromium sulfate form to the solution. Again the final solution was sonicated for 10 min at room temperature. Then Aqueous ammonia (NH_4OH) was added dropwise along with constant stirring until the pH of the solution reached to 9>10. After stirring, the obtained solution was aged at room temperature for 12 h. The precipitates, thus obtained were filtered and washed several times with ethanol and distilled water to remove impurities. The precipitates were dried in oven at 70°C for 36 h and annealed at 500°C for 6 h. The samples thus obtained were used for further characterization.

8.3.2 THIN FILM PREPARATION FOR OPTICAL SENSOR

Optical sensors were developed from the self-standing pure and doped Cr_2O_3 film by cutting 0.75 × 1.5 cm pieces. In order to promote the formation of thin films with good integrity, Cr_2O_3 pure and doped was dissolved in acetone as a solvent. The resulting solution of Cr_2O_3 NPs is sonicated for 15 min at room temperature then thin films were prepared by spin coating method at room temperature. The film made at room temperature was kept for 60 min to remove impurities/Moistures in the oven for 1 h 30 min. The fabricated pure and doped chromium oxide thin films in this device configuration work as an optical sensor.

8.3.3 GAS SENSING SETUP

Photoluminescence Ammonia Gas Sensing Study

Figure 8.2 shows the optical ammonia gas sensor setup, the gas chamber has an inlet to which a gas source is attached, an outlet with a gas collector and a thin-film holder at the bottom. This gas chamber was connected

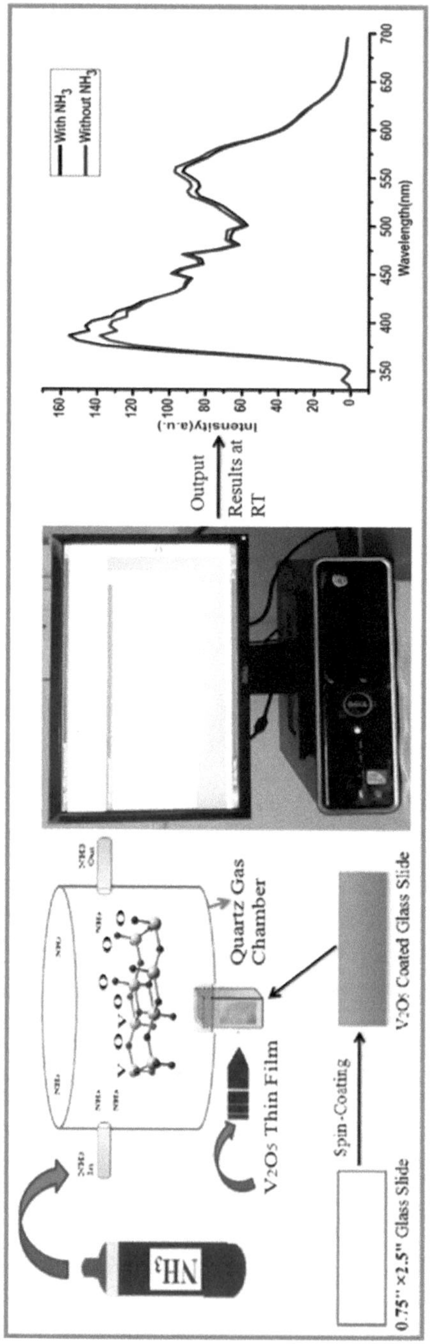

FIGURE 8.2 Photoluminescence ammonia gas sensing setup at room temperature.

to the F-7000 Hitachi system and which in turn was connected with the computer the spectrophotometer gives comparative data between sample in air and sample in ammonia gas for different ppm.

8.4 RESULTS AND DISCUSSION

8.4.1 X-RAY DIFFRACTION ANALYSIS

X-ray diffraction (XRD) patterns were obtained from Bruker D8 advance (Cu-Kα radiation, λ =1.5406 Å) diffractometer at room temperature. The crystal structure of all samples was identified by XRD in the 2θ range of 20°–70°. The structural phase analysis for as-synthesized Li-doped Cr_2O_3 NPs is shown in Figure 8.3. No extra peaks due to impurities were detected. The diffraction peaks were indexed in the Rhombohedral phase structure with space group $R\bar{3}c$ as described in the JCPDS card no. 3814-79. The hkl values corresponding to 2θ values are 24.55° (0112), 33.69° (104), 36.31° (110), 41.51° (113), 50.28° (024), 54.89° (116), 63.60° (214), and 65.28° (300). The strongest peaks for the single phase were observed at 33.69°θ value. The average crystallite sizes of pure and Li-doped Cr_2O_3 powder were determined by using the Scherrer–Debye formula.[39–44]

$$D = \frac{k\lambda}{\beta\cos\theta} \tag{8.2}$$

where k is constant (0.9), λ is the wavelength of the incident radiation (λ=1.5406 Å for Cu-Kα), β is the full-width half maxima (FWHM), and θ is Bragg diffraction angle. The average crystallite size was found to be about 34.97–35.68 nm, respectively (Fig. 8.4).

$$\delta = \frac{1}{D^2} \tag{8.3}$$

where D is the average crystallite size. The dislocation density decreased with increases crystallite size for 0.5–2.0 wt.% doped Cr_2O_3 NPs as Table 8.1. The radius of Cr^{3+} is 0.615 Å, which is larger than that of Li^{2+} (0.59 Å) at the same condition. Therefore, the substitution of Li^{2+} by Cr^{3+} induced the high intensity of diffraction peaks, confirming that Li^{2+} is incorporated into the Cr^{3+} lattice.

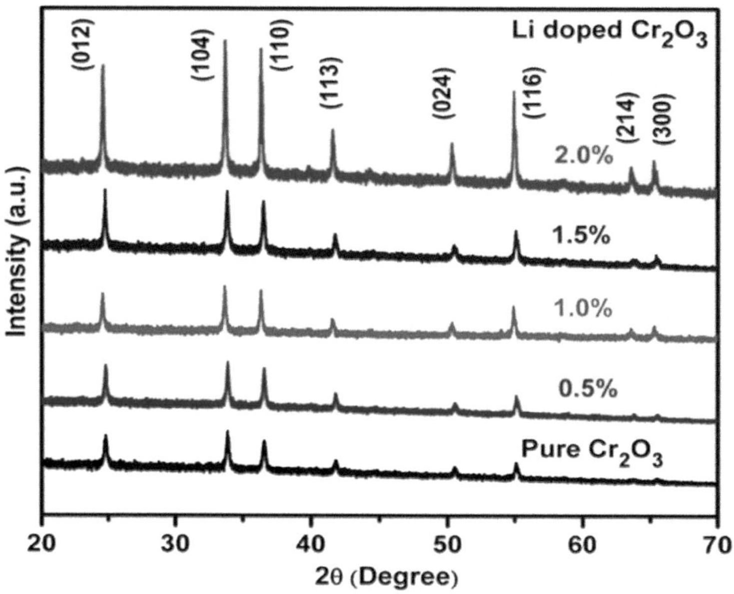

FIGURE 8.3 X-ray diffraction patterns of pure and Li (0.5–2.0 wt.%) doped Cr_2O_3 nanoparticles.

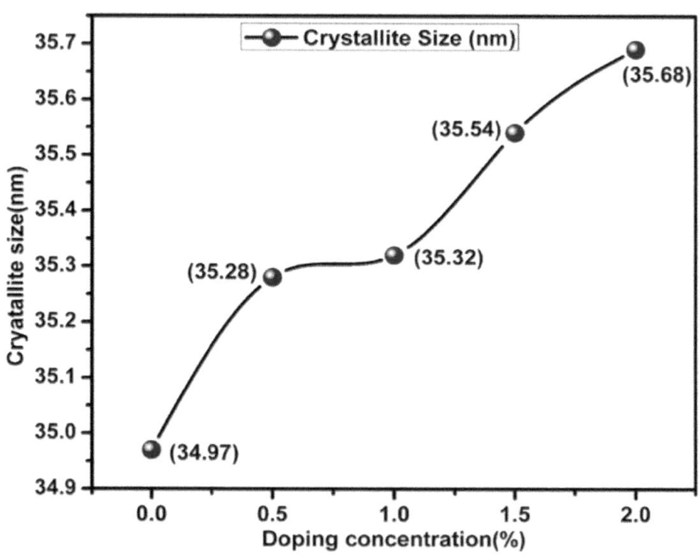

FIGURE 8.4 Graph between crystallite size *vs* Li doping concentration.

Figure 8.5 shows an expansion of the XRD patterns around the maximum peak of pure and doped Cr_2O_3 NPs. The dislocation density can be determined as[45–49] (Table 8.2).

TABLE 8.2 Crystallite Size from Debye–Scherrer Method and δ of Pure and Doped Cr_2O_3 Nanoparticles.

S. No.	Doping concentration (%)	Crystallite size D (nm)	Dislocation density δ (Lines/m²)
1	Pure Cr_2O_3	34.97	8.1773×10^{-4}
2	0.5 wt.% Li-doped Cr_2O_3	35.28	8.0410×10^{-4}
3	1.0 wt.% Li-doped Cr_2O_3	35.32	8.0160×10^{-4}
4	1.5 wt.% Li-doped Cr_2O_3	35.54	7.9171×10^{-4}
5	2.0 wt.% Li-doped Cr_2O_3	35.68	7.8551×10^{-4}

8.4.2 ATOMIC FORCE MICROSCOPY ANALYSIS

Figures 8.5a–d and 8.6a–d show the typical two- and three-dimensional atomic force microscopy (AFM) images (1 × 1μ) of the pure and doped Li-doped Cr_2O_3 NPs with concentration 0.5–2.0 wt.% thin film. Pure and doped thin films were made using spin coater on the quartz glass substrate at room temperature and then mild annealing process. Topography of the coatings produced at different conditions shows a significant difference in surface roughness (2D and 3D), mean width diameter, and typical surface features.

From the obtained AFM images, it was observed that with increasing Li doping concentration, the average grain size increase from 30.2 to 44.6 nm, the average roughness increase from 0.525 to 3.207 nm (2D) and 0.974 to 3.674 nm (3D) at 52.4 room humidity. The main characteristics of all thin films, such as grain size, surface roughness, mean width, and diameter are shown in Table 8.3. The increase in surface roughness of Li-doping with different concentration thin films indicates the probability of using in anti-reflection coating.[29]

The grain size distribution of the prepared pure and Li-doped Cr_2O_3 thin films was observed by examining the histograms of the height distribution from AFM. Figure 8.5 shows the characteristic histograms of doped and Pure chromium oxide NPs based thin films. With the increase in dopant concentration, the grain size increased from 30.2 to 44.6 nm. All

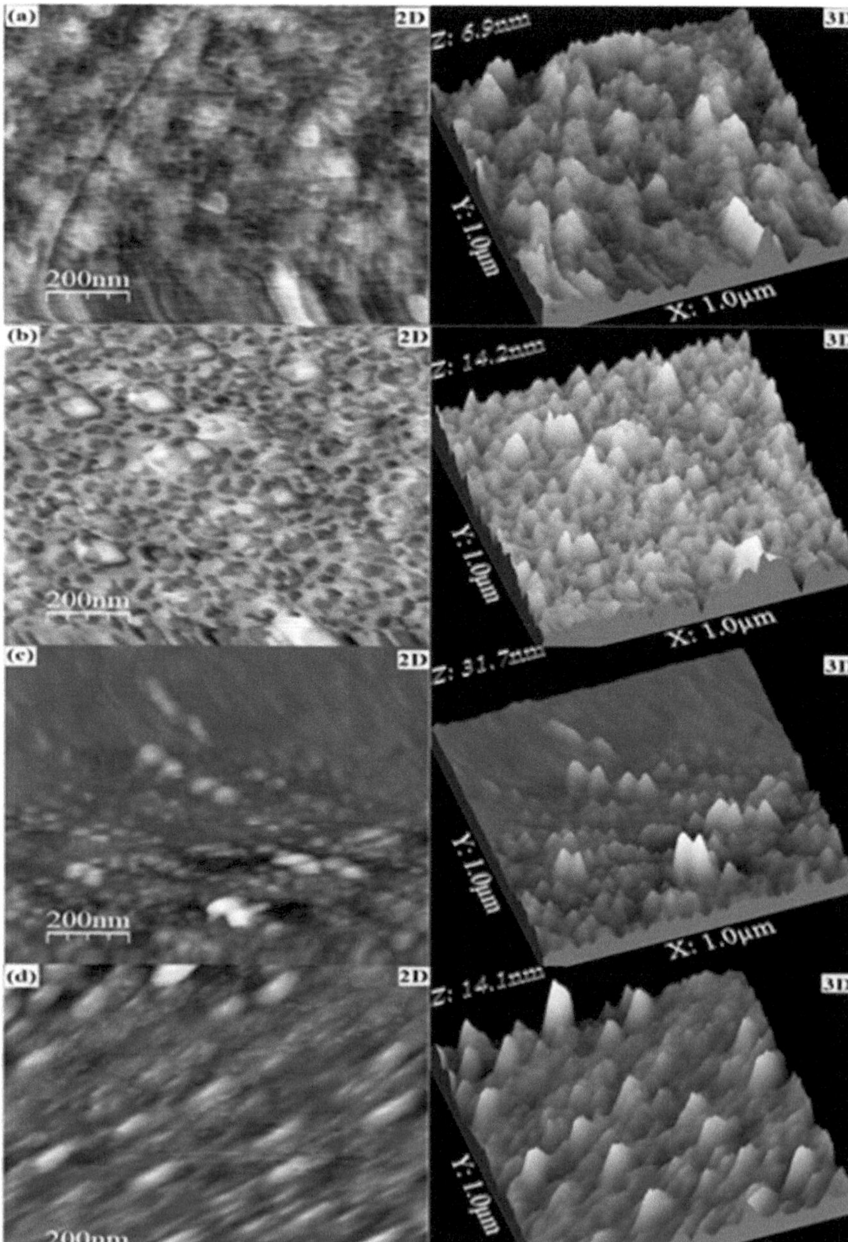

FIGURE 8.5 AFM images of pure and Li-doped Cr_2O_3 nanoparticles films on quartz substrate.

histogram results showed a unimodal behavior of distribution in which the average grain size calculated for the Li (0–2.0 wt.%) doped Cr_2O_3 was 30.2–44.6 nm (Table 8.3). The maximum shift in height from pure and doped chromium oxide NPs-based films predicts that the grains tend to grow in their *z*-dimension.[50]

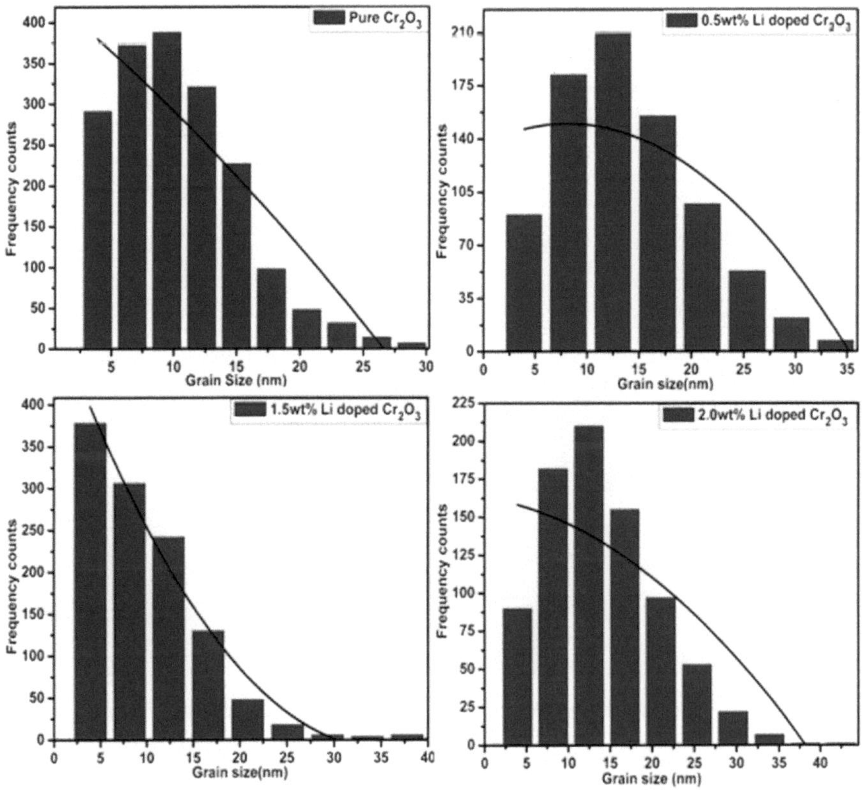

FIGURE 8.6 Histograms of the grain size distribution from AFM for the pure and doped Cr_2O_3 nanoparticles thin films.

8.4.3 TRANSMISSION ELECTRON MICROSCOPY ANALYSIS

Figure 8.7a shows a typical transmission electron microscopy (TEM) image of Cr_2O_3 NPs. This image shows particles size distribution of chromium oxide NPs. The particle sizes can be measured directly from the TEM images, which are, to a large extent, well-separated from one another throughout the field of the images. Figure 8.7b shows the SAED pattern of Cr_2O_3 NPs. From the TEM image results, we can observe a large quantity of irregular chromium oxide NPs and average particle size of 36 nm calculated by ImageJ software. Histogram analysis of the TEM image is illustrated in Figure 8.8c. The results obtained are in good agreement with the particle sizes estimated by XRD analysis using Debye–Scherrer's formula.[45,51]

TABLE 8.3 Atomic Force Microscopy Analysis Results Pure and Doped Cr_2O_3 Nanoparticles.

Doping concentration (%)	Grain size (nm)	Roughness (nm) 2D	Roughness (nm) 3D	Mean width (nm)	Diameter (nm)
Pure Cr_2O_3	30.2	0.525	0.974	15.8	15.8
0.5 wt.% Li-doped Cr_2O_3	36.7	1.264	1.184	16.9	16.9
1.5 wt.% Li-doped Cr_2O_3	41.4	2.965	3.278	27.5	27.5
2.0 wt.% Li-doped Cr_2O_3	44.6	3.207	3.674	30.6	30.6

FIGURE 8.7 (a) TEM images of pure Cr_2O_3 nanoparticles scale bar is 50 nm, (b) SAED pattern and (c) Histogram analysis of TEM image.

8.4.4 PHOTOLUMINESCENCE SPECTROSCOPY ANALYSIS

The PL spectrum of pure and Li-doped Cr_2O_3 NPs was recorded at room temperature. Figure 8.8 shows sharp PL emission spectra obtained from Cr_2O_3 NPs at room temperature. A maximum intensity of emission is obtained when the excitation wavelength 320 nm at 3.87 eV in photon energy is used. Nanomaterials have a higher density of crystalline defects, such as oxygen defects, stacking faults. In this case, the PL emission peak can be assigned to such crystal defects within the pure and doped Cr_2O_3

matrix. The PL (emission) spectra of all samples of Chromium oxide are as shown in Figure 8.8. Seven emission peaks were observed at 398 nm (violet region), 412 nm (violet region), 452, 469 and 482 nm (blue region), and 493 and 523 nm (green region) corresponding to approximate photon energy 3.11, 3.01, 2.74, 2.64, 2.57, 2.52, and 2.37 eV, respectively. A broad UV emission peak ~398 nm dominates in the spectra, attributed to the transition involving the $3d^3$ electrons of the Cr^{3+} ions.[52–54]

Moreover, oxygen vacancies create defect centers in the oxide nanomaterials and peak appears at lower energy, which increases PL emission by recombining electron that may be attached to oxygen vacancies. It is assumed that the emission wavelength of such material depends mainly on the particles size and morphology. When the particle size increases, the intensity of the visible emission also increase due to fast hole trap process.[55] Further higher doping amount causes a reduction in emission intensity due to concentration compromising effect.[56] The PL peak of pure and doped Cr_2O_3 at 523 nm due to oxygen vacancies, increases significantly with the Li doping as access electrons flood in the band because of substitution of the Cr^{3+} with Li^{+2}. In this case, the emission peak intensity decreased with increase in doping concentration because ionic radii of Cr^{3+} is 0.615 Å and Li^{2+} is 0.59 Å.

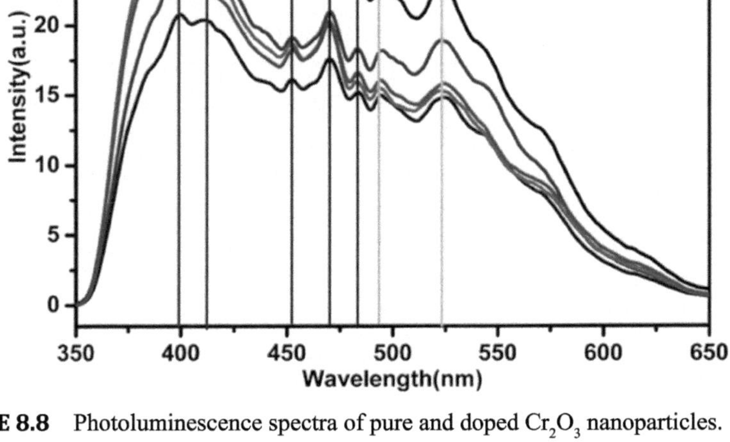

FIGURE 8.8 Photoluminescence spectra of pure and doped Cr_2O_3 nanoparticles.

Photoluminescence Ammonia Sensing Optical Properties of Pure and Li-doped Cr_2O_3 Nanoparticles

PL (Optical) gas sensing is of two types, the first is focused on optical adsorption, and the second is based on changes in active material absorption/emission when analyte gas reacts with it.[29,34] PL optical sensors have the advantages that can be easily miniaturized, non-contaminable, and do not require physical contact between the sensor and the optical detector. Most optical sensors work on the concept of reversible luminescence due to extinguish of a luminescent indicator luminophore's strength or excited-state lifetime.[57] PL is a nondestructive method for studying semiconductor properties both extrinsic and intrinsic. In n-type and p-type semiconductor materials, the PL technique can be used to evaluate bandgap, impurity levels, defect detection, and the recombination mechanism. Our sensor is part of the second group, in which the light emission corresponds to the charge transfer from oxygen to chromium ions, and chromium derivative complexes were observed at longer wavelengths.[34]

$$Cr^{3+} + Cr^{2+} \leftrightarrow Cr^{2+} + Cr^{3+} \tag{8.4}$$

or

$$Cr^{3+}–O^{2-} \xrightarrow{h\nu} Cr^{2+}–O^{-} \tag{8.5}$$

In case of pure and Li-doped Cr_2O_3 NPs based optical (PL) sensing of ammonia, the highly sensitive film deposited on a quartz substrate was inserted into a quartz gas chamber which was connected to ammonia source tube. All thin films were excited by 320 nm wavelength and corresponding emission spectra were noted with and without ammonia at room temperature. PL spectra as shown in Figure 8.8, exhibit a broad luminescence band. The morphology (shape), position, and strength of this band are all strongly influenced by grain size, roughness, and preparation conditions, as well as measurement conditions. Figure 8.9 shows the typical optical sensing response of pure and doped Cr_2O_3 NPs based thin films at different concentrations of NH_3 gas, at room temperature. In Figure 8.9 intensity variations are observed and compared to light emission in air and ammonia gas. An enhancement of PL intensity is recorded with ammonia and the opposite behavior is observed for the ambient. Thin films prepared for pure and Li (0.5–2.0 wt.%) doped Cr_2O_3 NPs exhibited linear increase in sensitivity as shown in Figure 8.9 (sensitivity vs doping concentration) and in Table 8.4. The device is studied for various PL sensor sensitivity parameters. The sensitivity of the sensor is affected by a number of factors, including the film surface, density of active centers, temperature, and sensor mass. Optical sensor sensitivity is defined as ac slope of the linear part of the calibration curve and the sensitivity of the PL sensors is directly proportional to the number of active atoms on the film surface.

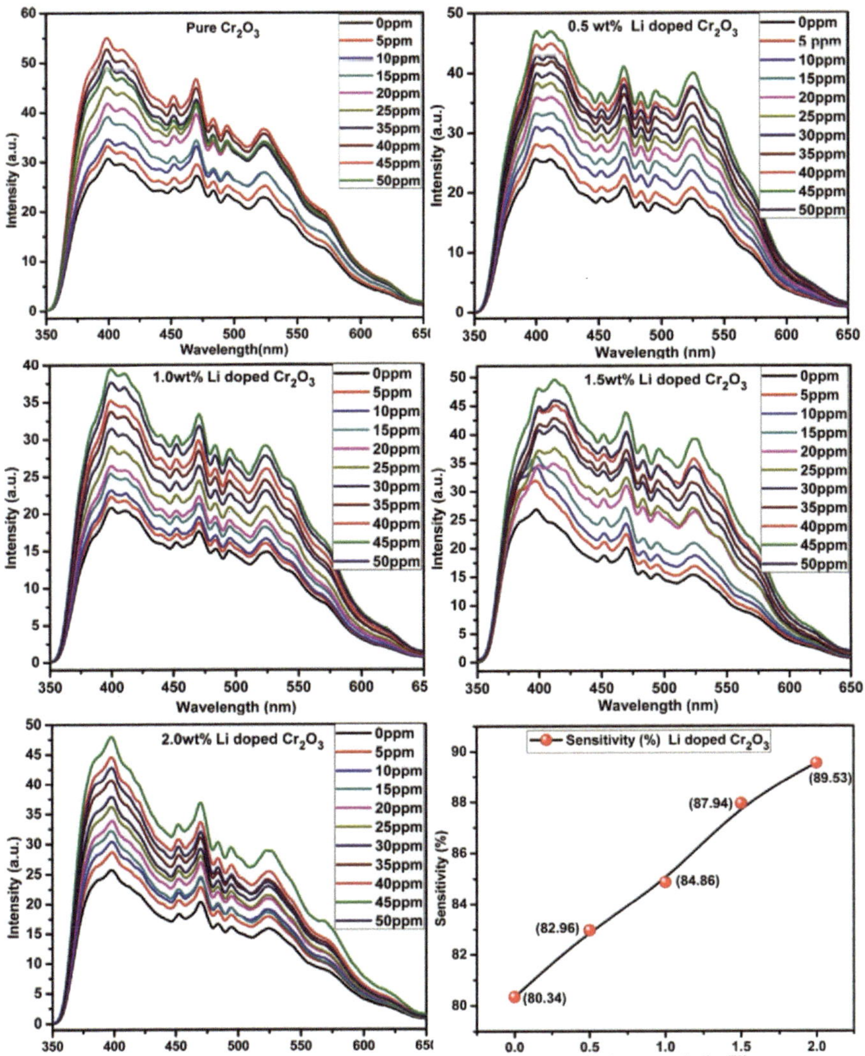

FIGURE 8.9 Photoluminescence ammonia gas sensing property of pure and doped Cr_2O_3 nanoparticles at room temperature and sensitivity vs doping concentration.

TABLE 8.4 Summary of Sensitivity Results of Pure and Li-doped Cr_2O_3 NPs for Ammonia Gas at Room Temperature.

S. No.	Doping concentration	NH_3 Gas (ppm)	Sensitivity (%)
1	Pure Cr_2O_3	45	80.34
2	0.5 wt.% Li-doped Cr_2O_3	45	82.96
3	1.0 wt.% Li-doped Cr_2O_3	45	84.86
4	1.5 wt.% Li-doped Cr_2O_3	45	87.94
5	2.0 wt.% Li-doped Cr_2O_3	45	89.53

From Figure 8.9 it follows that there is a remarkable difference in PL quenching response in NH_3 and air (ambient).

The sensing performance of pure and Li-doped Cr_2O_3 NPs was found and it was observed, as shown in Figure 8.9, that the sensitivity linearly increases with increasing doping concentration, respectively. Table 8.4 shows the summary of PL sensing for as fabricated ammonia gas sensor consisting of pure and Li-doped Cr_2O_3 NPs based thin films. The increase in sensitivity is due to less electron scattering with phonons as well as the structural defects in doped and pure Cr_2O_3 NPs.

8.5 CONCLUSION

Pure and Li-doped Cr_2O_3 NPs were successfully synthesized for ammonia gas optical sensing (PL) studies at room temperature. XRD, AFM, and TEM analysis confirmed the successful formation of chromium oxide NPs. Increase in crystallite size, grain size, mean width, and diameter with increasing doping concentration was observed. The highest optical sensitivity was found for 2.0 wt.% Li-doped Cr_2O_3 NPs that was 89.53% at 45 ppm concentration of ammonia at room temperature. Wavelength-dependent studies demonstrate that the fabricated doped Cr_2O_3 sensor is highly sensitive from visible to green region. The results of studied NPs show great opportunities for room temperature PL ammonia gas sensing applications.

ACKNOWLEDGMENTS

The authors are grateful to the Director, MANIT, Bhopal for providing facilities. The authors would like to owe sincere thanks to Director, UGC-DAE-CSR Indore Center (M.P.) for providing the XRD facility.

KEYWORDS

- **lithium**
- **Cr_2O_3 nanoparticles**
- **thin films**
- **optical ammonia sensor and sensitivity**

REFERENCES

1. Morrison, S. Roy; Mechanism of Semiconductor Gas Sensor Operation. *Sens. Actuat.* **1987,** *11* (3), 283–287.
2. Takeda, Y.; Kanno, R..; Tsuji, Y.; Yamamoto, O.; Taguch, H. Chromium Oxides as Cathodes for Lithium Cells. *Journal of Power Sources* **1983,** *9* (3), 325–328.
3. Suryawanshi, D. N.; Patil, D. R.; Patil, L. Fe2O3-Activated Cr_2O_3 Thick Films as Temperature Dependent Gas Sensors. *Sens. Actuat. B Chem.* **2008,** *134* (2), 579–584.
4. Aghaie-Khafri, M.; Kakaei Lafdani, M. H. A Novel Method to Synthesize Cr_2O_3 Nanopowders Using EDTA as a Chelating Agent. *Powder Technol.* **2012,** *222*, 152–159.
5. Abdullah, M. M.; Rajab, Fahd M.; Saleh M., Al-Abbas. Structural and Optical Characterization of Cr_2O_3 Nanostructures: Evaluation of Its Dielectric Properties. *Aip Adv.* **2014,** *4* (2), 027121.
6. Cao, H.; Qiu, X.; Liang, Y.; Zhao, M.; Zhu, Q. Sol-Gel Synthesis and Photoluminescence of P-type Semiconductor Cr_2O_3 Nanowires. *Applied physics letters.* **2006,** 88(24), 241112.
7. Ekwealor, A. B. C.; F. I. Ezema. Optical and Structural Properties Variations with Precursor Concentration for CrxOy Thin Films Synthesized in a Polymer Matrix by Chemical Bath Deposition Technique. *Digest J. Nanomater. Biostruct.* **2013,** *8* (1), 377–384.
8. Cho; Chul-Hyung; Ki-Cheol Park; Tae-Young Ma; Jeong-Gyoo Kim. Ammonia Gas-sensing Characteristics of Cr_2O_3 Thick Films. *Sens. Sci. Technol.* **2004,** *13* (6), 424–429.
9. Pei, Z.; Hongbin, X.; Zhang, Y. Preparation of Cr_2O_3 nanoparticles via C_2H_5OH Hydrothermal Reduction. *J. Alloys Comp.* **2009,** *468* (1–2), L5–L8. 468.
10. Hassen, A.; El Sayed, A. M.; Morsi, W. M.; El-Sayed, S. Influence of Cr_2O_3 Nanoparticles on the Physical Properties of Polyvinyl Alcohol. *J. Appl. Phys.* **2012,** *112*, 093525.
11. Makhlouf, Salah A.; Bakr, Zinab H.; Al-Attar, H.; Moustafa, M. S. Structural, Morphological and Electrical Properties of Cr_2O_3 Nanoparticles. *Mater. Sci. Eng.: B* **2013,** *178* (6), 337–343.
12. Fu, X.-Z.; Luo, X.-X.; Luo, J.-L.; Chuang, K. T.; Sanger, A. R.; Krzywicki, A. Ethane Dehydrogenation over Nano- Cr_2O_3 Anode Catalyst in Proton Ceramic Fuel Cell Reactors to Co-Produce Ethylene and Electricity. *J. Power Sour.* **2011,** *196* (3), 1036–1041.

13. Kim, D.-W.; Shin, S.-I.; Lee, J.-D.; Oh, S.-G.. Preparation of Chromia Nanoparticles by Precipitation–Gelation Reaction. *Mater. Lett.* **2004,** *58* (12–13), 1894–1898.
14. Pei, Z. Z.; Hong, B. X.; Zhang, Y. Preparation of Cr_2O_3 Nanoparticles via Hydrothermal Reduction of CH_3OH. In *Adv. Mater. Res.* **2012,** *463*, 760–763.
15. Balachandran, U.; Siegel, R. W.; Liao, Y. X.; Askew, T. R. Synthesis, Sintering, and Magnetic Properties of Nanophase Cr_2O_3. *Nanostruct. Mater.* **1995,** *5* (5), 505–512.
16. Vollath, D.; Szabó, D. V.; Willis, J. O. Magnetic Properties of Nanocrystalline Cr_2O_3 Synthesized in a Microwave Plasma. *Mater. Lett.* **1996,** *29* (4–6), 271–279.
17. Leilei, X.; Wang, Z.; Song, H.; Chou, L. Catalytic Dehydrogenation of Isobutane over Ordered Mesoporous Cr_2O_3–Al_2O_3 Composite Oxides. *Catal. Commun.* **2013,** *35*, 76–81.
18. Yang. M.; Men, Y.; Li, S.; Chen, G. Enhancement of Catalytic Activity over TiO_2-Modified Al_2O_3 and ZnO–Cr_2O_3 Composite Catalyst for Hydrogen Production via Dimethyl Ether Steam Reforming. *Appl. Catal. A Gen.* **2012,** *433*, 26–34.
19. Xia, P.; Zuo, S.; Liu, F.; Qi, C. Ceria Modified Crystalline Mesoporous Cr_2O_3 Based Nanocomposites Supported Metal Oxide for Benzene Complete Oxidation. *Catal. Commun.* **2013,** *41*, 91–95.
20. Zhang, J.; Qin, Z.; Zeng, D.; Xie, C. Metal-Oxide-Semiconductor Based Gas Sensors: Screening, Preparation, and Integration. *Phys. Chem. Chem. Phys.* **2017,** *19* (9), 6313–6329.
21. Wang, C.; Yin, L.; Zhang, L.; Xiang, D.; Gao, R. Metal Oxide Gas Sensors: Sensitivity and Influencing Factors. *Sensors* **2010,** *10* (3), 2088–2106.
22. Lin, T.; Lv, X.; Hu, Z.; Xu, A.; Feng, C. Semiconductor Metal Oxides as Chemoresistive Sensors for Detecting Volatile Organic Compounds. *Sensors* **2019,** *19* (2), 233.
23. Hahn, S. H.; Barsan, N.; Weimar, U.; Ejakov, S. G.; Visser, J. H.; Soltis, R. E. CO Sensing with SnO_2 Thick Film Sensors: Role of Oxygen and Water Vapour. *Thin Solid Films* **2003,** *436* (1), 17–24.
24. He, X.; Li, J.; Gao, X.; Wang, L. NO_2 Sensing Characteristics of WO_3 Thin Film Microgas Sensor. *Sens. Actuat. B Chem.* **2003,** *93* (1–3), 463–467.
25. Shankar, P.; Rayappan, J. B. B. Gas Sensing Mechanism of Metal Oxides: The Role of Ambient Atmosphere, Type of Semiconductor and Gases-A Review. *Sci. Lett. J.* **2015,** *4* (4), 126.
26. Tomchenko, A. A.; Harmer, G. P.; Marquis, B. T.; Allen, J. W. Semiconducting Metal Oxide Sensor Array for the Selective Detection of Combustion Gases. *Sens. Actuat. B Chem.* **2003,** *93* (1–3), 126–134.
27. Kanazawa, E.; Sakai, G.; Shimanoe, K.; Kanmura, Y.; Teraoka, Y.; Miura, N.; Yamazoe, N. Metal Oxide Semiconductor N_2O Sensor for Medical Use. *Sens. Actuat. B* **2001,** *77*, 72–77.
28. Korotcenkov, G. Metal Oxides for Solid-State Gas Sensors: What Determines Our Choice? *Mater. Sci. Eng. B* **2007,** *139*, 1–23.
29. Singh, N. Optical and Chemoresistive NH_3 Gas Sensing Studies on Sn^{4+} Doped Metal Oxide Nanoparticle. Thesis on National Institute of Technology, Bhopal, **2019,** 1–185.
30. Franke, M. E.; Koplin, T. J.; Simon, U. Metal and Metal Oxide Nanoparticles in Chemiresistors: Does the Nanoscale Matter? *Small* **2006,** *2*, 36–50.
31. Barsan, N.; Schweizer-Berberich, M.; Göpel, W. Fundamental and Practical Aspects in the Design of Nanoscaled SnO2 Gas Sensors: A Status Report. *Fresenius J. Anal. Chem.* **1999,** *365*, 287–304.

32. Singh, N.; Bamne, J.; Mishra, K. M.; Singh, N.; Haque, F. Z. Photoluminescence and Chemoresistive Gas Sensing: A Comparative Study Using V_2O_5 Nanostructures for NH_3. *Emerg. Trends Nanotechnol.* 279.
33. Mishra, M. K.; Singh, N.; Pandey, V.; Haque, F. Z. Synthesis of SnO_2 Nanoparticles and Its Application in Sensing Ammonia Gas through Photoluminescence. *J. Adv. Phys/* **2016,** *5* (1), 8–12.
34. Valerini, D.; Cretì, A.; Caricato, A. P.; Lomascolo, M.; Rella, R.; Martino, M. Optical Gas Sensing through Nanostructured ZnO Films with Different Morphologies. *Sens. Actuat. B Chem.* **2010,** *145* (1), 167–173.
35. Lakshmi, K.; Basha, S. S.; Rao, M. C. Optical and Photo-luminescent Studies on VO^{2+} Doped SnO_2 Thin Films. *RASAYAN J. Chem.* **2017,** *10* (2), 682–688.
36. Singh, N.; Syed, F.; Haque, F. Z. Ionic Liquid Controlled Growth of Zinc Oxide Nano Particles and Their Fluorescence Study in the Presence of NH_3 Gas. *Mater. Sci. Res. India*, **2014,** *11* (1), 27–34.
37. Vinita, P. Synthesis and Application of Metal Doped SnO_2 Nanostructures in Photoluminescence Based NH_3 Gas. *M.A.N.I.T., India*, Ph.D. Thesis, 2017.
38. Pandey, V.; Singh, N.; Malik, M. M.; Khan, F. Z. SnO_2, SnO_2: Mg and SnO_2: Ni Nanoparticles Based Luminescence Ammonia Sensors. *Mater. Focus* **2017,** *6*(3), 359–363.
39. Parra, M. R.; Haque, F. Z. Aqueous Chemical Route Synthesis and the Effect of Calcination Temperature on the Structural and Optical Properties of ZnO Nanoparticles. *J. Mater. Res. Technol.* **2014,** *3* (4), 363–369.
40. Siddiqui, H.; Parra, M. R.; Pandey, P.; Qureshi, M. S.; Haque, F. Z. Utility of Copper Oxide Nanoparticles (CuO-NPs) as Efficient Electron Donor Material in Bulk-Heterojunction Solar Cells with Enhanced Power Conversion Efficiency. *J. Sci. Adv. Mater. Device* **2020,** *5* (1), 104–110.
41. Li, Y.; Yao, J.; Uchaker, E.; Zhang, M.; Tian, J.; Liu, X.; Cao, G. Sn-Doped V_2O_5 Film with Enhanced Lithium-Ion Storage Performance. *J. Phys. Chem. C.* **2013,** *117* (45), 23507–23514.
42. McMurdie, H. F.; Morris, M. C.; Evans, E. H.; Paretzkin, B.; Wong-Ng, W.; Hubbard, C. R. Methods of Producing Standard X-Ray Diffraction Powder Patterns. *Powder Diffrac.* **1986,** *1* (1), 40–43.
43. Parra, M. R.; Haque, F. Z. Optical Study and Ruthenizer (II) N_3 Dye-Sensitized Solar Cell Application of ZnO Nanorod-Arrays Synthesized by Combine Two-Step Process. *Opt. Spectrom.* **2015,** *119*, 672.
44. Bamne, J.; Sharma, P. K.; Haque, F. Z. Effect of Solvent Mixing and Calcination Temperature on the Growth of TiO_2 Nanoparticle Prepared via Sol–Gel Method. *Mater. Focus.* **2018,** *7* (2), 232–241.
45. Bamne, J. Synthesis and Characterization of Rare Earth Materials Doped TiO_2 Nanoparticles for the Decomposition of Hazardous Substance Present in Waste Water; Barkatullah University Bhopal, India, Ph.D. Thesis, 2020; pp 1–193.
46. Parra, M. R.; Haque, F. Z. Poly (Ethylene Glycol) (PEG)-Assisted Shape-Controlled Synthesis of One-Dimensional ZnO Nanorods. *Optik.* **2015,** *126*, 1562–1566.
47. Singh, N.; Singh, N.; Mishra, K. M.; Haque, F. Z. Photoluminescence Properties of ZnO Micro/Nanostructures Capped with Various Surfactants. *J. Adv. Phys.* **2016,** *5* (2), 184–189.

48. Siddiqui, H; Qureshi, M. S.; Haque, F. Z. Surfactant Assisted Wet Chemical Synthesis of Copper Oxide (CuO)Nanostructures and Their Spectroscopic Analysis. *Optik.* **2016,** *127,* 2740–2747.
49. Rajeshwari, S; Kumar, J. S.; Rajendrakumar, R. T.; Ponpandian, N.; Thangadurai, P. Influence of Sn Ion Doping on the Photocatalytic Performance of V_2O_5 Nanorods Prepared by Hydrothermal Method. *Mater. Res. Exp.* **2018,** *5* (2), 025507.
50. Singh, N.; Umar, A.; Singh, N.; Fouad, H.; Alothman, O. Y.; Haque, F. Z. Highly Sensitive Optical Ammonia Gas Sensor Based on Sn Doped V_2O_5 Nanoparticles Materials. *Res. Bull.* **2018,** *108,* 266–274.
51. Kluensner, T.; Shen, Q.; Hlawacek, G.; Teichert, C.; Fateh, N.; Fontalvo, G. A.; Mitterer, C. Morphology Characterization and Friction Coefficient Determination of Sputtered V_2O_5 Films. *Thin Solid Films* **2010,** *519* (4), 1416–1420.
52. Parra, M. R.; Pandey, P.; Siddiqui, H.; Sudhakar, V.; Krishnamoorthy, K.; Haque, F. Z. Evolution of ZnO Nanostructures as Hexagonal Disk: Implementation as Photoanode Material and Efficiency Enhancement in Al: ZnO Based Dye Sensitized Solar Cells. *Appl. Surf.ace Sci.* **2019,** *470,* 1130–1138.
53. Singh, N.; Pandey, V.; Singh, N.; Malik, M. M.; que, F. Z. Application of TiO_2/SnO_2 Nanoparticles in Photoluminescence Based Fast Ammonia Gas Sensing. *J. Optics* **2017,** *46* (3), 199–203.
54. Suresh, R.; Giribabu, K.; Manigandan, R.; Munusamy, S.; Praveen Kumar, S.; Muthamizh, S.; Stephen, A.; Narayanan, V. Doping of Co into V_2O_5 Nanoparticles Enhances Photo Degradation of Methylene Blue. *J. Alloys Comp.* **2014,** *598,* 151–160.
55. Tang, H.; Prasad, K.; Sanjines, R.; Schmid, P. E.; Levy, F. Electrical and Optical Properties of TiO_2 Anatase Thin Films. *J. Appl. Phys.* **1994,** *75* (4), 2042–2047.
56. Tae, K. I.; Choi, A. S.; Jeong, J. W. Precise Control of a Correlated Color Temperature Tunable Luminaire for a Suitable Luminous Environment. *Build Environ.* **2012,** *57,* 302–312.
57. Mani, G. K.; Rayappan, J. B. B. Influence of Copper Doping on Structural, Optical and Sensing Properties of Spray Deposited Zinc Oxide Thin Films. *J. Alloy Compd.* **2014,** *582,* 414–419.

CHAPTER 9

Valorization of Waste Biomass as a Strategy to Alleviate Ecological Deficit: A Case Study on Waste Biomass Derived Stable Carbon

PRITAM DEY[1], RUPAK ROY[1], ARABINDA MUKHERJEE[2], PONNAGANTI SAI KRISHNA[2], RADHAGOVINDA KOIJAM[2], and SRIMANTA RAY[1*]

[1]*Department of Chemical Engineering, National Institute of Technology Agartala, Jirania 799046, Tripura, India*

[2]*Department of Civil Engineering, National Institute of Technology Agartala, Jirania 799046, Tripura, India*

[*]*Corresponding author. E-mail: rays.nita@gmail.com*

ABSTRACT

The growing population has caused increased resource consumption and higher waste generation, resulting in a widening ecological deficit. Therefore, the valorization and utilization of waste have become crucial to impede the growing ecological footprint. The waste biomass can be valorized into a variety of value-added products which can help minimize the ecological footprint. The stable carbon is one such value-added product derived from waste biomass. The nature of waste biomass utilized for obtaining stable carbon and biomass conversion techniques influence the overall characteristics of the stable carbon. A proper

Advanced Microscopy: A Strong Analytical Tool in Materials Science. Merin Sara Thomas, Jozef T. Haponiuk, Sabu Thomas, & Anne George (Eds.)

selection of biomass conversion technique and conditions hold the key to produce stable carbon for a specific application. A case study discussed in this chapter reviews stable carbon prepared from two different waste biomass—a crop residue and a vegetable peel. The case study showed that the vegetable peel derived stable carbon has good biosorptive capacities for aqueous metal ion and organic dye from aqueous medium. The waste-derived carbon product reported 70% and 95% sorption efficiencies for iron(III) ions and methylene blue dye, respectively. On the other hand, the stable carbon prepared from crop residue showed good potential as a soil additive and resulted in 29%, 250%, and 140% increase in the organic carbon content, above soil biomass and soil potassium respectively. The case study corroborates that valorization of waste biomass in form of stable carbon has great potential that can present a remedial solution to the widening ecological deficit and pave the way for a sustainable future.

9.1 INTRODUCTION

9.1.1 ECOLOGICAL FOOTPRINT

The ecological resources are considered as the wealth of the nation. The increase in the human population and growing human needs are putting tremendous pressure on ecological resources resulting in ecological deficits. The implications of ecological deficits in the long term can be devastating and may lead to the collapse of the ecosystem.[1] A recent study estimated that the present human consumption rate of ecological resources is 1.7 times more than the rate of ecological renewal,[2,3] which means 1.7 planet Earth are needed to meet the present need of the global human population. The National Footprint Accounts defines ecological footprint as "a measure of the area of biologically productive land and water an individual, population, or activity requires to produce all the resources it consumes and to absorb the waste it generates, using prevailing technology and resource management practices".[2] The present world-average ecological footprint has been computed as 2.8 global hectares per capita.[3] This number is on the rise across the globe, which makes it an alarming point. A country like India by holding a large share of the world's bio-capacity has low (0.8 global hectares per capita) compared to the global average. But, the population of India on the rise, projected to reach 1.7 billion by 2050.[4] The growing population will cause increased resource consumption and higher waste generation resulting in a widening ecological deficit.[1] The ecological footprint will likely rise due to increased waste generation. Thus, waste management is a crucial factor to minimize the ecological footprint and present a sustainable future.

9.1.2 WASTE MANAGEMENT

The rate of waste generation, solid waste generation, in particular, is on the rise across the globe. In addition to that, the per capita waste generation is also increasing year-by-year. The rise in population, growing economy, expanding urbanization, and enhancement in living standards are major contributors towards this rising trend of solid waste generation. The global solid waste generated rate was estimated as 0.74 kg and that for India was 0.57 kg per day per capita.[5] Waste generation in urban India was estimated to exceed 0.7 kg per person per day in 2025, approximately four to six times higher than in 2000.[6] The major categories of solid waste are listed in Table 9.1.

TABLE 9.1 Major Categories of Solid Wastes.[6,7]

Solid waste categories	Examples
Compostable waste	Food and kitchen waste, green waste (vegetables, flowers, leaves, fruits), and paper
Recyclable waste	Paper, glass, bottles, cans, metals, certain plastics, etc.
Inert waste matter	Dirt, debris.
Composite waste	Waste clothing, multilayered food packs, waste plastics such as toys
Domestic hazardous Waste and toxic waste	Waste medicine, e-waste, paints, chemicals, light bulbs, fluorescent tubes, spray cans, fertilizer or pesticide containers, and batteries.

The most common method of solid waste management in developing countries is dumping in landfills or predefined locations. Incineration is another solid waste management route. However, both dumping in uncon-structed landfill and incineration are linked with environmental and public health concerns thereby increase the ecological footprint.[8] Accordingly, utilization or recycling of solid waste holds great significance.

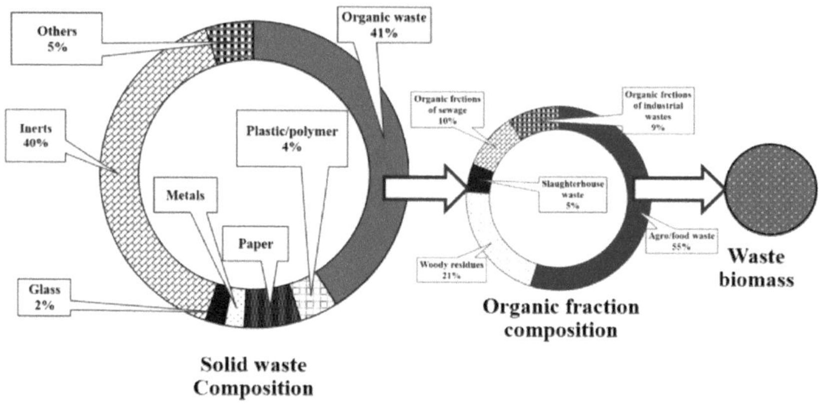

FIGURE 9.1 Contribution of waste biomass in the total solid waste in India.

The composition of solid waste varies across the globe. In developing countries like India, a good portion of solid waste is compostable waste or organic wastes[9] (Fig. 9.1). The organic fraction of solid waste includes biodegradable wastes from the kitchen, canteen, hotel, marketplace, and food-processing units, along with agricultural residues (Table 9.1). The agricultural residues in the organic fraction of solid waste, otherwise waste biomass, have great potential toward valorization into a variety of value-added products and holds the key toward minimizing the ecological footprint.

9.1.3 NATURE OF WASTE BIOMASS

Plant-based biomass is presently the largest available carbon source on earth accounting for about 450 billion tons of carbon.[10] Primarily, biomass is composed of lignin, cellulose, and hemicellulose. The composition of cellulose, hemicellulose, and lignin may vary from biomass to biomass depending upon age, geographical location, environmental conditions, and so on.[11] Biomass can be classified under three broad categories, namely woody biomass, nonwoody biomass, and agricultural residues. Table 9.2 displays a brief idea about the classification based on the composition of major structural components in biomass. Waste biomass encompasses among other things, vegetation, crops, as well as biosolids, animal, forestry, and agricultural residues; the organic fractions of municipal solid waste, and certain types of industrial wastes.[12] Figure 9.2 provides a brief idea about the general classification of waste biomass.

TABLE 9.2 Compositions of Various Types of Biomass.[11–16]

Biomass type	Examples	Lignin (%)	Cellulose (%)	Hemicellulose (%)
Woody biomass				
Hardwood	Mahogany, walnut, oak	16.5–52.4	25.2–51.1	6.8–32.9
Softwood	Pine, cedar, spruce	12.1–31.4	30.5–57.5	8.2–34.2
Nonwoody biomass				
Crops	Jowar, bajra, ragi	15.6–24.9	20.2–38.3	15.6–24.8
Grasses	*Miscanthus*, cogon, Napier	4.4–15.7	25.4–37.6	27.8–42.5
Agricultural residues				
Husks	Rice, mustard, soybean	5.4–54.8	28.5–43.9	0.2–35.6
Stalks /straws	Rice, wheat, maize	6.2–19.7	27.4–45.6	23.2–40.9
Vegetable peels	Potato, coffee, orange	5.6–15.9	9.7–48.2	7.2–42.1
Shells	Coconut, groundnut, walnut	17.1–52.3	25.5–48.6	18.9–34.7

Agricultural residues encompass a wide variety of biomass that is generated, which can be best classified under plant biomass residues, crop stacks or straws, and yard trimming residues.[12–15] Agricultural residues, food-based wastes, if not treated properly, may cause several environmental issues.[16] Since the millennia, efforts have shifted from waste remediation strategies to waste prevention ones, emphasizing the applications of the principles of *green chemistry*.[17] Now the focus is moving toward exploiting those wastes that are largely unavoidable, rather than disposing them off.

The appeal of waste biomass is largely due to its availability and its ability to be produced and consumed on a carbon-neutral basis.[18] A high percentage of elemental carbon, specifically in woody and agricultural biomass, makes them the most suitable for conversion into stable carbon products.[19] Thus, converting them into useful products is a wise course of action.

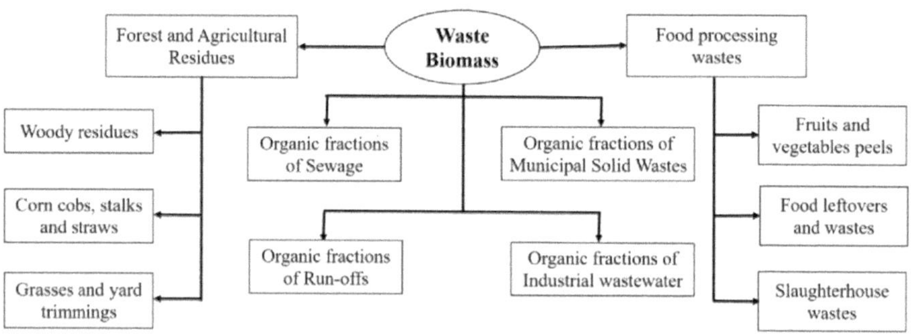

FIGURE 9.2 Different types of waste biomass.

9.1.4 WASTE BIOMASS CONVERSION AND VALORIZATION

The waste biomass is converted to desired products by various processes. The waste biomass can be transformed into fuel gas through partial combustion, converted to biogas through anaerobic digestion, converted to bioalcohol through biochemical processes, converted to biofuel, pyrolyzed into a bio-oil, or transformed into syngas for synthesizing chemicals and fuels.[11,16,20] Biomass waste-to-energy strategies encompass a wide variety of biomass types, conversion options, end-of-use applications, and technological means.[14,15] Two kinds of biomass conversion processes—thermochemical and biochemical conversion processes—are primarily employed.

The most communal technologies for biomass conversion are the thermochemical treatment processes which include incineration or combustion, pyrolysis, and gasification.[21,22] The hydrothermal conversion and biomass-to-liquid fuel are other thermal chemical treatment processes. On the other hand, biochemical treatment processes involve composting, fermentation, and bio-methanation/anaerobic digestion.[23,24] The mechanical attrition, under the influence of heat and catalysts of specific waste biomass like algal biomass, yields liquids fuels in the form of biodiesel.[20] Each of the processes mentioned above has its relative merits and demerits from the point of view of sustainability.

9.1.4.1 THERMOCHEMICAL CONVERSION PROCESSES

Thermochemical conversion processes, for example, combustion, gasification, and pyrolysis, convert biomass into solid (biochar), liquid (bio-oil), or gaseous fuel.[21] Combustion, the most conventional process, usually operates between 800°C and 1000°C and converts the accessible chemical energy into heat, leaving behind only ash and flue gases. In the gasification process, partial oxidation of carbonaceous contents of the biomass converts into gaseous fuel at an elevated temperature range of 700°C–900°C.[22]

The most popular thermochemical process is biomass pyrolysis. Depending on the heating rate and solid residence time, biomass pyrolysis can be divided into three main types including slow (conventional) pyrolysis, fast pyrolysis, and flash pyrolysis.[14,15] Slow pyrolysis, also known as carbonization, has been conventionally applied for the production of charcoal for centuries, where the temperature ranges between 200°C and 700°C.[16] Fast pyrolysis usually involves a 10°C–200°C/s rate of heating and 1–20 s of residence time.[14] The flash pyrolysis process is characterized by higher heating rates of 10^2°C—10^4°C/s and shorter residence times of less than a second, resulting in very high bio-oil yields.[15] A variety of process parameters such as the composition of the biomass, process temperature, heating rate, etc., can be optimized to yield different quantities of products with varying physical and chemical characteristics.[19] Table 9.3 lists several biomass feedstocks converted to desirable products by thermochemical routes.

9.1.4.2 BIOCHEMICAL CONVERSION PROCESSES

In biochemical treatment processes, microorganisms and specific enzymes transform the biomass into desired products through various biochemical

TABLE 9.3 Thermochemical Processes for Biomass Conversion.

Biomass feedstock	Conversion process	Significant outcomes	Reference
Pinewood and coconut fiber	Pyrolysis	Low-temperature pyrolysis in batch reactor between 200°C and 330°C to produce biochar. Highest HHV of 25.4 and 24.5 MJ/kg for pinewood and coconut fiber, respectively at 330°C. While the highest biochar yield at 200°C.	[16]
Pinewood, sal wood, and nut husk	Pyrolysis	Pyrolysis at a fixed temperature of 500°C in a six successive bed containing a fixed-bed reactor. The fourth bed gave the highest yields of bio-oil with HHV between 23.2 and 28.3 MJ/kg.	[18]
White ash, switchgrass, corn stover	Pyrolysis	Pyrolysis at a temperature between 300°C and 500°C in a fixed-bed reactor to produce bio-char, syngas, and bio-oil. Highest HHV of 33.5, 31.5, and 30.5 MJ/kg at 500°C for White ash, switchgrass, corn stover, respectively. High bio-oil yield at 300°C.	[25]
Rice straw and willow wood	Combustion	Combustion of pelletized straw and willow wood in a 4:1 ratio in a batch furnace. HHV ranges from 17 to 20 MJ/kg for different pellet proportions. Optimum biomass ratio was determined for lower CO and ash formation.	[26]
Corn cobs, vine pruning, and rice husk	Gasification	Gasification in a downdraft reactor (350 kW) with the three feeds. Plant's net efficiency of (21.1–21.6%), (20–22%), and (10–14%) for Corn cobs, Vine pruning, and rice husk, respectively.	[27]
Waste animal leather and poultry feather	Gasification	Co-gasification of the waste biomasses with wood pellets in two pilot plants of 2 and 3.5 MW capacities. A reduction in wood pellets was reported. Gasification produced less hazardous bottom ash for the two biomass than direct combustion.	[28]

conversion pathways.[19] Due to the recalcitrance nature of the lignin fraction of the biomass, the microbes are unable to degrade them into further simpler compounds.[23] The basic underlying mechanism of this type of technique involves hydrolysis of a complex lignocellulosic matrix into relatively simpler compounds so that microbes such as white rot, brown rot, and soft fungus may use them to produce value-added products. Breakdown of cellulose into glucose by enzymatic degradation is regarded as one of the major ways for biofuel production. Low-energy requirement and nonusage of hazardous materials are a few of the primary advantages of the biochemical conversion techniques. The major disadvantage of this conversion technique is that they require a relatively long time as compared with the other techniques of similar categories. Anaerobic digestion generally yields biogas, rich in methane, and sludge which can be used as manure; while aerobic processes are used to produce bioethanol.[23] According to Liang and McDonald,[31] anaerobic digestion is the best-suited waste to energy options for waste biomass. The conversion of various biomass through biochemical processes is presented in Table 9.4.

Easy handling, scaling up, and wide range of process parameters make biomass an acceptable material for different conversion technologies. The stable carbon is often produced as a product from various thermochemical conversion of waste biomass. Since the last decade, researchers have accordingly shifted their focus on stable carbon products as they find applications in various domains.[32,33] Also, the production of stable carbon-rich materials is seen as a possible approach to alleviate the carbon dioxide levels in the atmosphere, with their ability to store carbon in the soils.[33,34] Since the stable carbon-rich materials are highly recalcitrant in nature, they bear the potential to mitigate climate change by slowing down the movement of elemental carbon back to the atmosphere.[34] Thus, a case on stable carbon from biomass for sustainable applications holds good pertinence.

9.2 CASE STUDY: WASTE BIOMASS DERIVED STABLE CARBON—A SUSTAINABLE UTILIZATION APPROACH

9.2.1 STABLE CARBON FROM BIOMASS FEEDSTOCK

Stable carbon is the solid combustible residue of any carbon-rich material which remains upon heating at very high temperatures after the removal of moisture and volatile substances. More precisely, stable carbon is the

TABLE 9.4 Biochemical Processes for Biomass Conversion.

Biomass feedstock	Conversion process	Significant outcomes	References
Rice straw	Anaerobic digestion	Biogas yield and biodegradation of rice straw in a batch anaerobic digester under room temperature for 60 days. High biogas production at inoculum/substrate ratio of 0.8, with methane content over 70%.	[23]
Potato peel	Anaerobic digestion	Prefermentation potato peel waste and its lactic acid fermentation ratio were anaerobically digested for 40 days. Above 70% of methane yield was achieved before 20 days.	[29]
Pineapple peel	Fermentation	Alkaline pretreatment for 48 h followed by hydrolysis helped in enhancing ethanol production by fermentation by yeast. The highest ethanol yield of 8 g/L was recorded after 48 h.	[13]
Rice straw	Fermentation	Two microbes produced complementary lignin-degrading enzymes, laccase, and lignin peroxidase. 22.74 g/L of reducing sugar was produced under optimal conditions. The productivity of ethanol was found up to 2.17 g/(L h).	[30]
Potato peel	Bio-methanation	Stepwise digestion of peel waste mitigated acidogenesis at earlier stages and enhanced biomethane yield with an increase in methane percent from 46.47% to 60.4% and a reduction in total COD to about 94 %.	[24]
Waste activated sludge	Anaerobic digestion followed by biomethanation	Waste-activated sludge was chemically pretreated in alkaline and acidic conditions at varying temperatures and operating time. Increased COD during anaerobic digestion resulted in an increased methane generation—up to 43.6% under alkaline pretreatment.	[31]

amount of carbon in the percentage of the total weight of the solid. Stable carbon products are generally produced through the thermochemical breakdown of biomass at temperatures of 200°C–900°C under anoxic conditions.[35] In the production of stable carbon-rich material, however, the main product of interest is the solid residue. The removal of moisture along with the volatile fractions results in a surge in the comparative stable carbon content of the biomass residue. Weber and Quicker[36] confirmed that polymerized organics in vapors and gases may lead to secondary charring, further helping in the solid yield. Currently, a wide range of feedstocks, which include woody biomass (chips and wood pellets), tree barks, crop residues (straws, nutshells, and hulls), grasses of different families, organic wastes including distillers' grain, bagasse from the sugarcane industry, chicken litter, dairy manure, sewage sludge and paper sludge, is used throughout the world.[14,15,19]

Feedstocks used for obtaining stable carbon products (e.g., woody biomass, crop residues, and grasses) influence the overall characteristics of stable carbon, including concentrations of elemental constituents, density, porosity, and hardness.[12] Optimizing the stable carbon product for a specific application may require proper selection of biomass type as well as conversion techniques and conditions. Several articles that targeted the relationships between production conditions, characteristics, and potential end-users of the stable carbon product are present.[14,15] Besides the yield of stable carbon, its properties vary with different process parameters, viz., temperature, residence time, heating rate, particle size, etc. A detailed review was prepared by Tripathi et al.[19] Table 9.5 enlists different stable carbon products obtained from pyrolysis of different biomass.

A major portion of the crop's body is nonedible and is discarded as agricultural residue.[14,15] Straws comprise close to half of the cereal crops yield and are potentially more suitable for conversion to stable carbon form. The most desirable feedstocks for pyrolysis are the agricultural residues, as crop cultivation demands lesser area and farming period compared to woody biomass. A huge amount of agricultural residue is generated every day in India as it is an agriculture-dependent country and the agricultural residues are more than often discarded as waste. Dyani and Bhasker[15] have reviewed different types of agricultural wastes that are converted to stable carbon-rich materials through pyrolysis. Therefore, the feedstocks investigated were two very distinct agricultural residues: a crop residue and a vegetable peel.

TABLE 9.5 Different Stable Carbon Products Obtained from Pyrolysis of Different Biomass Feedstocks.[12,14,15,19]

Biomass feedstock	Pyrolysis temperature and time	Stable carbon (%)	Volatile (%)	Ash (%)
Woody biomass				
Pinewood	700C, 70 min	57.1	3.2	38.8
Oakwood	400°C–450°C	78.3	15.6	2.9
Saw dust	550°C	82.6	13.6	2.8
Nonwoody biomass				
Rapeseed plant	800°C	79.7	6.0	15.3
Cotton fiber	300°C, 20 min	28.4	60.6	11.0
Switch grass	500°C, 120 min	31.6	7.1	52.5
Agricultural residues				
Corn stover	450°C	28.7	12.7	58.0
Soybean stover	700°C, 100 min	67.7	14.7	17.2
Rice husk	500°C	–	–	42.2
Fescue straw	700°C	71.6	9.1	19.3
Cottonseed hull	800°C	69.5	11.4	9.2
Peanut shell	700°C, 100 min	58.1	32.7	8.9
Potato peel	700°C, 120 min	27.9	60.9	7.0

9.2.2 EFFECTS OF PYROLYSIS TEMPERATURE AND RESIDENCE TIME ON STABLE CARBON YIELD

Pyrolysis temperature and residence time are the most vital process conditions that dictate the yield and overall characteristics of stable carbon.[36] The effects of pyrolysis temperature and residence time on the characteristics of the stable carbon product were studied using two different waste biomass as the feedstock. The proximate analyses of the two biomasses were conducted as per ASTM standards (ASTM D1104-84 and E872-82). Also, the bulk densities of ground biomass were determined as per standards (ASTM D854-92). The theoretical higher heating values (HHVs) of the biomasses were determined using the correlations given by Dashti et al.[37] The pyrolysis of waste biomass was performed in a muffle furnace (Simeco Pvt. Ltd., WB, India) without the continuous flow of inert gas to make the process cost-effective. The pyrolysis temperature was varied from 250°C to 650°C, with 100°C increment and the residence time was varied between 1 h and 4 h, with an increment of 1 h.

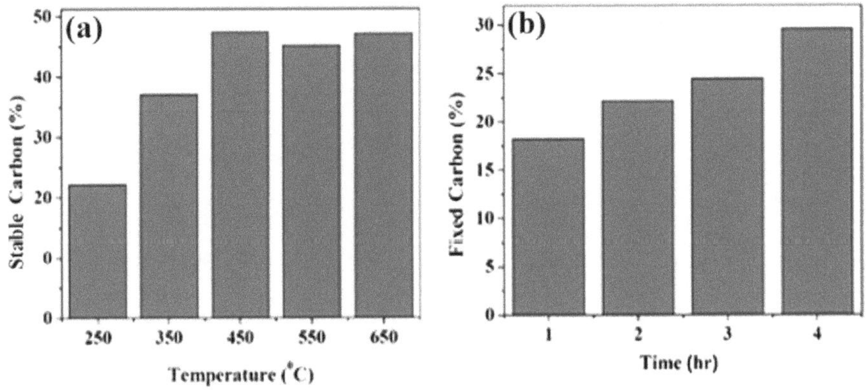

FIGURE 9.3 Comparative yields of stable carbon at varying (a) temperature and (b) residence time.

The quality of stable carbon is largely influenced by its fixed carbon content.[36] It was inferred that as the temperature of pyrolysis increased, the yield of stable carbon increased accordingly till a temperature of 450°C, above which the yield of stable carbon remains somehow unaffected (Fig. 9.3a). Similar results were obtained by others[38] who had concluded that with an increased temperature of pyrolysis, a rise in C while a fall in H and O contents takes place. A resultant lower molar H/C and O/C ratios, therefore indicating the occurrence of dehydration and deoxygenation of the biomass.[38] Additionally, beyond 4 h of continuous pyrolysis, there was no significant change in the amount of stable carbon present in the solid residue (Fig. 9.3b). Weber and Quicker[36] concluded that by increasing the residence time, an increased amount of stable carbon products can be obtained from pyrolysis. Moreover, they've also stated that the stable carbon yield is restricted once the volatile fractions of the biomass have been completely exhausted during the process. Thus, it can be inferred that 4 h of continuous heating results in the complete loss of volatile matter thereby producing highest stable carbon yield.

9.2.3 EFFECTS OF PYROLYSIS TEMPERATURE AND RESIDENCE TIME ON STABLE CARBON PROPERTIES

Table 9.6 enlists different physical properties of the two waste biomass under investigation before and after pyrolysis. As evident from the table, the stable

TABLE 9.6 Physical Characterization of Two Agricultural Residues.

Biomass	Moisture (%)	Volatile (%)	Ash (%)	Stable carbon (%)	HHV (MJ/Kg)	Bulk density (g/cm3)	SSA (m2/g)
Before pyrolysis and grounded							
Vegetable peel	11.43	59.17	14.25	15.15	14.47	0.541	0.75
Crop residue	14.82	54.71	11.23	19.24	15.24	0.341	1.27
After pyrolysis at 250°C for 2 h							
Vegetable peel	10.27	51.12	17.06	21.55	15.46	0.396	1.57
Crop residue	9.36	45.39	19.78	25.47	15.93	0.175	4.88
After pyrolysis at 550°C for 2 h							
Vegetable peel	11.67	13.67	29.54	45.12	17.86	0.225	9.91
Crop residue	2.71	13.80	53.05	30.44	17.38	0.132	32.87

carbon content increased as the pyrolysis temperature was increased. It was because, at higher temperatures, the more volatile matter was removed from the waste biomass. The ash content of the stable carbon material is largely dependent on the type of biomass waste. The ash content is known to rise with an increasing pyrolysis temperature because, after the removal of moisture and volatile fractions, a big chunk of ash remains in the pyrolysis residue.[36] The ash content was found to increase drastically with temperature.

Any biomass, irrespective of its source, primarily consists of cellulose, hemicellulose, and lignin in specific percentages. During the thermal treatment processes, these organic compounds behave differently from one another and directly impact the quantity of the product and its physical and chemical characteristics.[36] At temperatures between 220°C and 300°C, the thermal decomposition of hemicellulose present in the agricultural residues starts.[38] Cellulose, being thermally more stable, is decomposed between 280°C and 400°C.[36] Lignin consists of a variety of functional groups therefore have a wider range of thermal degradation, between 220°C and 900°C.[38] Thus, the bulk density of solids decreases as the pyrolysis temperature is increased, due to the consequent disintegration of the structural constituents of the biomass.

The gross calorific value or HHV of pyrolyzed solid is very much comparable to those of lignite and usually falls within the range of 15–23 MJ/kg,[15,37] thus making them an interesting choice as alternate solid fuels for power generation. At 250°C, there was a little increase in the HHV of the two waste biomasses. However, the HHV of the two waste biomass increased as the pyrolysis temperature was increased to 550°C, which was 14–24% higher than their respective feedstocks. The HHV of the vegetable peel stable carbon was higher than those of the crop waste stable carbon, because of the higher stable carbon content and a lower ash content of the former. Moreover, HHV increased with the increase in the pyrolysis temperature, in accordance with the study by Qin et al.[39]

As evident from Table 9.6, an increment in the surface area occurred with a rise in the pyrolysis temperature. The stable carbon-rich materials showed a little increment in its specific surface area (SSA) upon the pyrolysis temperature of 250°C. But, interestingly, at the pyrolysis temperature of 550°C, there was a drastic increase in the SSA of the pyrolyzed material from 1.6 to 9.9 m^2/g for vegetable peel waste and 4.9–32.9 m^2/g for the crop residue. The stable carbon produced at temperatures above 700°C usually possesses lower surface area, due to the deformation or obstruction of micropores in the structure.[40]

9.2.4 EFFECTS OF PYROLYSIS TEMPERATURE AND RESIDENCE TIME ON STABLE CARBON MORPHOLOGY

Pyrolysis temperature and residence time play important roles in the morphology and surface structural integrity of the stable carbon.[36] The solid products of pyrolysis were characterized using scanning electron microscopy (SEM) (NOVA SEM 450, FEI, BB, Netherlands) to visualize the morphology of the solids.

Figure 9.4 illustrates the SEM micrographs of various stable carbon products of the two agricultural residues pyrolyzed at three different temperatures. Visual inspection showed microstructure differences between the various stable carbon products for the pyrolysis temperature. Distinct macropores are observable in the stable carbon product of vegetable peel pyrolysis (Fig. 9.4a–c). It is evident that as the pyrolysis temperature was increased, porosity (both macro and meso) of the pyrolyzed material was also increased. Figure 9.4d,e suggests that upon increasing the pyrolysis temperature of the crop residue, the surface integrity was lost considerably.

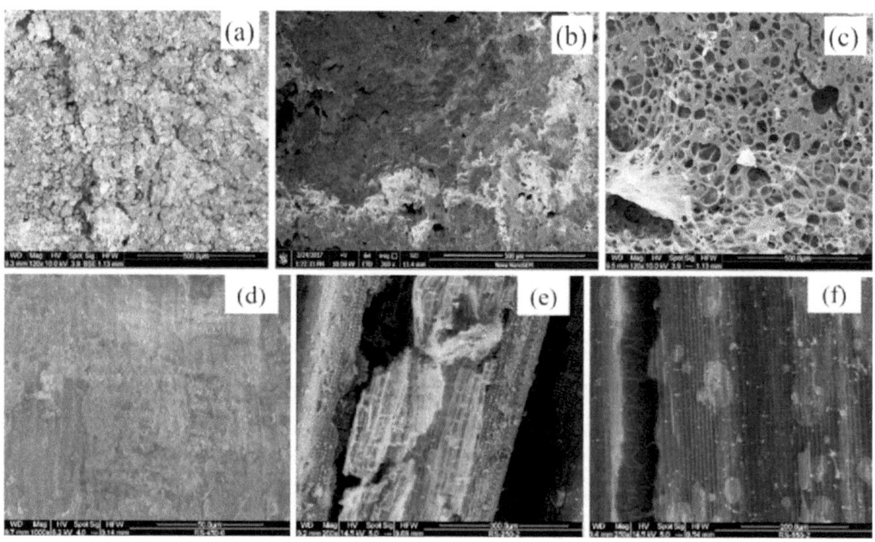

FIGURE 9.4 SEM images of stable carbon materials under different pyrolysis temperatures of (a) 250°C, (b) 350°C, and (c) 550°C for vegetable peel and (d) 250°C, (e) 350°C, and (f) 550°C for crop residue.

Also, X-ray crystallography (XRD) (D8 Advance, Bruker, US) was performed for the identification of the phase and structure of the adsorbent using Cu kα radiation of 40 KV at diffraction angles from 20° to 80°. The XRD patterns of the stable carbon products of agricultural residue pyrolyzed at different temperatures are shown in Figure 9.5. The sharp peak at 28.15° is due to the crystal plane index C(002).[25] This C(002) plane is due to the parallel and azimuthal orientation of the aromatic, partially carbonized lamellae. Sharper peaks are indicative of a higher degree of orientation.[34] Similarly, another sharp peak at 42.18° in both the solids is due to the crystal plane index of C(100). This C(100) peak is due to the condensed aromatic carbonized planes.[34] Moreover, small peaks at 26.59° and 67.84° conforms to SiO_2 (quartz) (JCPDS card no. 46-1045),[41] at 50.54° to $Ca(OH)_2$ (JCPDS card no. 01-073-5492),[42] whereas, peak at 65.14° indicate the presence of MnO_2 (JCPDS Card no. 44-0141).[43] Thus, increasing C(002) and C(100) peaks depict a considerable degree of crystallinity of carbon in the stable carbon products with increasing pyrolysis temperature.

A variety of feedstock is acceptable for pyrolysis, but the variable characteristics of the biomass feed are responsible for the inconsistency in the yield and characteristics of the final product. As a result, the scale-up from the laboratory setup to pilot plants turn out to be quite challenging.[15] Thus, to make use of the pyrolyzed biomass, it is necessary to understand the various applications of such materials.

9.3 UTILIZATION OF STABLE CARBON PRODUCTS

The applications of stable carbon, derived from the pyrolysis of biomass, are diversified. Stable carbon can be used as a solid fuel for thermal power generation, in metallurgy, in agronomy and animal husbandry, as a building material, and in the biomedical field—to name a few applications. With the idea to lower the greenhouse gas emissions, stable carbon has gained immense popularity over the years as a substitute for fossil-based carbon in several of these applications.[25,37] Stable carbon product can be used as a fuel. In such applications, stable carbon is carbon neutral, hence can also be used as a carbon sink.[37] Stable carbon from biomass feedstock is an excellent fertilizer, as it improves soil texture,[33] retains and slowly releases nutrients and water into the soil, and acts as a support for beneficial organisms required for plant growth.[34] Stable carbon has shown its potential as a good adsorbent of heavy metals and organic pollutants.[32,44] Thus, stable carbon obtained from waste biomass was utilized for various applications in the fields of adsorption, soil nutrient management, and catalysis, as shown in Table 9.7.

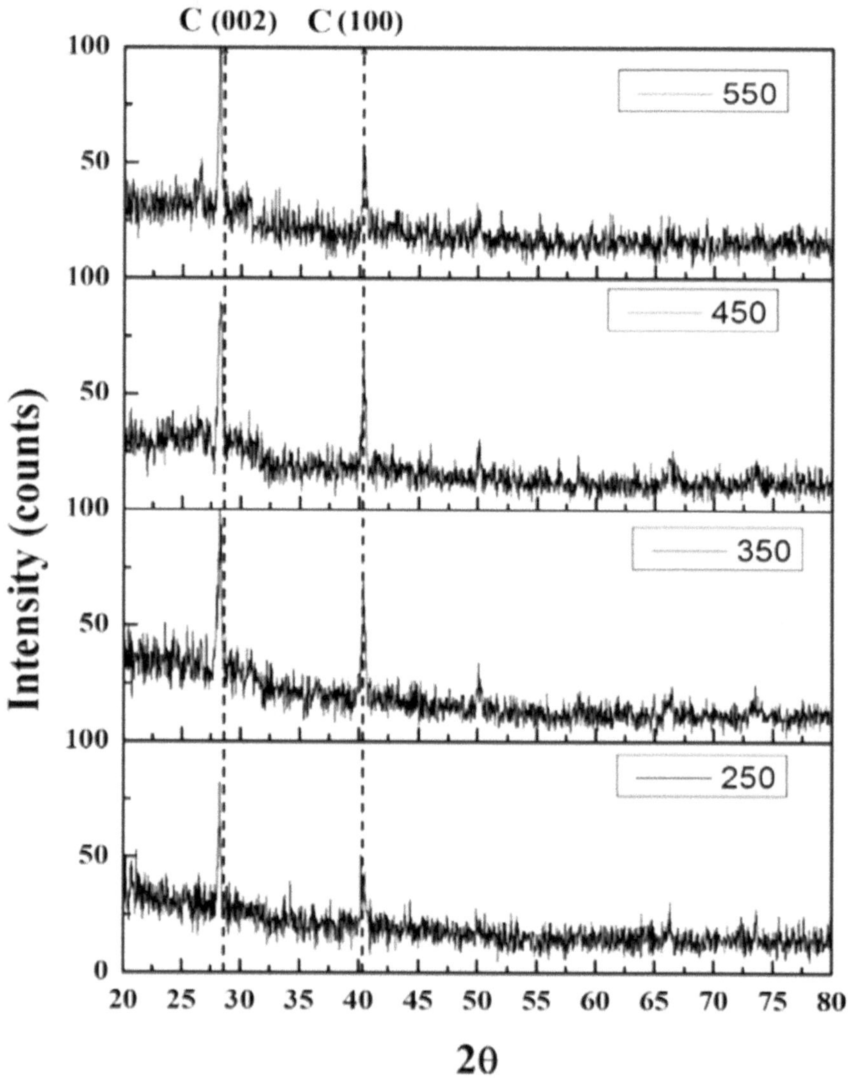

FIGURE 9.5 XRD pattern of the waste vegetable peel along with stable carbon products prepared different pyrolysis temperatures (250°C–550°C) with a residence time of 2 h.

TABLE 9.7 Applications of Stable Carbon Products from Different Biomass Feedstocks.

Biomass feedstock	Carbon (%)	Significant outcomes	References
As fuel (bio-oil and solid biofuel)			
Cotton fiber	15.0–28.4 SC	Increase in the combustion property of biofuel (increased thermal conversion efficiency and reduced pollutant emissions) and mass loss at the high-temperature zone with increasing pyrolysis temperature	[16]
Pinewood	18.6–47.9 SC		
Rice straw	27.9 SC	Pretreatment with sulfuric acid enhanced the energy recovery potential by of waste rice straw by 15.67% as compared to untreated rice straw.	[38]
Pine nutshell	74.6 SC	Increased HHV of stable carbon at 550°C and also at 1.0 MPa pressure—31.9 and 33.9 MJ/kg, respectively.	[39]
In soil fertility, crop growth, and carbon sequestration			
Rise husk	82.6 SC	Increase in soil pH, conductivity, organic matter, organic carbon content, and cation exchange capacity. Also a significant reduction in CO_2 emission from the soil with the addition of stable carbon products in soil.	[34]
Corn stover	76.6 SC		
Rice husk	NA	Increased soil concentrations of nitrogen, phosphorus, potassium, and enhanced soil microbial growth—leading to a higher yield of paddy plants.	[45]
Douglas fir	90.0 SC	Increased soil pH, organic matter content, phosphorus, potassium, sulfur contents, and the shoot and root biomass of wheat was observed with stable carbon application	[46]
Biosorption of heavy metals from water			
Pine nutshell	74.6 SC	Lead(II) The adsorption capacity of 273.3 mg Pb(II)/g of stable carbon product, followed Freundlich isotherm and pseudo-second-order kinetics	[39]
Miscanthus	58.3–88.6 SC	Cadmium(II) The adsorption capacity of 12.9 mg Cd(II)/g of stable carbon product, followed Freundlich isotherm and pseudo-second-order kinetics	[47]
Cottonwood	NA	Arsenic(V) The adsorption capacity of 3.2 mg As(V)/g of iron oxide nanoparticles modified stable carbon product, followed Langmuir isotherm and pseudo-second-order kinetics	[48]

TABLE 9.7 *(Continued)*

Biomass feedstock	Carbon (%)	Significant outcomes		References
Walnut shell	NA	Chromium(VI)	The adsorption capacity of 124.4 mg Cr(VI)/g of Chitosan modified stable carbon product, followed Freundlich isotherm and pseudo-second-order kinetics	[49]
Mustard waste and soy waste	NA	Cobalt(II)	The adsorption capacity of 24.2 mg Co(II)/g and 16.6 mg Co(II)/g for mustard waste and soy waste stable carbons, respectively. Adsorption followed Langmuir isotherm and pseudo-second-order kinetics	[50]
Biosorption of organic contaminants from water				
Corn stover	52.7 SC	Fluoride	The adsorption capacity of the magnetic stable carbon was 4.1 mg/g, followed Langmuir isotherm and pseudo-first-order kinetics	[51]
Corn straw	42.9 C	Atrazine	The adsorption capacity of sodium sulfide treated stable carbon was 2.69 mg/g, followed Langmuir isotherm and pseudo-first-order kinetics. 76% removal was achieved.	[52]
Korean cabbage leaf, rice straw, and wood chips	NA	Crystal violet	The adsorption capacities of Korean cabbage, rice straw, and wood chips were 1304, 620, and 196 mg/g, respectively. All three followed Langmuir isotherm and pseudo-second-order adsorption kinetics	[53]
Waste fiberboard	65.7 SC	Tetracycline	The adsorption capacity of stable carbon was 6.37 mg/g, followed Langmuir isotherm and pseudo-first-order adsorption kinetics. 68.6% removal was achieved.	[54]
Litchi peel	NA	Malachite green Congo red	Malachite green and Congo red adsorption capacity of 2468 and 404 mg/g, respectively, and followed Elovich kinetic model. Malachite green adsorption was best fitted in Freundlich model, whereas, Congo red followed Temkin model	[55]

NA = Not available; C = as elemental carbon; SC = as stable carbon.

9.3.1 STABLE CARBON IN BIOSORPTION OF POLLUTANTS

In the past few years, around the world, the potential of stable carbon materials as an adsorbent for removing water contaminants from aqueous solution was investigated.[32,49,55] The adsorption of organic pollutants from an aqueous medium on stable carbon-rich material is possible because of its microporous morphology and a high surface area.[32] Table 9.7 enlists a few such adsorption studies carried out onto sable carbon products produced from different feedstocks and for various metal ions and organic contaminants in water. Batch adsorption studies were conducted for the removal of a metal ion (Fe(III)) and an organic pollutant—Methylene Blue (MB) using the stable carbon-rich solid obtained from the pyrolysis of vegetable peel at both 250°C and 550°C temperatures. Different concentrations of Fe(III) and MB were prepared in deionized water for various biosorbent doses. The pH of the synthetic solutions was not altered and samples were taken out at different time intervals. The concentrations of Fe(III) and MB were determined using a UV-Vis spectrophotometer (Lambda-35, Perkin Elmer, Singapore).

Huang et al.[49] have linked the occurrence of chemisorption with the plentiful existence of charged species on the surface of stable carbon-based biosorbents. Stable carbon-rich material was found to be an excellent biosorbent, as evident from Figures 9.6 and 9.7. The removal efficiency increased with the increase in the concentration of ions for a fixed adsorbent dose and then it kept on decreasing. It occurred because of the availability of a large adsorbent surface at lower concentrations, which kept on exhausting as more and more ions were adsorbed from the solution on the adsorbent surface.[39,52]

The rate of adsorption of the soluble pollutants at the beginning of the reaction was high because of the presence of a large number of adsorption sites to the ions in solution. This may be attributed to an increase in the driving force of the concentration gradient with the increase in the initial concentration of pollutants in the solution.[39,54] With passing time, the number of adsorption sites occupied in adsorbent surface increases, and the number of empty places for placement of pollutants are reduced, thus reducing the rate of absorption. Upon exhaustion of the adsorption sites on the adsorbent's surface, the diffusion of adsorbate into the pores control the rate of its sorption.[48,55] Upon kinetic experiments, it was found that equilibrium adsorption was achieved in 60 min for both Fe(III) and in 30 min for MB, after which the removal became somewhat constant.

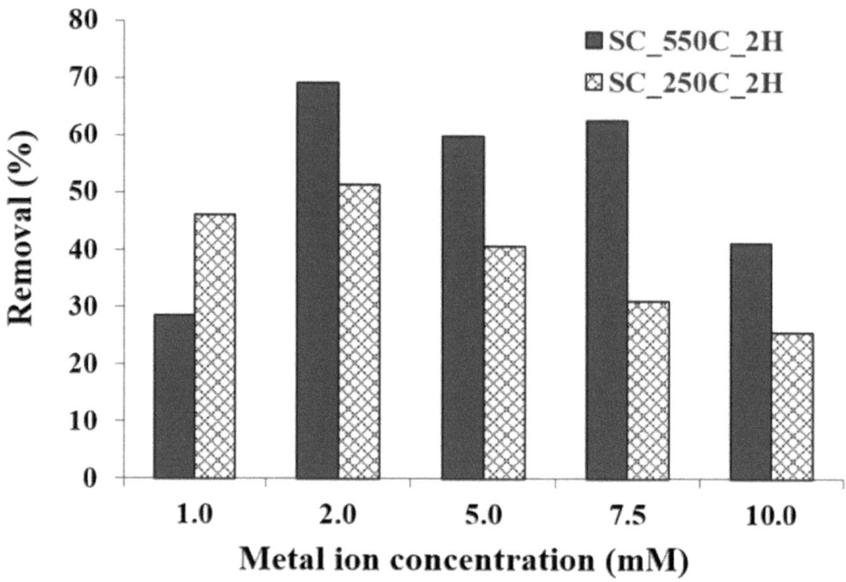

FIGURE 9.6 Removal efficiency of stable carbon obtained from waste vegetable peel for different Fe^{3+} concentrations.

The stable carbon produced from vegetable peel pyrolysis at 550°C demonstrated an adsorption capacity of 72.2 mg Fe(III)/g, removing close to 70% of the Fe(III) ions from the solution in just half an hour for an initial concentration of 2.0 mM (Fig. 9.7a). But, for the stable carbon-rich material produced at 250°C, the equilibrium adsorption capacity was mg/g along with 51% removal efficiency. Although, at a low iron concentration of 1.0 mM, stable carbon material pyrolyzed at 250°C was able to demonstrate better removal efficiency than the 550°C material, in general, waste vegetable peel pyrolyzed at 550°C has shown good potential in removing Fe(III) ions. The adsorption of Fe(III) onto stable carbon material's surface followed pseudo-second-order kinetics and was best described with Langmuir's isotherm model. Such results were also obtained by others who had used pyrolyzed biomass, with high stable carbon content, in adsorptive removal of metal ions from water.[48,50]

Also, the stable carbon-rich product, produced at a pyrolysis temperature of 250°C, effectively removed MB to the extent of 95%. This biosorbent was better than other stable carbon products produced from different biomass feedstocks, as listed in Table 6. As the pyrolysis temperature was increased, adsorption capacity decreased linearly, which is evident from Figure 9.7b. The MB adsorption on stable carbon products from crop waste followed Langmuir adsorption and pseudo-second-order kinetics, with a maximum adsorption capacity of 38.02 mg/g. Mohan et al.[51] also reported that the Langmuir model had a good description of the organic pollutant adsorption on biosorbents based on the correlation coefficients. Moreover, a similar model was proposed for the adsorption of organic pollutants on stable carbon products from different sources.[52,53]

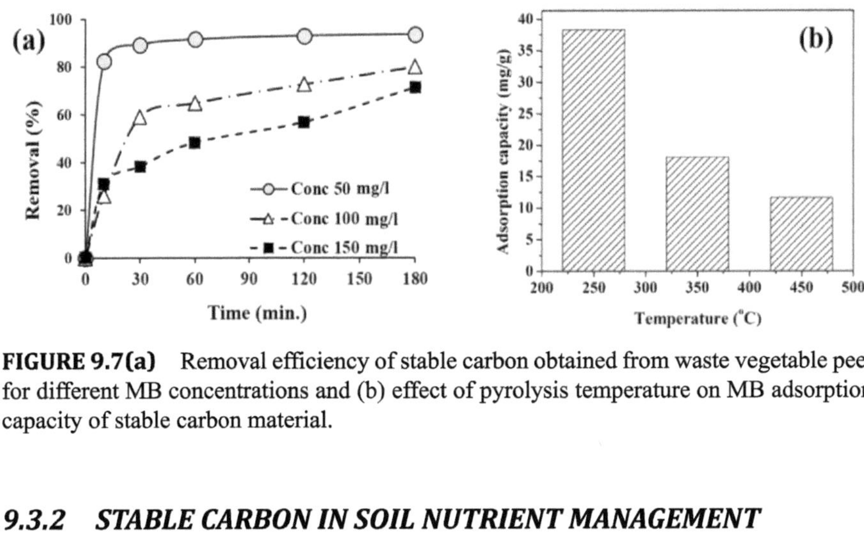

FIGURE 9.7(a) Removal efficiency of stable carbon obtained from waste vegetable peel for different MB concentrations and (b) effect of pyrolysis temperature on MB adsorption capacity of stable carbon material.

9.3.2 STABLE CARBON IN SOIL NUTRIENT MANAGEMENT

Stable carbon-rich products can achieve enormous impact in the agricultural world due to its ease of application. Modes of application include: mixing with manures, fertilizers and seeds, top-dressing (on the soil surface), and uniform soil mixing, with or without plow deep banding, mixing with compost, and layering at planted edges.[34] Growth in the agricultural yields in soils, which are infertile or degrading, can be beneficial for the third world countries. The stable carbon is generally is stored in a recalcitrant form that can improve the soil's capacity to retain moisture and various nutrients, typically resulting in increased plant growth.[34] It has a positive priming effect on soil and a significant increase in the germination of seed and growth in plants has been observed for the soil with the addition of stable carbon-rich materials.[45] Its addition to soil has shown a greater increase of exchangeable cations of calcium, potassium, magnesium, and sodium in the soil, due to their negatively charged reactive surface which allows for cations to be electrostatically bounded and available for exchange with the plant roots.[34,46]

Stable carbon has been investigated for its application in soil nutrient management, specifically in evaluating its ability to retain moisture and organic carbon, soil biomass improvement, and potassium availability in soil. Amending soil with stable carbon products has led to a significant increase in the percentage of organic carbon in the soil after 180 days (Fig. 9.8a). Similar results were obtained by.[34] Soil mixed with stable carbon product pyrolyzed at 250°C showed higher organic carbon content on before and after 180 days of the growing period, as compared to 550°C pyrolyzed product.

The above soil biomass content also increased by 3.5 times for 250°C stable carbon product and 3.0 times for 550°C stable carbon product, relative to that for the control soil after 180 days (Fig. 9.8b). The percentage available potassium in topsoil decreases with the growth of the plant. However, after 180 days, the available potassium was 12–14 times higher in the stable carbon-containing soil than the control soil (Fig. 9.8c).

Upon amending the toil soil with the stable carbon-rich products, the initial and final potassium content in the soil were significantly higher than those for control soil. Since stable carbon-rich materials are characteristically stable in soil, they could very easily sequester far more elemental carbon in the soil than any other biomass with analogous carbon contents.[46] Thus, stable carbon material can play an important role in harnessing the carbon cycle.

9.4 CONCLUSION

The valorization of waste biomass as stable carbon was explored in this study as a means to alleviate the widening ecological deficit. The waste biomass derived stable carbon being carbon negative can minimize ecological footprint due to growing consumption. Two different waste biomass—a vegetable peel and a crop residue—were converted to stable carbon by the thermochemical route of pyrolysis. The stable carbon

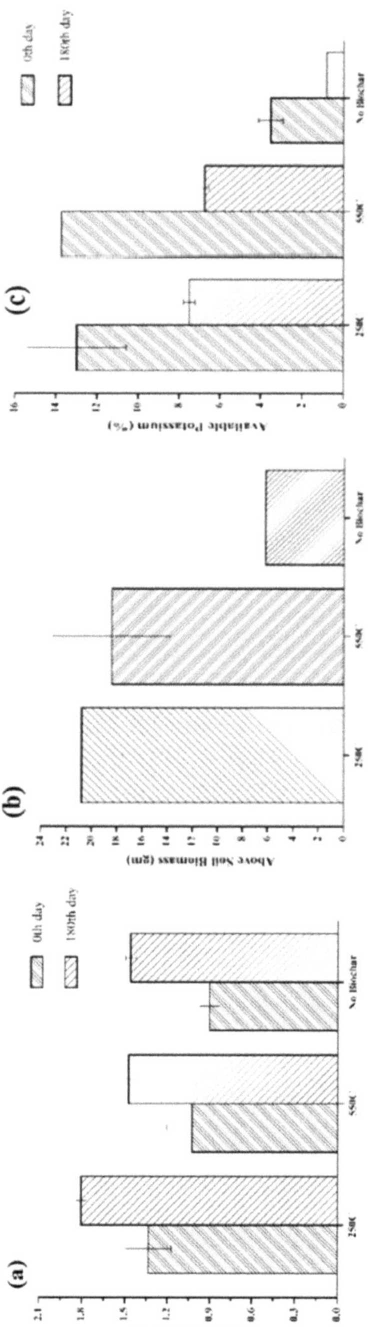

FIGURE 9.8 Investigations on (a) percentage of organic carbon, (b) biomass, and (c) available potassium in soil using stable carbon materials of two pyrolysis temperature.

content was found to rely on two process parameters—temperature and residence time. Increasing both the process parameters led to an increase in the stable carbon yield till 550°C and 2 h, beyond which not much change was observed. The HHV and SSA increased with a rise in the pyrolysis temperature, because of the removal of O and H and a decrease in H/C content in the carbon product. More than a 10-fold increase in SSA of the solids made them better suited as a sorbent of metal ions or organic pollutants in aqueous solution. The vegetable peel derived stable carbon showed good biosorption capacities for iron (Fe(III)) and methylene blue (MB) dye (organic dye) in synthetic water. Stable carbon produced at pyrolysis temperature of 550°C was the better adsorbent for Fe(III) ion adsorption, with 70% removal efficiency. While, the stable carbon produced at a temperature of 250°C was better, with more than 95% removal efficiency, for organic dye removal. Both the adsorption processes followed pseudo-second-order kinetics and were best fit with Langmuir isotherm, suggesting monolayer formation on the solid surface. The maximum adsorption capacities were 72.2 mg Fe(III)/g and 38.02 mg dye/g for the stable carbon. The stable carbon product prepared from crop residue at 550°C was added to the soil for nutrient management. The addition of stable carbon in the soil resulted in a 29%, 250%, and 140% increase in the organic carbon content, above soil biomass and soil potassium respectively. Hence, the waste biomass derived stable carbon was proved as a good additive for increasing soil fertility and maximizing crop growth. Therefore, utilization of waste biomass by a simpler conversion process can solve the problems of water contamination with heavy metals and organic pollutants as well as improve soil nutrient management for better crop productivity. The great potential of waste biomass derived stable carbon can be envisaged to facilitate the valorization of waste biomass and thereby impede the increment of the ecological footprint.

ACKNOWLEDGMENT

The authors thank the Central Research Facility at the National Institute of Technology Agartala and Rubber Board of Tripura for technical support.

KEYWORDS

- **ecological footprint**
- **waste biomass**
- **stable carbon**
- **biosorption**
- **soil nutrient management**

REFERENCES

1. Global Footprint Network Report. 2017-04-17 (accessed Oct 5, 2020).
2. National Footprint Accounts Report, 2017. data.footprintnetwork.org (accessed Oct 5, 2020).
3. Lin, D.; Hanscom, L.; Murthy, A.; Galli, A.; Evans, M.; Neill, E.; Mancini, M. S.; Martindill, J.; Medouar, F-Z.; Huang, S.; Wackernagel, M. Ecological Footprint Accounting for Countries: Updates and Results of the National Footprint Accounts, 2012–2018. *Resources* **2018,** *7* (3). https://doi.org/10.3390/resources7030058.
4. Hackett, C. By **2050,** India to Have World's Largest Populations of Hindus and Muslims. *Pew Research Center*. April 21, 2015. http://pewrsr.ch/1HPGwo8 (accessed Oct 5, 2020).
5. Kaza, S.; Yao, L. C.; Bhada-Tata, P.; Van Woerden, F. *What a Waste 2.0: A Global Snapshot of Solid Waste Management to 2050. Urban Development*; World Bank: Washington, DC, 2018.
6. Kumar, S.; Smith, S. R.; Fowler, G.; Velis, C.; Kumar, S. J.; Arya, S. R.; Kumar, R.; Cheeseman, C. Challenges and Opportunities Associated with Waste Management in India. *R.. Soc. Open Sci.* **2017,** *4*, 1607–1664.
7. Joshi, R.; Ahmed, S. Status and Challenges of Municipal Solid Waste Management in India: A Review. *Cogent. Environ. Sci.* **2016,** *2*, 1139–1143.
8. Ahmed, Z.; Wang, Z.; Mahmood, F.; Hafeez, M.; Ali, N. Does Globalization Increase the Ecological Footprint? Empirical Evidence from Malaysia. *Environ. Sci. Pollut. Res.* **2019,** *26* (18), 18565–18582.
9. Nizami, A. S.; Rehan, M.; Waqas, M.; Naqvi, M.; Ouda, O. K. M.; Shahzad, K.; Miandad, R.; Khan, M. Z.; Syamsiro, M.; Ismail, I. M. I.; Pant, D. Waste Biorefineries: Enabling Circular Economies in Developing Countries. *Bioresour. Technol.* **2017,** *241*, 1101–1117.
10. Bar-On, Y. M.; Phillips, R.; Milo, R. The Biomass Distribution on Earth. *Proc. Natl. Acad. Sci.* **2018,** *115* (25), 6506–6511.
11. Wu, Z.; Li, Y.; Xu, D.; Meng, H. Co-Pyrolysis of Lignocellulosic Biomass with Low-Quality Coal: Optimal Design and Synergistic Effect from Gaseous Products Distribution. *Fuel.* **2019,** *236*, 43–54.

12. Cai, J.; He, Y.; Yu, X.; Banks, S. W.; Yang, Y.; Zhang, X.; Yu, Y.; Liu, R.; Bridgwater, A. V. Review of Physicochemical Properties and Analytical Characterization of Lignocellulosic Biomass. *Renew. Sustain. Energy Rev.* **2017,** *76*, 309–322.
13. Casabar, J. T.; Unpaprom, Y.; Ramaraj, R. Fermentation of Pineapple Fruit Peel Wastes for Bioethanol Production. *Biomass Conver. Biorefin.* **2019,** *9* (4), 761–765.
14. Kan, T.; Strezov, V.; Evans, T. J. Lignocellulosic Biomass Pyrolysis: A Review of Product Properties and Effects of Pyrolysis Parameters. *Renew. Sustain. Energy Rev.* **2016,** *57*, 1126–1140.
15. Dhyani, V; Bhaskar, T. A Comprehensive Review on the Pyrolysis of Lignocellulosic Biomass. *Renew. Energy.* **2018,** *129*, 695–726.
16. Liu, Z.; Han, G. Production of Solid Fuel Biochar from Waste Biomass by Low Temperature Pyrolysis. *Fuel* **2015,** *158*, 159–165.
17. Sheldon, R. A. Green Chemistry and Resource Efficiency: Towards a Green Economy. *Green Chem.* **2016,** *18* (11), 3180–3183.
18. Mishra, R. K.; Mohanty, K. Pyrolysis of Three Waste Biomass: Effect of Biomass Bed Thickness and Distance between Successive Beds on Pyrolytic Products Yield and Properties. *Renew. Energy.* **2019,** *141*, 549–558.
19. Tripathi, M.; Sahu, J. N.; Ganesan, P. Effect of Process Parameters on Production of Biochar from Biomass Waste through Pyrolysis: A Review. *Renew. Sustain. Energy Rev.* **2016,** *55*, 467–481.
20. Goh, B. H. H.; Ong, H. C.; Cheah, M. Y.; Chen, W. H.; Yu, K. L.; Mahlia, T. M. I. Sustainability of Direct Biodiesel Synthesis from Microalgae Biomass: A Critical Review. *Renew. Sustain. Energy Rev.* **2019,** *107*, 59–74.
21. Rago, Y. P.; Mohee, R.; Surroop, D. A Review of Thermochemical Technologies for the Conversion of Waste Biomass to Biofuel and Energy in Developing Countries. In *The Nexus: Energy, Environment and Climate Change*; Filho, W. L., Surroop, D., Eds.; Springer: Cham, 2018; pp 127–143.
22. Molino, A.; Larocca, V.; Chianese, S.; Musmarra, D. Biofuels Production by Biomass Gasification: A Review. *Energies* **2018,** *11* (4), 811.
23. Candia-García, C.; Delgadillo-Mirquez, L.; Hernandez, M. Biodegradation of Rice Straw under Anaerobic Digestion. *Environ. Technol. Innov.* **2018,** *10*, 215–222.
24. Jacob, S.; Chintagunta, A. D.; Banerjee, R. Selective Digestion of Industrial Potato Wastes for Efficient Biomethanation: A Sustainable Solution for Safe Environmental Disposal. *Int. J. Environ. Sci. Technol.* **2016,** *13* (10), 2363–2374.
25. Chen, T.; Liu, R.; Scott, N. R.; Characterization of Energy Carriers Obtained from the Pyrolysis of White Ash, Switchgrass and Corn Stover—Biochar, Syngas and Bio-Oil. *Fuel Process. Technol.* **2016,** *142*, 124–134.
26. Greinert, A.; Mrówczyńska, M.; Grech, R.; Szefner, W. The Use of Plant Biomass Pellets for Energy Production by Combustion in Dedicated Furnaces. *Energies* **2020,** *13* (2), 463.
27. Biagini, E.; Barontini, F.; Tognotti, L. Gasification of Agricultural Residues in a Demonstrative Plant: Vine Pruning and Rice Husks. *Bioresour. Technol.* **2015,** *194*, 36–42.
28. Dudynski, M. Gasification of Selected Biomass Waste for Energy Production and Chemicals Recovery. *Chem. Eng. Trans.* **2018,** *65*, 391–396.

29. Liang, S.; McDonald, A. G. Anaerobic Digestion of Pre-Fermented Potato Peel Wastes for Methane Production. *Waste Manage.* **2015,** *46*, 197–200.
30. Wu, X.; Zhang, J.; Xu, E.; Liu, Y.; Cheng, Y.; Addy, M.; Ruan, R. Microbial Hydrolysis and Fermentation of Rice Straw for Ethanol Production. *Fuel.* **2016,** *180*, 679–686.
31. Tulun, Ş.; Bilgin, M. Enhancement of Anaerobic Digestion of Waste Activated Sludge by Chemical Pretreatment. *Fuel* **2019,** *254*, 115671.
32. Tan, X.; Liu, Y.; Zeng, G.; Wang, X.; Hua, X.; Gu, Y.; Yang, Z. Application of Biochar for the Removal of Pollutants from Aqueous Solutions. *Chemosphere* **2015,** *125*, 70–85.
33. Wu, S.; He, H.; Inthapanya, X.; Yang, C.; Lu, L.; Zeng, G.; Han, Z. Role of Biochar on Composting of Organic Wastes and Remediation of Contaminated Soils—A Review. *Environ. Sci. Pollut. Res.* **2017,** *24* (20), 16560–16577.
34. Mohan, D.; Abhishek, K.; Sarswat, A.; Patel, M.; Singha, P.; Pittman Jr., C. U. Biochar Production and Applications in Soil Fertility and Carbon Sequestration—A Sustainable Solution to Crop-Residue Burning in India. *RSC Adv.* **2018,** *8*, 508–520.
35. Leng, L.; Huang, H.; Li, H.; Li, J.; Zhou, W. Biochar Stability Assessment Methods: A Review. *Sci. Total Environ.* **2019,** *647*, 210–222.
36. Weber, K.; Quicker, P. Properties of Biochar. *Fuel* **2018,** *217*, 240–261.
37. Dashti, A.; Noushabadi, A. S.; Raji, M.; Razmi, A.; Ceylan, S.; Mohammadi, A. H. Estimation of Biomass Higher Heating Value (HHV) Based on the Proximate Analysis: Smart Modeling and Correlation. *Fuel* **2019,** *257*, 115931.
38. Roy, R.; Ray, S. Effect of Various Pretreatments on Energy Recovery from Waste Biomass. *Energy Sourc. A. Recov. Util. Environ. Eff.* **2019**. https://doi.org/10.1080/15567036.2019.1680767.
39. Qin, L.; Wu, Y.; Hou, Z.; Jiang, E. Influence of Biomass Components, Temperature and Pressure on the Pyrolysis Behavior and Biochar Properties of Pine Nut Shells. *Bioresour. Technol.* **2020,** *313*, 123682.
40. Hassan, M.; Liu, Y.; Naidu, R.; Parikh, S. J.; Du, J.; Qi, F.; Willett, I. R. Influences of Feedstock Sources and Pyrolysis Temperature on the Properties of Biochar and Functionality as Adsorbents: A Meta-Analysis. *Sci. Total Environ.* **2020,** 140714.
41. Jin, C.; Jin, L.; Guo, M. X.; Liu, P.; Zhang, J. N.; Bian, S. W. MnO_2 Nanotubes Assembled on Conductive Graphene/Polyester Composite Fabric as a Three-Dimensional Porous Textile Electrode for Flexible Electrochemical Capacitors. *J. Colloid Interf. Sc.* **2017,** *508*, 426–434.
42. Xu, X.; Hu, X.; Ding, Z.; Chen, Y. Effects of Copyrolysis of Sludge with Calcium Carbonate and Calcium Hydrogen Phosphate on Chemical Stability of Carbon and Release of Toxic Elements in the Resultant Biochars. *Chemosphere* **2017,** *189*, 76–85.
43. Madrid, J.A; Lanzón, M. Synthesis and Morphological Examination of High-Purity Ca(OH)2 Nanoparticles Suitable to Consolidate Porous Surfaces. *Appl. Surf. Sci.* **2017,** *424*, 2–8.
44. Li, H.; Dong, X.; da Siva, E. B.; de Oliviera, LM.; Chen, Y.; Ma L. Q. Mechanisms of Metal Sorption by Biochars: Biochar Characteristics and Modifications. *Chemosphere* **2017,** *178*, 466–478.
45. Singh, C.; Tiwari, S.; Gupta, V. K.; Singh, J. S. The Effect of Rice Husk Biochar on Soil Nutrient Status, Microbial Biomass and Paddy Productivity of Nutrient Poor Agriculture Soils. *Catena.* **2018,** *171*, 485–493.

46. Bista, P.; Ghimire, R.; Machado, S; Pritchett, L. Biochar Effects on Soil Properties and Wheat Biomass Vary with Fertility Management. *Agronomy* **2019,** *9* (10), 623.
47. Kim, W. K.; Shim, T.; Kim, Y. S.; Hyun, S.; Ryu, C.; Park, Y.K;, Jung, J. Characterization of Cadmium Removal from Aqueous Solution by Biochar Produced from a Giant Miscanthus at Different Pyrolytic Temperatures. *Bioresour. Technol.* **2012,** *138*, 266–270.
48. Zhang, M.; Gao, B.; Varnoosfaderani, S.; Hebard, A.; Yao, Y.; Inyang, M. Preparation and Characterization of a Novel Magnetic Biochar for Arsenic Removal. *Bioresour. Technol.* **2013,** *130*, 457–462.
49. Huang, X.; Liu, Y. G.; Liu, S.; Tan, X. F.; Ding, Y.; Zeng, G.; Zhou, Y.; Zhang, M.; Wang, S.; Zheng, B. Effective Removal of Cr(VI) Using Alpha-Cyclodextrin–Chitosan Modified Biochars with Adsorption/Reduction Bifuctional Roles. *RSC Adv.* **2016,** *6* (1), 94–104.
50. Lucaci, A. R.; Bulgariu, D.; Ahmad, I.; Lisă, G.; Mocanu, A. M.; Bulgariu, L. Potential Use of Biochar from Various Waste Biomass as Biosorbent in Co(II) Removal Processes. *Water*. **2019,** *11* (8), 1565.
51. Mohan, D.; Kumar, S.; Srivastava, A. Fluoride Removal from Ground Water Using Magnetic and Non-Magnetic Corn Stover Biochars. *Ecol. Eng.* **2014,** *73*, 798–808.
52. Tan, G.; Sun, W.; Xu, Y.; Wang, H.; Xu, N. Sorption of Mercury (II) and Atrazine by Biochar, Modified Biochars and Biochar Based Activated Carbon in Aqueous Solution. *Bioresour. Technol.* **2016,** *211*, 727–735.
53. Sewu, D. D.; Boakye, P.; Woo, S. H. Highly Efficient Adsorption of Cationic Dye by Biochar Produced with Korean Cabbage Waste. *Bioresour. Technol.* **2017,** *224*, 206–213.
54. Xu, D.; Gao, Y.; Lin, Z.; Gao, W.; Zhang, H.; Karnowo, K.; Hu, X.; Sun, H.; Syed-Hassan, S. S. A.; Zhang, S. Application of Biochar Derived from Pyrolysis of Waste Fiberboard on Tetracycline Adsorption in Aqueous Solution. *Front. Chem.* **2019,** *7*, 1–12.
55. Wu, J.; Yang, J.; Feng, P.; Huang, G.; Xu, C.; Lin, B. High-Efficiency Removal of Dyes from Wastewater by Fully Recycling Litchi Peel Biochar. *Chemosphere* **2020,** 125734.

CHAPTER 10

Using Microscopy to Unravel the Hallmarks of Cancer Cells

SHAMEE BHATTACHARJEE* and DEBA PRASAD MANDAL

Department of Zoology, West Bengal State University, Berunanpukuria, Malikapur, Kolkata 700126, India

**Corresponding author. E-mail: shamee.zoology@wbsu.ac.in*

ABSTRACT

Visualizing cells, cellular macromolecules, organelles, and even cellular processes with the aid of microscopes facilitate the study and analysis of finer details of individual cells by researchers. The field of cancer research and diagnosis has been conventionally aided by microscopic techniques. Even to this day, light microscopy is an indispensable tool in cancer diagnosis and staging, which is heavily dependent on histopatholgical and cytopathological assessment of tumor tissues. With the advent of various optical, fluorescence, and electron microscopic techniques, it is possible to study the hallmark features of cancer such as apoptosis, angiogenesis, epithelial-to-mesenchymal transition (EMT), invasion, and metastasis. Advances in confocal microscopy have helped scientists specify localization of molecules and events more precisely in cells.

Cancer-biology laboratories across the globe regularly perform molecular and morphological characterization of cancer cells using high-resolution electron microscopy and versatility of optical miscroscopic techniques. By enhancing the contrast in unstained cells, phase Contrast and Differential Interphase contrast (DIC) miscroscopy aid in the detection

Advanced Microscopy: A Strong Analytical Tool in Materials Science. Merin Sara Thomas, Jozef T. Haponiuk, Sabu Thomas, & Anne George (Eds.)

of gross morphological changes associated with diverse types of cell deaths including apoptosis, necrosis, and autophagy. Scanning electron microscopy (SEM) is also used to examine apoptotic cells. Similarly, morphological features associated with tumor angiogenesis, such as vessel sprouting, tortuous vessels, etc., can be observed by SEM. The mesenchymal phenotype of cells, pseudopodial extension characterizing metastatic cancer cells can also be visualized by phase contrast and DIC microscopy. Moreover, information regarding the location and quantity of protein marker expression associated with these hallmark features of cancer cells is facilitated by immunofluorescence microscopy.

With the advent of new and more powerful microscopes with improved technology, scientists can now visualize cancer cells in 3D and 4D, leading to unprecedented access to cellular and histological structure and function. A combination of imaging technologies assisted by recent advances in software and photography represents cutting edge in microscopy which is instrumental in improving cancer treatment and diagnosis.

> *"By the help of microscopes, there is nothing so small, as to escape our inquiry; hence there is a new visible world discovered to the understanding."*
> — Robert Hooke

10.1 INTRODUCTION

To peep into the invisible world, either too far away or too minute to be seen with the unaided eye, has always fascinated mankind. Over the years we have learnt to apply properties of light to visualize and magnify images of the hitherto invisible world. One such technique which has facilitated the discovery of the invisible microbial world as well as that of the fine cellular details is microscopy, thus, making it an indispensable tool in biology. Therefore, as rightly pointed out by Sir Robert Hooke in the above statement, microscopes have provided us a window to look into and discover the full complexity of the natural world.[1] The first compound microscope was invented by Hans Janssen and his son Zaccharias in 1590s.[2] Since very early times, microscopes have played a vital role in understanding the structure and function of various life forms. The term cell was coined by Robert Hooke in the 1650s when he viewed a cork tissue under the microscope and made his observations public in Micrographia in 1665.[3] Therefore, it would not be an exaggeration to state that the discovery of cell as the structural and functional unit of life was made possible due to the invention of microscopes. Subsequently, it was Antonie van Leeuwenhoek who actually observed bacteria, protozoa, sperm cells, blood cells, and many others under the microscope.[4] Thus, with the aid of microscopes, biologists could peek into the hitherto invisible world of various life forms.

Microscopes have evolved extensively over the years. With the advent of sophisticated microscopes, biologists are now visualizing a wide spectrum of objects, ranging from whole organisms to individual cells to subcellular structures to cellular macromolecules, under the microscope. In addition, observing cellular activity and molecular interactions in live cells with ever more temporal and spatial resolution has lead to unprecedented detailing of cellular structure and function. These constantly evolving microscopic techniques have revolutionized cell biology.

A thorough knowledge of cell biology is absolutely essential to understand cancer and the complex molecular mechanisms that underlie its development. Deregulation in cellular signaling endows cancer cells with several attributes which are acquired during the multistep process of carcinogenesis. Eight such attributes have been identified by Hanahan and Weinberg as the *hallmark features of cancer*.[5] These include continued signal for cell proliferation, escaping growth suppressors, resisting cell death, attaining replicative immortality, induction of angiogenesis, invasion and metastasis, metabolic reprogramming and evading immune response.[5] The study of these essential characteristics of cancer cells is facilitated by the evolution of various microscopy techniques thereby unraveling the fundamental and dynamic aspects of various hallmarks of cancer. Moreover, microscopy also plays a very crucial and vital role in cancer diagnosis and therapy. This chapter intends to discuss the various basic microscopic techniques used to obtain information about cell, cellular organelles, macromolecules, molecular interactions and dynamics which has greatly facilitated cancer research.

10.2 COMMONLY USED MICROSCOPIC TECHNIQUES IN CELL BIOLOGY

Optical microscopy or light microscopy is by far the most widely used microscopic techniques in cell biology research. Conventionally, light microscopy is divided into two main types: Brightfield and Fluorescence.

In brightfield microscopy, light from an incandescent source passes through the cell to the objective lens and the observation of distinct cellular parts relies on the contrast ensuing from absorption of visible spectrum of the electromagnetic radiation by the various cellular components. For many biological samples, especially those which are unstained, contrast is so poor that it becomes impossible to visualize the specimen under the microscope. Various staining techniques have been developed to increase the contrast in these samples. Prior to staining, specimens for light microscopy have to be fixed with alcohol or formaldehyde. Therefore, staining essentially kills the specimen and is not suitable for imaging live specimens. For live specimens, therefore, various contrast-enhancing techniques have been developed to improve specimen visibility without altering their structure or function. Two such techniques are routinely used in laboratories: Phase contrast and DIC. Phase contrast microscopy, first developed by Fritz Zernike, makes use of the differences in refractive indices of cellular constituents which causes phase differences in light. It functions by converting the phase differences of light into amplitude differences which imparts contrast to unstained biological samples.[6] In DIC microscopes, gradients in optical path lengths are converted into amplitude differences which increase contrast in resulting image. The contrast in the image increases with the increasing differences in optical path. Wollaston or Normaski compound prisms are employed in DIC microscopes to generate wave pairs which experience little variations in optical path difference while navigating through the specimen due to differences in thickness and/or refractive indices. This difference in optical path is converted to an amplitude difference in the final image.[4,7]

Fluorescence microscopy is a powerful technique in the field of biomedical sciences because a wide range of molecular and physiological events occurring within live or fixed cells can be studied using it. The phenomenon of fluorescence was first reported by Sir Frederik William Herschel in 1845 when he noted for the first time that a colorless solution of quinine shows blur color when viewed under certain wavelength of sunlight.[8] Later, in 1852, it was George Stokes who explained fluorescence in detail.[9] The first fluorescence microscopes were devised by the companies Carl Zeiss and Carl Reichart at the beginning of the twentieth century.[10] In fluorescence microscopy, fluorophores are used to tag molecules of interest within either fixed or live cells. One frequent application of fluorescence microscopy is to observe the distribution of proteins within the cells by using antibodies tagged with fluorescent labels. In live cells, proteins can be visualized by fusing the protein of interest with green fluorescent protein (GFP) of jelly fish. Confocal microscopy is a superior form of conventional wide-field fluorescent microscopy in which the emitted fluorescence from the specimen is passed through a confocal pinhole aperture that eliminates out-of-focus light creating a sharper image with a superior lateral and transverse optical resolution. Fluorescence microscopy has been constantly involving to include many elaborate techniques in its tool box such as FRAP (Fluorescence recovery after photobleaching), FRET (Forster resonance energy transfer), and FLIM (Fluorescence lifetime imaging microscopy) which enable visualization of dynamic cellular events and molecular interactions within cells. FRAP is a quantitative technique used to measure the dynamics or mobility of molecules in the timescale of seconds to minutes. In FRAP, a specific region of the sample fluorescence is deliberately bleached by using high-intensity laser followed by studying the exchange and redistribution of the fluorescent and bleached molecules in the region under low light illumination.[10] FLIM, a method to measure the duration up till which a fluorophore remains excited, that is, the lifetime of its fluorescence, produces image based on the differential decay rates of fluorescence emitted from individual fluorophore in a sample.[11] FLIM is used to measure dynamic signaling events, protein–protein interactions, ionic concentrations, etc. within live cells. FRET is the transmission of energy from one fluorophore to another in close proximity. Lifetime of fluorescence is linearly reduced with increasing FRET efficiency. Thus, FLIM provides a convenient and quantitative way to measure FRET.[12] FRET is used to visualize protein–protein interaction such as receptor dimerization, ligand receptor complex formation, colocalization, etc. between closely spaced (2–10 nm) protein molecules.[13] However, deep tissue imaging, photobleaching, and UV fluorophore imaging which have always been main problems of fluorescence microcopy could be overcome to some extent with the advent of multiphoton microscopy.[14] Multiphoton microscopy is used for imaging live specimens from molecular level to whole organism over time scales ranging from a few microseconds to days or weeks because of its improved depth penetrations and reduced photodamage.[15]

The most vital aspect of a microscope is its resolving power or resolution which is directly proportional to (1) the refractive index of the fluid between the sample and objective and (2) the angular aperture of the objective. Another important determinant of a microscope's resolution is the wavelength of the incident light which is inversely proportional to it. Therefore, given the range of visible spectrum, 400–700 nm, the highest resolution obtained using a light microscope is 200 nm. This limit of resolution can be improved significantly using electron microscopes which use electron beams with a wavelength of only 0.005 nm to focus onto electromagnetic lenses thereby achieving a much higher resolution. In practice, a resolution of 0.2 nm can be attained with an electron microscope which, in case of biological samples gets limited to around 2 nm due to problems in specimen preparation, lack of inherent contrast, etc.[16] However, this is about a 100-fold improvement over the resolving power of light microscopes, biologists can visualize cellular structures over a range of dimensions from atomic resolution for cellular macromolecules to the range of a few nanometers in case of cellular organelles with the help of electron microscopes. The two types of electron microscopy which are extensively used to analyze cellular components are scanning and transmission electron microscopes. In transmission electron microscopy (TEM), electrons are passed through the specimen to obtain an image whereas in scanning electron microscopy (SEM), a focused electrons beam scans the specimen surface. Therefore, information provided by SEM is usually restricted to surface characteristics of cells and TEM aids in studying internal structures of a cell such as subcellular organelles and macromolecules.

Another technique to analyze biological surfaces at atomic scale resolution is scanning probe microscopy (SPM). Scanning Tunneling Microscope (STM) is the first of this type to be applied to visualize biomolecules such as DNA. However, samples for STM must be electrically conductive which limits its application for biological specimen.[17] On the other hand, another SPM technique, Atomic Force Microscopy (AFM) can be used for obtaining 3D images of both conductive and nonconductive surfaces at atomic or sub-atomic resolution and is being widely applied for biological systems.[18]

Despite the low resolution of light microscopes in comparison to electron or scanning probe microscopes, it is still the most widely used tool to study biological structure and function because it allows 4D (x-y-z- t) imaging of biological specimens in hydrated state.[19]

The various forms of microscopy discussed in the above section allow researchers to peek into and study the intricate details of the cells. Microscopy is an essential tool to understand carcinogenesis at the cellular level and to elucidate the structural and functional variations between cancer cells and normal cells. As mentioned earlier, studying the hallmark attributes of cancer cells is important to comprehend the complexities of tumor biology. Each of the hallmark attribute of cancer cells and its mechanistic pathway can be characterized using appropriate microscopic techniques. Here, in the next few sections, we discuss the various microscopic techniques that are routinely used in cancer research laboratories and clinic to unravel the cellular and molecular aspects of hallmarks of cancer.

10.3 VISUALIZING APOPTOSIS

The ability of malignant cells to evade apoptosis is a hallmark feature of cancer and hence induction of apoptosis is considered to be an effective anticancer strategy.[20] Moreover, resistance to anti-cancer therapy is also correlated with resistance to apoptosis. Therefore evaluation of apoptosis is an important criterion to assess the effectiveness of anticancer drugs. The various morphological changes in the cell occurring during apoptosis have been characterized by light and electron microscopy.[21] In 1972 Kerr et al.[22] demonstrated apoptosis in several tissues by light microscopy and TEM. Observations made by the group under TEM established the characteristic ultrastructural hallmark features of apoptosis such as pyknosis, karyorrhexis, etc.[23] SEM is used to study surface alterations associated with apoptosis such as membrane blebbing, reduction in microvilli and formation of apoptotic bodies[24,25] (Fig. 10.1).

However, due to the high cost and technical expertise associated with performing electron microscopy, light microscopy is the method which is routinely used to detect apoptosis (Fig. 10.2).

Mainly two dyes, Hematoxylene and Giemsa, are used to stain cells for light microscopic evaluation of apoptosis. On histologic examination with hematoxylin and eosin stain, the cytoplasm of the apoptotic cell appears dark eosinophilic with purple nuclear chromatin fragment near the nuclear lamina.[26]

Mitochondrial dysfunction orchestrates the induction of apoptosis. Indeed, the release of apoptogenic factors in the cytoplasm occurs due to the loss of the mitochondrial transmembrane potential ($\Delta\psi_m$) which is stimulated by opening of the mitochondrial permeability transition pore.[27] Relative changes in the $\Delta\psi_m$ can be monitored microscopically by using

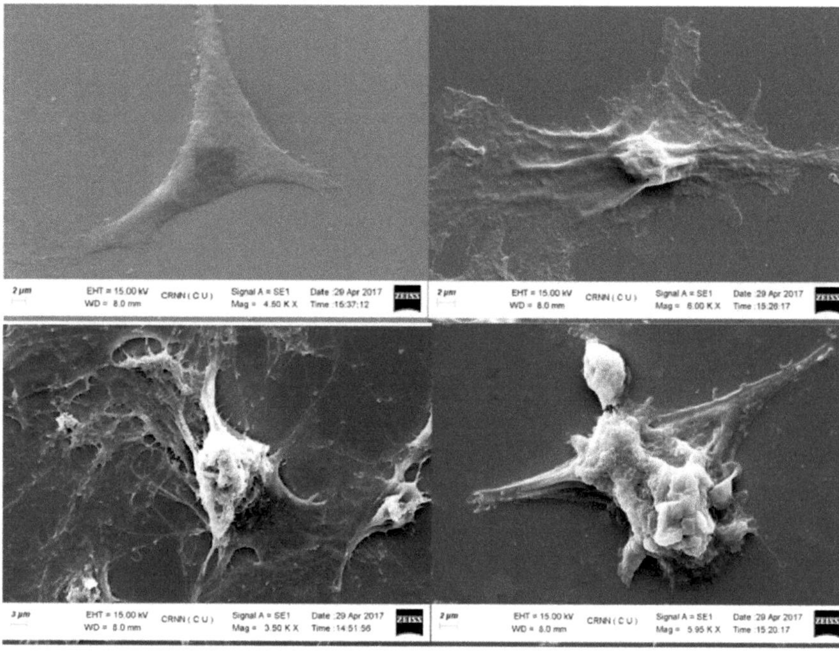

FIGURE 10.1 SEM image of a cancer cell undergoing apoptosis.

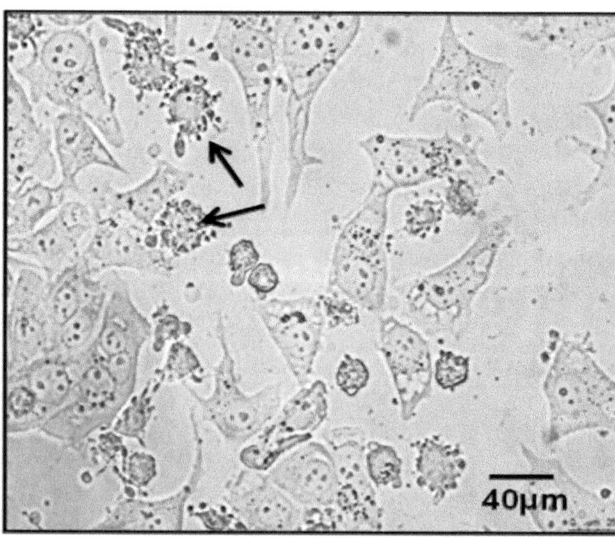

FIGURE 10.2 Visualizing apoptosis under an inverted light microscope (40x). Apoptotic bodies are indicated by arrow heads.

lipophilic fluorescent dyes such 5,5′,6,6′-tetrachloro- 1,1′,3,3′-tetraethyl-benzimi- dazolylcarbocyanine iodide, commonly known as JC1, which forms J- aggregates exhibiting red fluorescence at the high concentrations accumulated in intact healthy mitochondria. In apoptotic cells having depolarized mitochondrial membrane, JC1 cannot form aggregates and remain as monomers showing green fluorescence.[28] Apoptosis can also be established by analyzing molecular events such as nuclear DNA fragmentation using terminal deoxynucleotidyl transferase-UTP nick-end labeling (TUNEL) assay and single gel electrophoresis assay or comet assay (Fig. 10.3).

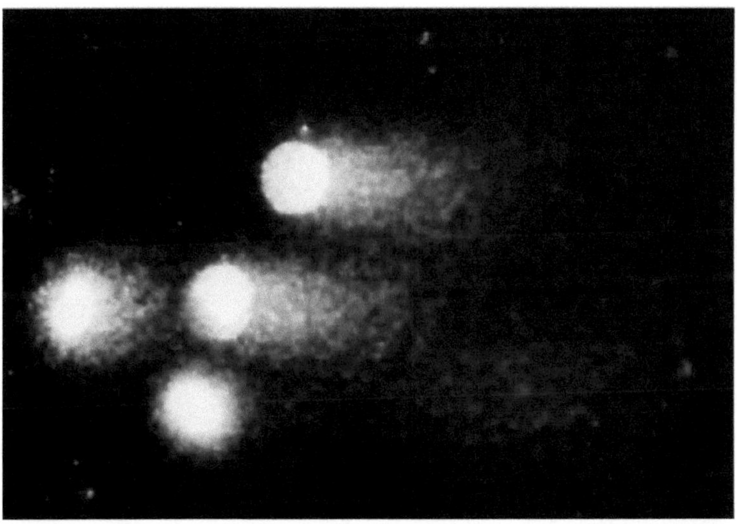

FIGURE 10.3 Fragmented DNA (Comets) visualized under a fluorescent microscope in Comet Assay (40x).

Acridine orange/ethidium bromide (AO/EB) double fluorescent staining is also applied to detect apoptotic cells under a fluorescent microscope[29] (Fig. 10.4).

10.4 VISUALIZING ANGIOGENESIS

Judah Folkman had postulated that tumor tissues cannot grow beyond 1–2 mm without the formation of new vasculature.[30] Since then, tumor angiogenesis has become an important criterion among cancer researchers and clinicians to assess tumor growth and aggressiveness. The development of new blood vessels from pre-existing vasculature is termed as angiogenesis. There are several *in vitro* and *in vivo* angiogenic models to study the effect of anti- or proangiogenic factors. The *in vitro* assays largely focus on endothelial cell proliferation, differentiation, and migration.[31] A very simple way to measure the migration of endothelial cells is by scraping a part of the confluent endothelial cell monolayer thus creating a 'wound'[32] (Fig. 10.5). The degree of endothelial cell migration to that *wounded* area is then measured at various intervals by time-lapse phase contrast microscopy.[33]

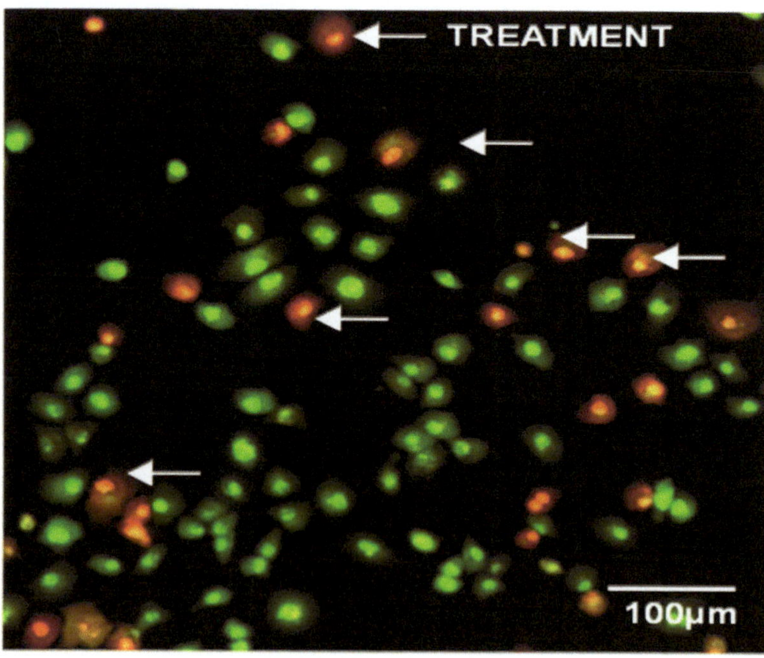

FIGURE 10.4 Fluorescence microscopic images (10x) of cells stained with dual AO and EB. Viable cells appear green and apoptotic cells stain yellow to orange (indicated by white arrows).

Endothelial cell proliferation is measured by counting cells after staining them with a vital dye in a hemocytometer under a light microscope. Formation of capillary-like tubules by endothelial cells is representative of later stages of angiogenesis. In tube formation assay, endothelial cells are stimulated to differentiate by culturing endothelial cells on gel matrix such as collagen or matrigel, which stimulate formation of tube-like structures.[34] Images of the vessel formation can be obtained through phase contrast microscopy. Fluorescent images can also be obtained if calcein AM is added to visualize the tube network through fluorescence microscopy.[35]

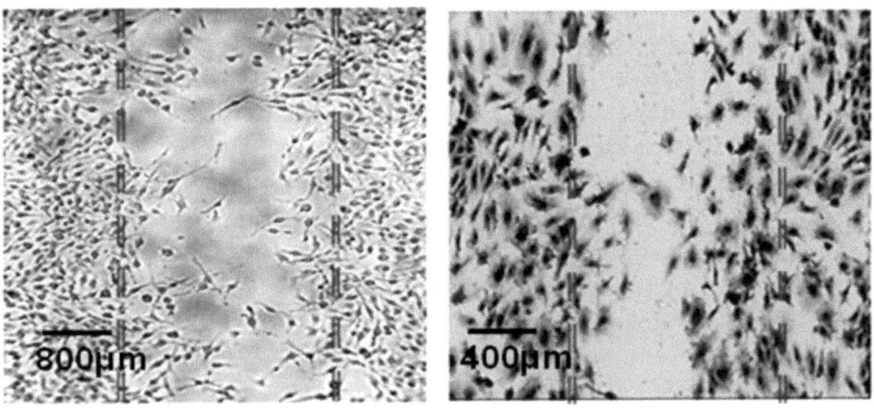

FIGURE 10.5 Endothelial cell migration into the *wounded area* assessed under a light microscope (10x).

Specific morphological features such as formation of cellular junctions between endothelial cells and of lumen, structural alterations such as appearance of cellular strings and their development into mono-layer strands, network formation, and 3D organization of endothelial cells during tube formation can be effectively monitored by electron microscopy.[36]

Organotypic assays such as rat aorta and chick aortic ring assays[37,38] are also widely used to study angiogenesis. In these methods, rat aorta or chick aortic arches are cut into segments and placed on matrigel, a substratum containing extracellular matrix (ECM) constituents. Over the next few days, outgrowth of endothelial cells from the explants to form vessel-like structures are monitored by DIC or phase contrast microscopy.[39]

Understanding the process of angiogenesis and evaluation of pro- and anti-angiogenic drugs in preclinical models have received a boost by the advances in imaging techniques. Quantification of the number and spacing of blood vessels, measurement of blood flow, assessment of the vascular permeability and identification of vessel wall abnormality have all been made possible by vascular imaging. Till date, measurement of microvessel density (MVD) by manual counting of blood vessels, a technique introduced by Weidner et al., is the most popular method to study angiogenesis.[40] In various types of cancer, MVD is considered to be an independent prognostic factor.[41] Images gathered by light microscope can be used to quantify angiogenesis by applying a number of techniques, for instance: (1) Counting the blood vessels manually,[42] (2) analyzing the density of vessel area and length and that of fractal dimensions[43], (3) determining of endpoints such as length density of radially arranged capillaries.[44]

Various *in vivo* models have been developed for quantification of angiogenesis. These include the corneal pocket angiogenesis assay, matrigel plug assay, and chick choriallantoic membrane (CAM) assay. Easy accessibility, low cost, and better visualization of the rich vascular network of CAM assay make it a robust model for research.[45] As imaging is done at multiple time points during the CAM assay, the changes in blood vessels can be quantified in each step thus providing a complete picture of the entire process of angiogenesis.[46]

10.5 VISUALIZING EPITHELIAL TO MESENCHYMAL TRANSITION AND METASTASIS

Metastasis is the primary cause of all cancer-related deaths. Although it is considered to be a late event in carcinogenesis, the acquirements of traits that facilitate their dissemination from the primary tumor and migrate to a secondary organ, commence at earlier stages. Histological techniques provide useful information regarding colonization of metastatic cells to secondary organs and lymph nodes. However, this provides only a static image which does not reflect the complexity and dynamicity of the entire metastatic cascade.[47]

EMT is a morphological change in which epithelial cells detach from their neighbouring cells and become migratory. This process is considered to play a vital role in tumor metastasis. EMT entails concerted cellular and molecular changes resulting in loss of cellular adhesion, polarity, and cytoskeletal integrity with simultaneous expression of mesenchymal markers, spindle-shaped cells, and attainment of invasive and migratory potential.[48] This morphological transition can be easily demonstrated by phase contrast or SEM imaging of cancer cells. Epithelial phenotype would exhibit a rounded cobble stone-like appearance while mesenchymal phenotype would be more elongated due to lamellipodial extensions (Fig. 10.6).

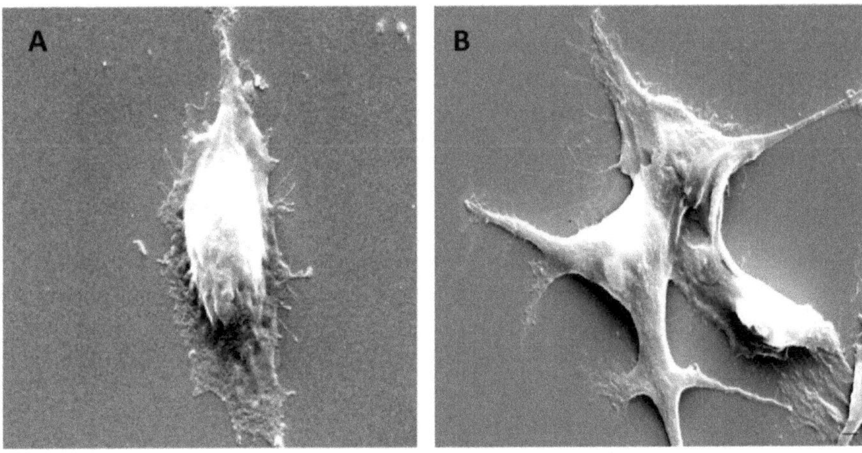

FIGURE 10.6 SEM images of cancer cell showing (A) Epithelial phenotype (B) Mesenchymal Phenotype.

The ability to migrate is a vital property of live cells and is important for cancer metastasis. This property of cancer cells assumes immense significance in cancer research because metastasis of cancer cells to distant sites is the primary cause of mortality in cancer patients. The dissemination of cancer cells to distant sites is facilitated by invasion of ECM at the primary site, intravasation, migration to distant sites, and ultimately extravasation to form secondary colonies.[49–51] The study of the events of metastatic cascade is facilitated by several methods such as the scratch wound method and the transwell migration/invasion assays.[49] Observing and measuring the distance covered by cells thereby closing the wound over time under an inverted light microscope gives an estimate of the degree of cell migration. Invasion assays in a Boyden's chamber is also facilitated by light microscopy. After cell migration in the lower Boyden's chamber, the migrated cells are stained with crystal violet, and when observed under the microscope, appear as purple stained.

Intravital microscopy (IVM) has evolved over the years to view the activities of disseminating cancer cells at subcellular resolution. High-resolution IVM techniques such as spinning disk confocal microscopy and multiphoton microscopy are being widely used to image tumors *in vivo*.

Due to high resolution acquisition, better depth penetration and reduced photobleacing, multiphoton IVM technique is the technique preferred by tumor biologists. Moreover, cellular and molecular interactions during metastasis can be monitored by using multicolor IVM in which different cell types are genetically labeled using green fluorescent proteins and its variants.[52] Additionally, tumor cells labeled with fluorescent proteins can be imaged simultaneously with cells tagged with fluorescent dyes. However, a major disadvantage of such multiphoton microscopy is spectral overlap between the fluorescent proteins and dyes. To overcome this problem, as well as to tackle autofluorescence FLIM which measures fluorescence life time (described in earlier sections) is used. Molecular mechanism underlying metastasis such as protein–protein interaction, activation of proteins, etc., can be measured by FRET.

10.6 TUMOR HISTOPATHOLOGY AND CYTOLOGY

The diagnosis of cancer still largely depends on skilled examination of tissues and/or cells under the microscope. Carcinogenesis is a multistep process involving gradual accumulation of mutation resulting in selection of cells with progressively greater capacity to proliferate, invade, and metastasize. As the cancer cells undergoes multistep progression of tumor development, they acquire the hallmarks traits of cancer, as mentioned above.[5] Tumor histopathology and/or cytology provides evidence for this gradual progression of cancer from benign through malignant stages.[53] Both these techniques rely on the observation of tissue sections and cells, respectively, under the microscope. Histopathological assessment of tumor tissues is the gold standard for the clinical diagnosis and staging of cancer.[54] Cancer treatment as well as survival prediction is largely governed by staging survival.[55]

For histological preparations, tissue or biopsy samples are collected, fixed, and then supported by freezing or embedding in a solid mold (e.g., paraffin wax) to facilitate thin sectioning for microscopic observation.[56] The thickness of tissue sections for light microscopy is approx. 5 μm and for electron microscopy it is 80–100 nm. To visualize the tissue sections with a light microscope, contrast enhancement is done by staining the sections with dyes. Hematoxylene and Eosin is the routine stain for light microscopy. Hematoxylene stains the nuclei blue and eosin stains the cytosol and connective tissues pink (Fig. 10.7).

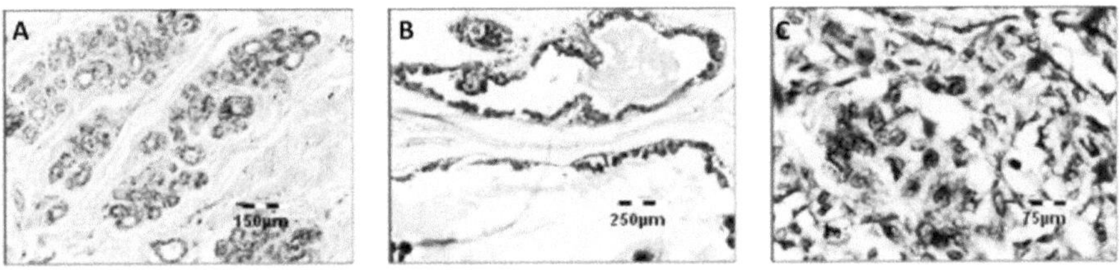

FIGURE 10.7 H-E stained microscopic images of histopathological sections of breast tissue showing (A) Dysplastic changes (B) Carcinoma *in situ* (C) Poorly differentiated invasive duct carcinoma.

Cytological methods have contributed immensely to the detection and diagnosis of various forms of cancer. Depending on the site of the lesion, some commonly used methods of sampling cells from patients are Fine needle aspiration cytology (FNAC), swab, and scraping. Exfoliative cytology particularly that of the uterine cervix has been highly successful as a screening test.[57] Cervical epithelial cells are stained with Papnicolaou stain which provides excellent nuclear and cytoplasmic details upon microscopical observation.[58]

In order to analyze subcellular components or study molecular expressions in histological sections or cytological preparations, the objects of interest can be tagged with a fluorophore and fluorescence can be detected under a widefield or confocal fluorescent microscope. This technique is known as immunofluorescence.[59]

10.7 STUDY OF MOLECULAR MARKERS

A large number of molecular alterations are associated with the development of cancer.[60] With the help of fluorescence microscopy, very low concentrations of cellular molecules can be detected and quantified over a range of spatial and temporal dimensions. Detection of such molecular markers and their interactions provides a detailed understanding of the dysregulated signaling pathways in cancer resulting in the manifestation of the so called hallmark features of cancer cells. In other words, each of the hallmark features of cancer is concomitant with the expression of an assortment of markers. Therefore, it is quite reasonable to analyze the marker expressions in tumor or blood samples in order to identify their predictive, diagnostic, prognostic or therapeutic significance. Apoptosis occurring spontaneously or induced by therapies can be studied at the molecular level by analyzing the various markers of the extrinsic or intrinsic pathways of apoptosis, for example, initiator and effector caspases, Bcl-2 family proteins, cytochrome-c, PARP-1 etc.[20] Markers for angiogenesis include VEGF and its receptors Flt-1and KDR, angiopoietin, endothelial marker CD31 etc.[61]

Disappearance of epithelial features during EMT is marked by the expression of mesenchymal markers such as Vimentin, N-Cadherin, and β-Catenin.[62] Immunofluorescence microscopy is a valuable tool to study the localization as well as expression level of all these molecular markers in fixed or frozen tissue sections, cell smears, or cultures cells. In this technique, the molecules of interest or antigens are detected by binding of antigen-specific fluorochrome-tagged antibody. In direct immunofluorescence, the fluorochrome is bound directly to the antigen-specific primary antibody whereas in indirect fluorescence, the fluorochrome is attached to an anti-immunoglobulin which recognizes the primary antibody.[63] The emitted fluorescence from the fluorochrome is detected optically with a fluorescence microscope (widefield or confocal) (Fig. 10.8).

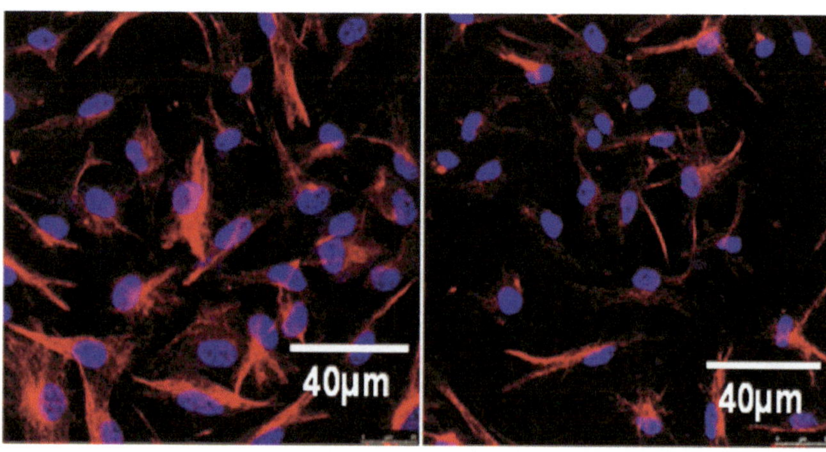

FIGURE 10.8 Representative fluorescent microscopic images of Phycoerythrin (PE) (a fluorochrome)-tagged vimentin expression in cancer cells. Nucleus is stained with DAPI.

10.8 DISCUSSION

It would not be an overstatement to mention that progress in the field of cell biology goes hand in hand with that of microscopy. Ever since its first realization, microscopy has contributed immensely to the understanding of steady state organization of cells as well as the dynamic aspects of cellular structure and function. Despite the incorporation of molecular biology techniques, advances in microscopy are heralding new avenues for understanding the structure and function of transformed cells. Apart from the several microscopic techniques described in the above sections, some more recent revolution in light microscopy include the development of super resolution microscopy which has changed the limit of optical resolution from approximately 250 to 10 nm[64] and the development of light sheet microscopy which optically sections tissues or organisms labeled with a fluorophore suitable for 3D reconstruction.[65] Recently, live multicolor imaging of various immune cells using multiphoton microscopy has led to a deeper understanding of communication between immune cells.[66] It is notable that apart from the hallmarks of cancer described in the preceding sections, evasion of the immune system by the tumor cells is now recognized as yet another emerging hallmark of cancer.[5] Therefore, now with the growing importance of immune-therapies in cancer treatment, analyzing the immune cell-cancer cell interaction in the tumor microenvironment has become a priority and is being facilitated by intravital multiphoton microscopy.[67]

However, there are many challenges in using advanced microscopic techniques in cancer research which limit their wide application, for example, cost of purchase, maintenance of the equipment, requirement of trained operator, specialized tissue processing etc. But despite these limitations, it is true that visualizing cell, subcellular components and cellular events leads to such an understanding of the cellular activity and its constituents in a manner that can never be done by molecular biology or biochemical techniques.

10.9 CONCLUSION

Microscopy and imaging play a prominent role in cancer research and diagnosis. Starting from basic morphological and histopathological analysis, microscopic techniques are constantly evolving to give rise to superior techniques such as multiplex imaging, image digitalization, intravital imaging which facilitates more elaborate dynamic and molecular information regarding tumor cells and the tumor microenvironment. Tissue or cell imaging ex-vivo is routinely done in research and clinical laboratories.

Tumor histothology still remains the gold standard for classification, grading and staging of tumors. On the other end of the 'spectrum', in vivo imaging, such as that using multiphoton microscopy, although not easily accessible to researchers, provide dynamic information enabling scientists unprecedented access to the working of cancer cells and their communication with the tumor microenvironment. Overall, a plethora of microscopy techniques, including high resolution electron microscopy, multiplex immunohistochemistry, or IVM, combined with various types of present day sophisticated imaging systems provides a thorough understanding of the mechanisms underlying carcinogenesis, drug delivery to tissues, diagnosis and staging of cancer. It can be envisaged that with the development of newer fluorochromes and continuing improvements in imaging and analysis techniques, microscopy will further shed light into the spatiotemporal processes in tumors and will continue to transform basic and clinical cancer research.

ACKNOWLEDGMENT

The authors thank research scholars Mr Arnab Sarkar, Dr Kartick Patra, Mr Ashikur Rahaman, and Miss Priyanka Pal for providing the microscopic images that has been included in the chapter. Authors acknowledge and thank DST- FIST(SR/FST/LSI-585/2014 (C) dated 28th December 2015) New Delhi, Govt. of India, and DBT-BOOST, Govt. of West Bengal for providing instrumental support.

KEYWORDS

- **cancer**
- **light microscope**
- **fluorescence microscope**
- **apoptosis**
- **angiogenesis**
- **epithelial to mesenchymal transition (EMT)**
- **metastasis**

REFERENCES

1. Hooke, R. *Micrographia*, 1665.
2. Ball, C. S. The Early History of the Compound Microscope. *Bios* **1966,** 51–60.
3. Hooke, R. *Micrographia: Or, Some Physiological Descriptions of Minute Bodies Made by Magnifying Glasses with Observations and Inquiries Thereupon*; John Martyn, 1667.
4. Wollman, A. J.; Nudd, R.; Hedlund, E. G.; Leake, M. C. From Animaculum to Single Molecules: 300 Years of the Light Microscope. *Open Biol.* **2015,** *5*, 150019.
5. Hanahan, D.; Weinberg, R. A. Hallmarks of Cancer: The Next Generation. *Cell* **2011,** *144*, 646– 674.
6. Zernike, F. Phase Contrast, a New Method for the Microscopic Observation of Transparent Objects. *Physica* **1942,** *9*, 686–698.
7. Nomarski, G.; Weill, A. R. Application of Polarized Two-Wave Interference Methods to Metallography. *Rev. Met. Paris*. **1955,** *52*, 121–134.
8. Herschel, J. F. W. IV. On a Case of Superficial Colour Presented by a Homogeneous Liquid Internally Colourless. *Phil. Trans. R. Soc.* **1845,** *135*, 143–145.
9. Stokes, G. G. On the Change of Refrangibility of Light. *Phil. Trans. R. Soc.***1852,** *142*, 463–562.
10. Renz, M. Fluorescence Microscopy—A Historical and Technical Perspective. *Cytometry Part A* **2013,** *83*, 767–779.
11. Korczyński, J.; Włodarczyk, J. Fluorescence Lifetime Imaging Microscopy (FLIM) in Biological and Medical Research. *PostepyBiochem*. **2009,** *55*, 434–440.
12. Bunt, G.; Wouters, F. S. Visualization of Molecular Activities Inside Living Cells with Fluorescent Labels. *Int. Rev. Cytol.* **2004,** 237, 205–277.
13. Rajoria, S.; Zhao, L.; Intes, X.; Barroso, M. FLIM-FRET for Cancer Applications. *Curr. Mol. Imaging.* **2014,** 3, 144–161.
14. Periasamy, A.; Skoglund, P.; Noakes, C.; Keller, R. An Evaluation of Two-Photon Excitation Versus Confocal and Digital Deconvolution Fluorescence Microscopy Imaging in Xenopus Morphogenesis. *Microsc. Res. Tech.* **1999,** *47*, 172–181.
15. Larson, A. Multiphoton Microscopy. *Nature Photon* **2011,** *5*, 1.
16. Cooper, G. M. *The Cell*, 2nd ed.; Sinauer Associates, 2000.
17. Bhushan, B.; Fuchs, H.; Tomitori, M. *Applied Scanning Probe Methods IX: Characterization: Nanoscience and Technology*; Springer: Berlin, New York, 2008.
18. Horber, J. K. H.; Miles, M. J. Scanning Probe Evolution in Biology. *Science* **2003,** *302*, 1002–1005.
19. Diaspro, A.; Bianchini, P.; Vicidomini, G.; Faretta, M.; Ramoino, P.; Usai, C. Multi-Photon Excitation Microscopy. *Biomed. Eng. Online* **2006,** *5*, 36.
20. Carneiro, B. A., El-Deiry, W. S. Targeting Apoptosis in Cancer Therapy. *Nat. Rev. Clin. Oncol.* **2020,** *17*, 395–417.
21. Hacker, G. The Morphology of Apoptosis. *Cell Tissue Res*. **2000,** *301*, 5–17.
22. Kerr, J. F. R.; Wyllie, A. H.; Currie, A. R. Apoptosis: A Basic Biological Phenomenon with Wide- Ranging Implications in Tissue Kinetics. *Br. J. Cancer*. **1972,** *26*, 239–257.
23. Taatjes, D. J.; Sobel, B. E.; Budd, R. C. Morphological and Cytochemical Determination of Cell Death by Apoptosis. *Histochem. Cell Biol.* **2008,** 129, 33-43.

24. Abdel Wahab, S. I.; Abdul, A. B.; Alzubairi, A. S.; Mohamed Elhassan, M.; Mohan, S. In Vitro Ultramorphological Assessment of Apoptosis Induced by Zerumbone on (HeLa). *J. Biomed. Biotechnol.* **2009**.
25. Xulu, K. R.; Hosie, M. J. HAART Induces Cell Death in a Cervical Cancer Cell Line, HCS-2: A Scanning Electron Microscopy Study. *J. Microsc. Ultrastruct.* **2017,** *5*, 39–48.
26. Elmore, S. Apoptosis: A Review of Programmed Cell Death. *Toxicol. Pathol.* **2007,** *35*, 495–516. 27. Cui, Y.; Pan, M.; Ma, J.; Song, X.; Cao, W.; Zhang, P. Recent Progress in the Use of Mitochondrial Membrane Permeability Transition Pore in Mitochondrial Dysfunction-Related Disease Therapies. *Mol. Cell Biochem.* **2020.**
27. Elefantova, K.; Lakatos, B.; Kubickova, J.; Sulova, Z.; Breier, A. Detection of the Mitochondrial Membrane Potential by the Cationic Dye JC-1 in L1210 Cells with Massive Overexpression of the Plasma Membrane ABCB1 Drug Transporter. *Int. J. Mol. Sci.* **2018,** *19*, 1985.
28. Devi, L. R.; Raza, M. K.; Musib, D.; Roy, M. New Selenonapthaquinone-Based Copper (II) Complexes as the Next-Generation Photochemotherapeutic Agents. *Anticancer Agents Med. Chem.* **2020.**
29. Folkman, J. Tumor Angiogenesis: Therapeutic Implications. *N. Engl. J. Med.***1971,** *285*, 1182–1186.
30. Staton, C. A.; Reed, M. W.; Brown, N. J. A Critical Analysis of Current in Vitro and in Vivo Angiogenesis Assays. *Int. J. Exp. Pathol.* **2009,** *90*, 195–221.
31. Sarkar, A.; Rahaman, A.; Biswas, I.; Mukherjee, G.; Chatterjee, S.; Bhattacharjee, S.; Mandal, D. P. TGFβ Mediated LINC00273 Upregulation Sponges mir200a-3p and Promotes Invasion and Metastasis by Activating ZEB1. *J. Cell Physiol.* **2020,** *235*, 7159–7172.
32. Pepper, M. S.; Belin, D.; Montesano, R.; Orci, L.; Vassalli, J. D. Transforming Growth Factor-Beta 1 Modulates Basic Fibroblast Growth Factor-Induced Proteolytic and Angiogenic Properties of Endothelial Cells in Vitro. *J. Cell Biol.* **1990,** *111*, 743–755.
33. Lawley, T. J.; Kubota, Y. Induction of Morphologic Differentiation of Endothelial Cells in Culture. *J. Invest. Dermatol.* **1989,** *93*, 59–61.
34. DeCicco-Skinner, K. L.; Henry, G. H.; Cataisson, C.; Tabib, T.; Gwilliam, J. C.; Watson, N. J.; Bullwinkle, E. M.; Falkenburg, L.; O'Neill, R. C.; Morin, A. Endothelial Cell Tube Formation Assay for the in Vitro Study of Angiogenesis. *J. Vis. Exp.* **2014,** *91*, e51312.
35. Hirschberg, R. M.; Sachtleben, M.; Plendl, J. Electron Microscopy of Cultured Angiogenic Endothelial Cells. *Microsc. Res. Tech.* **2005,** *67*, 248–259.
36. Liu, C.; He, L.; Wang, J.; Wang, Q.; Sun, C.; Li, Y.; Jia, K.; Wang, J.; Xu, T.; Ming, R. Anti- Angiogenic Effect of Shikonin in Rheumatoid Arthritis by Downregulating PI3K/AKT and MAPKs Signaling Pathways. *J. Ethnopharmacol.* **2020,** 113039.
37. Katakia, Y. T.; Duddu, S.; Nithya, S.; Kumar, P.; Rahman, F.; Kumaramanickavel, G.; Chatterjee, S. Ex Vivo Model for Studying Endothelial Tip Cells: Revisiting the Classical Aortic-Ring Assay. *Microvasc. Res.* **2020,** *128*, 103939.
38. Auerbach, R.; Lewis, R.; Shinners, B.; Kubai, L.; Akhtar, N. Angiogenesis Assays: A Critical Overview. *Clin. Chem.* **2003,** *49*, 32–40.
39. Weidner, N.; Semple, J. P.; Welch, W. R.; Folkman, J. Tumor Angiogenesis and Metastasis—Correlation in Invasive Breast Carcinoma. *N. Engl. J. Med.* **1991,** *324*, 1–8.

40. Kather, J. N.; Marx, A.; Reyes-Aldasoro, C. C.; Schad, L. R.; Zöllner, F. G.; Weis, C. A. Continuous Representation of Tumor Microvessel Density and Detection of Angiogenic Hotspots in Histological Whole-Slide Images. *Oncotarget* **2015,** *6*, 19163–19176.
41. Célérier, J.; Cruz, A.; Lamandé, N.; Gasc, J.-M.; Corvol, P. Angiotensinogen and Its Cleaved Derivatives Inhibit Angiogenesis. *Hypertension* **2002,** *39*, 224–228.
42. Parsons-Wingerter, P.; Elliott, K. E.; Farr, A. G.; Radhakrishnan, K.; Clark, J. I.; Sage, E. H. Generational Analysis Reveals That TGF-β1 Inhibits the Rate of Angiogenesis in Vivo by Selective Decrease in the Number of New Vessels. *Microvasc. Res.* **2000,** *59*, 221–232.
43. Tímár, J.; Döme, B.; Fazekas, K.; Janovics, Á.; Paku, S. Angiogenesis-Dependent Diseases and Angiogenesis Therapy. *Pathol. Oncol. Res.* **2001,** *7*, 85–94.
44. Ribatti, D.; Annese, T.; Tamma, R. The Use of the Chick Embryo CAM Assay in the Study of Angiogenic Activiy of Biomaterials. *Microvasc. Res.***2020,** 104026.
45. Auerbach, R.; Popp, B. L.; Kiley, L. M.; Gilligan, B. J. Angiogenesis Assays: Problems, Pitfalls and Potential. In *Proceedings of the Therapeutic Implications and Mechanisms of Angiogenesis—Inhibitors and Stimulators*; IBC Biomedical Library Series: Westborough, MA Conferences, 1996.
46. Beerling, E.; Ritsma, L.; Vrisekoop, N.; Derksen, P. W.; van Rheenen, J. Intravital Microscopy: New Insights into Metastasis of Tumors. *J. Cell Sci.* **2011,** *124*, 299–310.
47. Tsai, J. H.; Yang, J. Epithelial–Mesenchymal Plasticity in Carcinoma Metastasis. *Genes Dev.* **2013,** *27*, 2192–2206.
48. Albini, A. Tumor and Endothelial Cell Invasion of Basement Membranes. *Pathol. Oncol. Res.* **1998,** *4*, 230–241.
49. Fares, J.; Fares, M. Y.; Khachfe, H. H.; Salhab, H. A.; Fares, Y. Molecular Principles of Metastasis: A Hallmark of Cancer Revisited. *Signal Transduct Target Ther.* **2020,** *5*, 1–17.
50. Welch, D. R.; Hurst, D. R. Defining the Hallmarks of Metastasis. *Cancer Res.* **2019,** *79*, 3011–3027.
51. Perrin, L.; Bayarmagnai, B.; Gligorijevic, B. Frontiers in Intravital Multiphoton Microscopy of canceR. *Cancer Rep.* **2020,** *3*, e1192.
52. Schedin, P.; Elias, A. Multistep Tumorigenesis and the Micro Environment. *Breast Cancer Res.* **2004,** *6*, 93–101.
53. Fu, Y.; Jung, A. W.; Torne, R. V.; Gonzalez, S.; Vöhringer, H.; Shmatko, A.; Yates, L. R.; Jimenez-Linan, M.; Moore, L.; Gerstung, M. Pan-Cancer Computational Histopathology Reveals Mutations, Tumor Composition and Prognosis. *Nat. Cancer.* **2020,** *1*, 800–810.
54. Moawad, E. Y. Clinical and Pathological Staging of the Cancer at the Nanoscale. *Cancer Nanotechnol.* **2012,** *3*, 37–46.
55. Chandler, D. E.; Roberson, R. W. *Bioimaging: Current Concepts in Light and Electron Microscopy*; Jones & Bartlett Publishers, 2009.
56. Boddington, M. M.; Spriggs, A. I. Cytological Diagnosis of Cancer: Its Uses and Limitations. *Br. Med. J.* **1965,** *1*, 1523–1529.
57. Roe, C. J.; Hanley, K. Z. Updates in Cervical Cytology: The 90-Year-Long Journey from Battle Creek to Today. *Surg. Pathol. Clin.* **2018,** *11*, 589–599.

58. Ramjiawan, B.; Ariano, R. E.; Mantsch, H. H.; Maiti, P.; Jackson, M. Immunofluorescence Imaging as a Tool for Studying the Pharmacokinetics of a Human Monoclonal Single Chain Fragment Antibody. *IEEE Trans. Med. Imag.* **2002,** *21*, 1317–1323.
59. Nair, M.; Sandhu, S. S.; Sharma, A. K. Cancer Molecular Markers: A Guide to Cancer Detection and Management. *Semin. Cancer Biol.* **2018,** *52*, 39–55.
60. Shih, S. C.; Robinson, G. S.; Perruzzi, C. A.; Calvo, A.; Desai, K.; Green, J. E.; Ali, I. U.; Smith, L. E.; Senger, D. R. Molecular Profiling of Angiogenesis Markers. *Am. J. Pathol.* **2002,** *161*, 35–41.
61. Zhu, G. J.; Song, P. P.; Zhou, H.; Shen, X. H.; Wang, J. G.; Ma, X. F.; Gu, Y. J.; Liu, D. D.; Feng, A. N.; Qian, X. Y. Role of Epithelial-Mesenchymal Transition Markers E-Cadherin, N-Cadherin, β-Catenin and ZEB2 in Laryngeal Squamous Cell Carcinoma. *Oncol. Lett.* **2018,** *15*, 3472–3481.
62. Maclachlan, N. J.; Dubovi, E. J.; Barthold, S. W.; Swayne, D. E.; Winton, J. R. *Fenner's Veterinary Virology*, 5th ed.; Elsevier Inc., 2016.
63. Galbraith, C. G.; Galbraith, J. A. Super-Resolution Microscopy at a Glance. *J. Cell. Sci.* **2011,** *124*, 1607–1611.
64. Santi, P. A. Light Sheet Fluorescence Microscopy: A Review. *J. Histochem. Cytochem.* **2011,** *59*, 129–138.
65. Okada, T.; Takahashi, S.; Ishida, A.; Ishigame, H. In Vivo Multiphoton Imaging of Immune Cell Dynamics. *Pflugers Arch.* **2016,** *468*, 1793–1801.
66. Torcellan, T.; Stolp, J.; Chtanova, T. In Vivo Imaging Sheds Light on Immune Cell Migration and Function in Cancer. *Front. Immunol.* **2017,** *8*, 309.

CHAPTER 11

Mechanical and Microstructural Characterization of Cement Mortar with Nanosilica Additions

MAINAK GHOSAL* and ARUN KUMAR CHAKRABORTY

Indian Institute of Engineering Science & Technology, Shibpur, India

**Corresponding author. E-mail: mainakghosal2010@gmail.com*

ABSTRACT

Cement the only standard and binding material in concrete construction, also accounts for its major strength development. A typical process of cement-hydration (chemical reaction between cement and water) completes 90–95% in the first 28 days while the remaining hydration takes place throughout the year after which it attains a fixed hydration level with negligible increase in latter periods. The standard testing protocol for cement concrete suggests testing of various behavioral properties of cement and concrete at 28 days. Microstructure is the small-scale structure of a material, defined as the structure of a prepared surface of material as revealed by a microscope. This article presents the mechanical and microstructural characterization on the hardened properties of basic cement mortar of a suitable water–cement (w/c) ratio of 0.4 (based on cement's consistency) and sand (river): cement = 3:1 which were initially mixed and then modified with nanosilica (nS) [in proportions ranging 0%, 0.5%, 0.75%, 1%, 1.25%, 1.5% @ cement wt.] Several authors have tried microstructural analysis of cement concrete but without the presence of nanomaterials. Here, three important characterization tools, viz., (1) mechanical strength testing, (2) X-ray diffraction technique,

Advanced Microscopy: A Strong Analytical Tool in Materials Science. Merin Sara Thomas, Jozef T. Haponiuk, Sabu Thomas, & Anne George (Eds.)

and (3) field emission scanning electron microscopy with energy-dispersive X-ray spectroscopy were used to study the effect of nanomaterial (viz. nanosilica) additions on hydration of cement both for short (28 days) and long terms(365 days) including phase-dispersion studies.

11.1 INTRODUCTION

The mechanical behavior of cement concrete materials depends to a great extent on the cement paste hydration phenomena that are effective on a microscale and nanoscale. The size of the calcium silicate hydrate (C-S-H) phase, the primary component responsible for strength and other properties in cementitious systems, lies in the few nanometers range[9] as shown in Figure 11.1. The microstructure of concrete as shown in Figure 11.1 is a three-phase system with (1) aggregates, (2) transition zone, and (3) bulk cement paste. Chemically, the microstructure of concrete is also a three-phase system consisting of C-S-H (calcium silicate hydrate), C-H (calcium hydroxide) (Portlandite) and C-A-Ŝ-H (calcium sulfo-alumino hydrate (ettringite)). Here, the chemical abbreviations are— C=CaO (calcium oxide), S=SiO_2 (silicon dioxide), H=H_2O; Ŝ = SO_3 (sulfur trioxide), and A=Al_2O_3 (aluminum oxide).

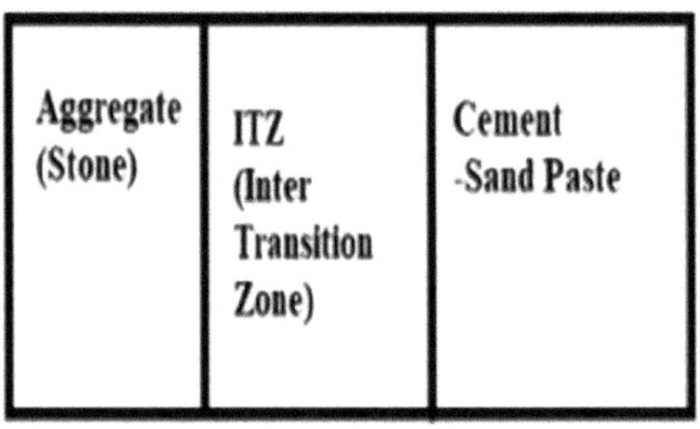

FIGURE 11.1 Schematic three-phase microstructure of cement concrete.

Concrete is weak in tension and tend to crack along with the intertransition zone planes between stone aggregates and cement–sand paste inherent in its microstructure. So, bad microstructure results in a poor concrete susceptible to factors responsible for the concrete deterioration, such as permeability, carbonation, chemical attack, alkali-aggregate reaction, and physical aggression, such as thermal shocks and abrasion. But concrete gains more strength as its microstructure develops with less and less voids and pores through the reaction of cement with water added externally. With time, these hydration products increase resulting in more strength though Indian and other codes like IS: 456-2000 which is a Code of Practice prescribes the standard time for strength test results, and it should be that the strength is achieved at 28 days. Addition of nanomaterials in optimized dosages helps in increasing these hydration products through the additional space gain on account of nanomaterials' nucleation power due to their very higher specific area. The incumbent nanomaterials also act as nucleating agents and the additional spaces created by their addition give rise to a dense region of calcium silicate hydrate (C-S-H) which could be differentiated as—the inner C-S-H and the outer calcium silicate hydrate (C-S-H) as shown in Figure 11.2. These novel contribution of nanomaterials results in a tight internal microstructure of concrete which offers practically few paths for external attacking agents as the increase in hydration products has invaded the spaces left out by normal hydration reaction. Literatures are abound regarding novel nanosilica contribution to microstructural via atomic force microscopy(AFM) and scanning electron microscopy (SEM) results[63] and strength development in cement concrete.[59,60–62,64]

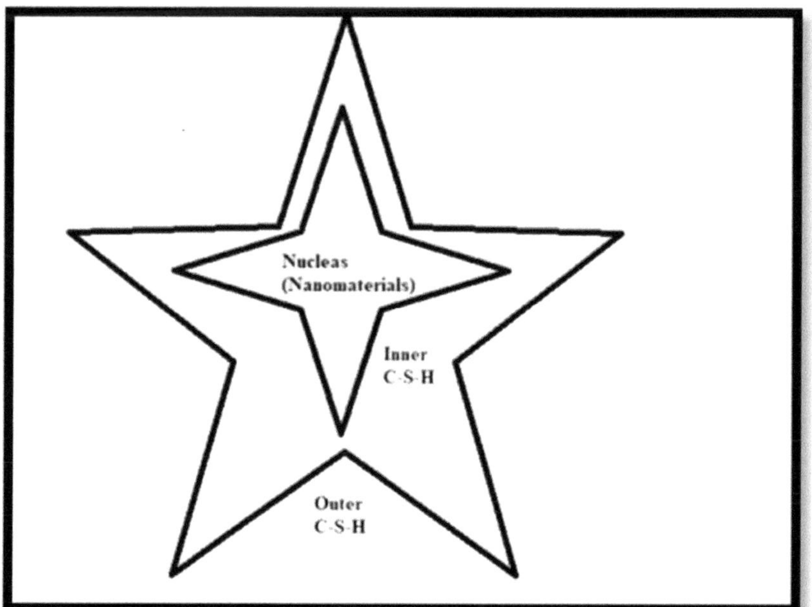

FIGURE 11.2 Schematic structure of nucleation.

11.2 REVIEW OF LITERATURES

Khanzadi et al.[1] have found that the influence of nanosilica particles on the mechanical properties and durability of concrete has been studied through the measurement of compressive and tensile strength, water absorption, and the depth of chloride penetration. The experimental results show that the mechanical properties measured, and the durability of the concrete mixed with the Nanoparticles were better than that of a plain concrete, also the SEM study of the microstructures showed that the nanoparticles filled the cement paste pores, and by reacting with calcium hydroxide crystals from calcium silicate hydration, decreased the size and amount of these crystals. Therefore, the results indicate that nanoscale silica behaves not only as a filler to improve microstructure, but also as an activator to promote pozzolanic reaction.

Sobolev et al.,[2] in their research have found that an increase in compressive strength of mortars with developed nanoparticles at early stages of hardening followed by the strength reduction at later age. Addition of super plasticizer was proposed to overcome this obstacle. Super plasticized mortars with selected nano-SiO_2 demonstrated a 15–20% increase of compressive strength, reaching up to 144.8 MPa at 90-day age. The developed high-performance cements demonstrate the 28-day compressive strength at the range of 93–115 MPa, which is higher than 72–89 MPa, the strength of reference cements.

Morsy et al.[3] have found in their work that CNT can improve the mechanical performance, durability, and the sustainability of concrete. When ordinary portland cement (OPC) was substituted by 6 wt.% of cement by nanometakaolin (NMK) and the carbon nanotube was added by ratios of 0.005, 0.02, 0.05, and 0.1 wt.% of cement, this combination reduces the amount of cement needed in concrete in order to obtain higher strength. It also reduces the construction periods and produce high-strength concrete with less curing time.

Givi et al.[4] have found that SiO_2 nanoparticles with two different sizes of 15 and 80 nm have been used as a partial cement replacement by 0.5, 1.0, 1.5, and 2.0 wt.%. It was concluded that concrete specimens containing SiO_2 particles with average diameter of 15 nm were harder than those containing 80 nm of SiO_2 particles at the initial days of curing. When nanoparticles are uniformly dispersed in the cement paste, it promotes cement hydration which is favorable for the strength of cement mortar.

Nanoparticles fill the cement pores, thus increasing the strength. Nano-Sio_2 can contribute to the hydration process to generate more C-S-H through reaction with $Ca(OH)_2$ Result shows that nano-Sio_2 particles of blended concrete have higher compressive, flexural, and tensile strength at all ages of moist curing.

Sadrmomtazi et al.[5] have found that nano-Sio_2 develops higher mechanical strength and lower water absorption when different amounts of nano-SiO_2 (0%, 1%, 3%, 5%, 7%and 9%) were incorporated into the ordinary cement mortar. Here, Nano-Sio_2 improves the bond strength of paste–aggregate interface and it was valuable for enhancing abrasion resistance and it accelerates the cement hydration due to its high activity.

Zaki et al.[6] have revealed that the well-known performance of concrete without nanoparticles was compared with that after the addition of nanoparticles for both fresh and hardened states. Nanosilica concrete requires additional amount of water or super plasticizer to maintain the same workability level. Nanocarbon tubes can improve the properties of concrete. Nanosilica addition provides significant increase in concrete compressive strength after 28-days up to 1 year and the optimum amount of nanosilica is 0.5% by weight of cementitious material.

Rana et al.[7] have reported that their article examines the potential areas where nanotechnology can benefit construction engineering. The data and information collected are from the current literature. The purpose is to point out clear cut direction among the nanotechnology development areas where the construction process would immediately harness nanotechnology, by specifying clear recommendations. The information would be beneficial to both construction engineering education and research. In this paper, a broad overview of the potential application of various nanotechnology developments in the construction engineering field is discussed, and the potential for further basic research that may lead to improved systems is evaluated. Nanotechnology has been concerned with developments in the fields of microelectronics, medicine, and materials sciences. The potential for application of many of the developments in the nanotechnology field in the area of construction engineering is growing.

Pacheco-Torgal et al.[8] have expressed that their paper reviews the current knowledge about nanotechnology and nanomaterials used by the construction industry. It covers the nanoscale analysis of Portland cement hydration products, the use of nanoparticles to increase the strength and durability of cementations composites, the photocatalytic capacity of nanomaterials and also nanotoxicity risks. Nanotechnology seems to hold the key that allows construction and building materials to replicate the features of natural systems improved until perfection during millions of years.

Sharif[11] focused on some of the nanoscopy techniques that are most frequently used in current research of cement-based nanostructured materials. In particular, SEM, transmission electron microscopy (TEM), AFM, scanning tunneling microscopy, tomography, scanning transmission X-ray microscopy, and laser scanning confocal microscopy are addressed. A number of case studies related to microscopic characterization of nanomaterials utilizing the aforementioned techniques from the published literature were discussed.

Li et al.[13] have experimentally studied the mechanical properties of nano-Fe_2O_3 and nano-SiO_2 cement mortars. The experimental results showed that the compressive and flexural strengths measured at the 7th and 28th day of the cement mortars mixed with the nanoparticles were higher than that of a plain cement mortar. The SEM study of the microstructures between the cement mortar mixed with the nanoparticles and the plain cement mortar showed that the nano-Fe_2O_3 and nano-SiO_2 filled up the pores and had reduced the $Ca(OH)_2$ compound among the hydrates. These mechanisms explained the supreme mechanical performance of the cement mortars with nanoparticles.

Reddy and Meena[59] used the combination of different by-products from various industries for replacing cement in concrete leads to saving in energy and natural resources. In this article, an attempt has been made to study the mechanical and water absorption properties of concrete incorporated with the combination of fly ash (FA), alccofine (ALC), and collodial nanosilica (CNS) at 7, 28, and 56 days curing period. Cement has been partially replaced by the combination of FA at 25%, ALC at 10%, and CNS at 0.5%, 1%, 2%, and 3% with water cement ratio of 0.43. The result indicates that the incorporation of combination of FA, ALC, and CNS can be very effective in the improvement of mechanical and water absorption properties of the concrete. The mix with a combination of 25% FA, 10% ALC, and 1% CNS is most effective in the improvement of mechanical and water absorption properties as compared with all other mixes.

Andrabi and Singh[60] studied the use of 215 nm nanosilica to increase the compressive strength of concrete. Experimental studies were performed at 0.4%, 0.7%, and 1% of BC. These tests have greatly improved the strength of the early pressure and the overall pressure strength of the concrete. As the percentage of nanosilica increases, an increase in strength is observed. Current research is the design of (M40 + M30) quality of concrete. The mixtures developed to obtain more rheological properties and solid properties were studied. In this study, the cement is replaced by silica smoke and nanosilica materials. In addition, it is planned to determine the number of predictions of cubes in the program. The current design of the probe is the strength of M40 and M30 quality of concrete. This thesis explores the standard optimization ratio for silica and fume to get the best compressive strength. Further, this has to check for Indian standard uses for developing the demand of quality users in India. Finally, this research leads us to create the console for evaluating the strength of concrete.

Elboghdadi et al.[61] aimed this experimental study to assess the combined effects of using nanosilica (NS) and steel fiber (SF) on the workability of fresh concrete and mechanical properties of hardened concrete. NS is used as partial cement replacement by 1, 1.5, 2, and 4 wt.%, and SF is used as volume substitution by 0.45, 0.9, and 1.35 %. Slump, compressive strength at 7 and 28 days of age, modulus of elasticity, and flexural strength are evaluated using different combinations between NS and SF. Thermogravimetric analysis (TGA) is provided for detecting changes in hydration material due to the addition of NS. Significant improvement in the mechanical properties of concrete on using NS and SF, addition of NS not only act as pozzolanic additive but also as pore filling leading to reduce the pore size structure of concrete. Optimum content of NS is 1.5–2 wt.% which improves the compressive strength for samples with various SF content. On the other hand, it can be stated that more than 2 wt.% of NS leads to decrease the compressive strength, as unreacted agglomerated nanoparticles existed when large amount of NS is used. Utilizing 2 wt. % NS with 0.9 % SF leads to doubling the modulus of elasticity compared with samples without either NS or SF. Finally, Flexural strength is improved to 87.7 % for samples of 2 wt.% NS and 1.35 % SF compared with samples without NS and SF.

Sidhardhan and Karthikeyan,[62] have reported the influence of nanosilica in mortar with the average diameter of 40 nm and also studied the pore-filling effect and its pozzolanic activity with cement toward the improvement of mechanical properties and durability aspects. The strength, durability, microstructure, and acoustical study of cement mortar incorporating NS in various percentages, such as 1%, 2%, 3%, 4%, and 5% as a partial replacement of cement are compared with the conventional mortar. The hardened properties of the mortar determined using compressive strength of all the specimens at the age of 7, 14 and 28 days. Durability properties have been carried out to determine the steel corrosion. The microstructure of the specimens has been determined by using XRD and SEM analysis. The results reveal that replacing nanosilica upto 2% shows enhanced strength, durability, and improved microstructure of the mortar specimens resulting in better performance as compared with conventional specimens. Thus, there is a scope for the development of crack-free concrete toward sustainable construction.

Boshehrian and Hossein[63] studied the mechanical properties (by compressive and flexural strength tests), durability (by water absorption test), and microstructural properties of interfacial transition zone (ITZ) (by SEM and AFM tests) of mortars applicable for the casting of ferrocement elements reinforced with nano-SiO_2 particles. The parameters of this study include the low replacement ratio of nano-SiO_2 particles with respect to cement in ordinary Portland cement (OPC) mortar mixture (including 1%, 2%, and 3%), water to binder ratio (including 0.35, 0.4, and 0.5), and also sand to binder ratio (including 2 and 2.5). The results have shown that the cement mortars containing nanoparticles have reasonably higher strength, low water absorption, and denser ITZ compared with those of the OPC ferrocement mortars. Furthermore, along with increasing the W/CM, the performance of silica nanoparticles has been reduced. Besides, using higher S/CM was followed by strength loss in both categories of mixtures including with and without silica nanoparticles. However, distinctive strengthening trend was not observed in mixtures with different S/CM (by holding the other variables constant such as silica nanoparticles).

Teja et al.,[64] in their work, have utilized the combination of different mineral admixtures, such as FA, ALC, and NS which are by-products from industries and possess pozzolanic nature and these are environmental friendly materials in construction industry. The environmental impact can be minimized by making use of many industrial by-products in a sustainable manner. The evaluation of strength and durability study was performed using concrete with FA, ALC, and NS as a partial replacement of OPC. The studied parameters were compressive strength, split tensile strength, and flexural strength, as well as the durability study that involves water absorption and sorptivity. The combination of FA 25%, ALC 10%, and NS 0%, 0.5%, 1%, 2%, and 3% was used in partial substitution of cement at by weight. The results show that among different NS content, high performance concrete containing FA–AL–NS in the ration of 25–10–1% exhibited the best results for the strength and durability property tests at 7, 28 days for strength and at 28, 90 days for durability properties.

11.3 MATERIALS AND TESTING PROGRAM

The materials used in testing were cement—OPC (43 Grade), Fine Aggregate (FA)–River sand conforming to Zone II of IS: 383-1970, potable water and nanomaterials (viz. nanosilica (nS) with particle size not exceeding 40 nm. The specific properties of nanosilica are given in Table 11.1.

TABLE 11.1 Specific Properties of Nanosilica Used.

Item	Description
Active nano content	35–40%
Particle size	5–40 nm
Ph (20°C)	9–10
Specific gravity	1.08–1.32

Our Testing Program included Compression Testing of cement mortars with and without nanosilica at both short and long terms along with X-Ray diffraction testing and SEM Testing with and without energy dispersive X-ray spectroscopy at 28 & 365 days respectively. But the other tests, such as TEM, AFM, scanning probe microscopy (SPM), scanning tunneling microscopy(STM), and tomography have been briefly outlined as they form an integral part in the characterization of nanomaterials.

(a) Compression Testing for Cement Mortar

Mortar cubes of 70.7 × 70.7 × 70.7 mm size were casted as per the Indian Standards (IS:4031). Nanosilica were added in various proportions ranging from 0%, 0.5%, 0.75%, 1.0%, 1.25%, and 1.5% with respect to cement weight, keeping the w/c ratio fixed at 0.4. The cubes were then ordinary cured under water and tested at 7, 28, 90, 180, and 365 days in a compression testing machine(CTM). The CTM generally consists of the following parts, namely, (1) Load Frame is usually consisting of two strong supports for the machine;(2) Load cell—A force transducer which converts the tensile or compressive force of the CTM into means of measuring the load, and (3) Output device—A means of providing the test result like dial or digital displays and chart recorders. Many newer machines have a computer interface for analysis and printing. The optimized proportion of the nanomaterials were found which gave maximum strength at 28 days.

(b) X-Ray Diffraction Test

X-ray diffraction techniques are widely used for determining the identity and structure of crystalline materials by a X-ray diffractometer. It consists of a source of radiation, a monochromator to choose the wavelength, slits to adjust the shape of the beam, a sample and a detector. Graph is drawn between 2θ and counts. Thus, the quantitative and qualitative measures of the sample have been done by using this test. The X-ray diffraction (XRD) technique is used to measure the atomic spacing between lattice layers in a crystal. This can be calculated from the X-ray diffraction test results using the Braggs Law equation which is given as follows: $n\lambda = 2d \sin\theta$, where, λ(lambda) = wavelength of the incidental X-ray beam, d = the atomic spacing between the layers, θ = glancing angle or angle of incidents, n is taken as unity.

(c) Scanning Electron Microscopic Test

SEM analyzer is an instrument used for the examination and analysis of the microstructural characteristics of the solid sample. For research concerning nanomaterials, it is necessary to observe the morphology and composition of samples. At present, SEM is heavily used in the pursuit of the further understanding of nanomaterials, the combination of higher magnification, larger depth of field, greater resolution, and compositional and crystallographic information makes the SEM one of the most heavily used instruments in research areas and industries.[14,15] The SEM provides images which provide information on

- Topography: The surface features of an object and its texture.
- Morphology: The shape and size of the particles making up the object.
- Composition: The elements and compounds that the object is composed of and the relative amounts of them.

- Crystallographic Information: How the grains are arranged in the object.

In SEM, electron beam is produced by an electron gun and is accelerated by anode voltage downward to the sample target. The cathode and Wehnelt electrode are floating on a negative potential from kV to several tens kV, while anode is grounded. The grounded anode accelerates emitted electrons from cathode (thermoionic or field emission).[16] The scanning generator powers x and y coil for both the electron beam probe and electron beam of cathode ray tube (CRT). These two scans are synchronized. The secondary electron signal collected by a detector is processed, amplified, and supplied to the CRT to modulate the intensity of electron beam for the image formation.[17]

Morphological characteristic is found out here by using the instrument Hitachi S-3000H. Appearance of the specimen image is gained by this test and it gives the direct result of the large depth of field. At greater depth of field, much more information about the specimen was found out. This instrument can give the morphological structure of the specimen for various magnifications. Here, cement mortar with and without nanomaterial additions were tested for SEM results at 28 and 365 days, respectively. Also energy dispersive X-ray spectroscopy (EDS or EDX) which is a chemical microanalysis technique was used in conjunction with SEM at 365 days. The EDS technique detects X-rays emitted from the sample during bombardment by an electron beam to characterize the elemental composition of the analyzed volume and to create element composition maps over a much broader raster area.

(d) Transmission Electron Microscopy

TEM allows a qualitative evaluation of the internal structure and spatial distribution of the various phases through direct visualization. In scientific research, now TEM is being widely used by the researchers for its outstanding performance in high-resolution imaging capability. This enables the user to examine fine detail—even as small as a single column of atoms, which is thousand times smaller than the smallest resolvable object in a light microscope. With the development of TEM, the associated technique of scanning TEM (STEM) was re-investigated with the development of field emission gun and adding a high quality objective lens.[18] Using this design, Crewe demonstrated the ability to image atoms using annular dark-field imaging. Crewe and coworkers at the University of Chicago built a STEM and were able to visualize single heavy atoms on thin carbon substrates.[19] In 2008, Meyer et al. described the direct visualization of light atoms such as carbon and even hydrogen using TEM and a clean single-layer graphene substrate.[20]

TEM consists of an assembly of electromagnetic lenses arranged in a vertical column with an electron source and fluorescent screen at opposing end of the column. In most cases, the microscope has electron source (gun) at top of the column. Electrons produced at the gun are accelerated down the column (which is evacuated to lower pressure than 10^{-6}Torr; HRTEM using field emission gun requires at least 10^{-9}Torr in the region of electron emission) toward the screen by a high voltage, typically between 100 and 300 kV.[21] The electrons emerging from the lower side of the specimen are focused by the objective lens, which surrounds the specimen. In the lower focal plane of this lens, a diffraction pattern of the specimen is formed. At a position below the lower focal plane, an inverted image of the specimen is created. The sole purpose of the subsequent post-specimen lenses is to magnify one of these two plane images. By varying the currents in these latter lenses, either a magnified diffraction pattern or a magnified image of the specimen can be obtained on the fluorescent screen. A camera, normally located under moveable fluorescent screen, is used for making a permanent record of the diffraction pattern or image. The final image magnification can be varied by changing the currents of the post-specimen lenses from typical 10^2–10^6 times.[22]

TEM has now become a part and parcel in nanoscale characterization of materials in the field of cancer research, virology, materials science, as well as pollution, nanotechnology, semiconductor and construction materials research. A few studies carried out by some researchers and their findings will be discussed.

Nanomaterials are being considered for various uses in the construction and related infrastructure industries. To achieve environmentally responsible nanotechnology in construction, it is important to consider the lifecycle impacts of nanomaterials on the health of construction workers and dwellers, as well as unintended environmental effects at all stages of manufacturing, construction, use, demolition, and disposal. The size distributions of the nanomaterials can be visualized with TEM.[23–27] Commercially available colloidal nano-SiO_2 (CNS) with a mean particle size of 20 nm and a solid content of 30% was used in a research.[28] Morphologic studies indicate that

CNS particles are generally round in shape. Although most of the particles are well-dispersed, agglomerations are found. The dark spot in the TEM image indicates that some seriously agglomerated clusters may be seen in CNS.

Through TEM, the interaction between the different components of a mixture can be clearly monitored. Cement kiln dust (CKD), a by-product of the cement industry, contains significant amounts of alkali, free lime, chloride and sulfate. In a study of CKD, the morphology of calcium aluminosilicate hydrate (C-A-S-H) gel present in CKD-based fly ash binders, the microstructure was examined under a TEM.[29] Fibrillar type morphology for C-A-S-H is clearly evident in TEM micrographs. In TEM examination, structural difference is distinguishable around the incumbent particle, as seen in several researches.

TEM can also help to identify crystalline phases and morphological and structural characteristics[30] with the TEM nanographs in bright field mode and selected area of electron diffraction patterns (SAEDP) for cement paste exhibited at an age of 90 days show specific phases.

Latest TEM developments with real-time observations of materials interactions at high resolution in their native fluid environment have attracted increasing attention across physical and biological science disciplines.[31] Recently, a microfabricated cell with electron transparent membranes was used to contain the fluid (liquid or gas) in the high-vacuum environment of the microscope for the in situ transformation observation.[31]

(e) Scanning Probe Microscopy

SPM is a family of techniques that measures surface topography and properties on the atomic scale. There are an ever increasing number of papers devoted to technical advances and applications of SPM. The invention of the scanning tunneling microscope (STM), the atomic force microscope (AFM) and other SPMs, such as the scanning near-field optical microscope and the scanning ion conductance microscope, has really enabled the pursuit of more ambitious goals in nanotechnology. Many scanning probe microscopes can image several interactions simultaneously. To form images, scanning probe microscopes raster scan the tip over the surface. At discrete points in the raster scan, a value is recorded (which value depends on the type of SPM and the mode of operation–constant interaction mode or constant height mode). These recorded values are displayed as a heat map to produce the final STM images, usually using a black and white or an orange color scale. The nature of an SPM probe tip depends entirely on the type of SPM being used. Rendering software is necessary to produce images and such software is produced and embedded by instrument manufacturers. Now, there is a growing interest to use SPMs to study the nanostructural characteristics of construction materials.

(f) Atomic Force Microscopy

The AFM is a scanning probe imaging instrument that provides spatial resolution of a fraction of a nanometer (depending on specimen type and surface roughness) with a sharpened probe, typically less than 10 nm in diameter.[32–34] The AFM is one of the foremost tools for imaging, measuring, and manipulating matter at the nanoscale. The information is gathered by "feeling" the surface with a mechanical probe. Piezoelectric elements that facilitate tiny but accurate and precise movements on (electronic) command enable the very precise scanning. There are also some variations in scanning. According to the nature of the tip motion, AFM operation is usually described as one of the three modes named contact mode/static mode, tapping mode/intermittent contact/vibrating mode, and non-contact mode.

To investigate the use of hybrid recycled powder as a supplementary cementing material, the pozzolanic property of hybrid powder is comprehensively studied here by utilizing advanced tools including AFM.[35] In order to study the effect of hybrid recycled powder on cement paste by AFM, a representative region containing the hybrid recycled particles was identified for the nanoscale probe. A representative region containing one hybrid particle and one fine aggregate (sand), which was denoted as Region I, was selected. For comparison, another region containing no hybrid recycled particles and no fine aggregates, denoted as Region II, was also selected to represent the normal cement paste. In Region I, the height varied from +2.2 to −2.2 μm. It was found that the RMS roughness of Region I was 595 nm. The surface topography of Region II was shown in the paper. In this region, the main constituent was C–S–H gel. Based on the quantitative measurement, the RMS roughness of Region II was 470 nm.

In another study, the early age mechanical response of the cement mortar was modified with graphene oxide using AFM.[36] The nanocomposite was prepared with OPC containing 3 wt.% of the graphene oxide (GO3).

The topography of the samples used for the analysis of their mechanical properties was named as, C –reference sample and D – nanocomposite. The nanocomposite exhibits the distribution of Young's Modulus in the range of 5–20 GPa. The results showed a high increase in the value of Young's modulus of composites containing GO sheets.

Lateral force microscopy (LFM) is a derivative of AFM. In LFM, the tip is maintained in contact with the sample surface and the "twisting of the cantilever" is measured in addition to the vertical deflection (i.e., the approach that is more common for AFM).[37–39] AFM and LFM techniques were used on mortar samples to distinguish between CSH, CH, and unhydrated cement particles in a research.[40] The LFM technique appears to be more sensitive to topographic changes than the conventional AFM and it can more clearly distinguish between the different phases at high magnification (low scan range) as discussed in the paper.

Joined AFM –scanning electrochemical microscopy (AFM–SECM) is a influential evolving technology capable of providing simultaneous topographical and electrochemical imaging at the sample surface. The integration of electrodes into AFM–SECM probes facilitates obtaining enhanced information during measurements of relevant building materials.

(g) Scanning Tunneling Microscopy

A STM is an instrument for imaging surfaces at the atomic level. For an STM, good resolution is considered to be 0.1 nm lateral resolution and 0.01 nm depth resolution[41–44] through which individual atoms within materials can be routinely imaged and manipulated. The process of scanning tunneling microscopy is based on the concept of quantum tunneling. In this technique, a fine probe tip is scanned over the conducting surface of the material to be examined with the help of a piezoelectric crystal at a distance of 0.5–1 nm, and the resulting current or the position of the tip required to maintain a constant tunneling current is monitored. When a conducting tip is brought very near to the surface to be examined, a bias (voltage difference) applied between the two can allow electrons to tunnel through the vacuum between them. The resulting tunneling current is a function of tip position, applied voltage, and the local density of states (LDOS) of the sample.[45] Information acquired by monitoring the current as the tip's position scans across the surface is usually displayed in image form. The use of STM in nanocharacterization is very much advantageous as it can be used not only in ultrahigh vacuum but also in air, water, and various other liquid or gas ambient, and at temperatures ranging from near-zero Kelvin to a few hundred degrees Celsius.[45]

STM provides a probe for the morphology and electrical properties at atomic resolution in all three dimensions. Atomically resolved STM topographs of graphene on SiO_2 substrates have been reported[46–50] from which the height of graphene ripples was determined to be 3–5 Å. Xu et al. reported on the scanning tunneling microscopy study of a new class of corrugations in exfoliated monolayer graphene sheets, with wrinkles of ~10 nm in width and ~3 nm in height.[51] Wrinkle-like structures have been seen before in TEM images of suspended graphene sheets,[52] and in STM and high-resolution SEM images of graphene grown on conducting substrates,[53,54] but their structure and properties have not been carefully characterized. Through STS, they further demonstrated that the wrinkles have lower electrical conductance when compared with other regions of graphene and are characterized by the presence of mid-gap states, which are in agreement with recent theoretical predictions.

It was surprising that atomically resolved topographs reveal very different structures for the wrinkles in comparison with other parts of graphene. A triangular pattern is observed over the entire graphene wrinkle, and the distance between adjacent bright spots is ~2.5 Å, indicating that the honeycomb six-fold symmetry of the graphene lattice is broken, and only three of the six carbon atoms in each hexagonal ring is observed, as the paper says. In comparison, the topograph taken on the same graphene sheet adjacent to the wrinkle reveals the honeycomb structure that is consistently observed on the "flat" (by "flat" meant only the ~4 Å ripples are present) parts of the monolayer graphene sheets investigated in their study.[51] In another example of STM study, gold films on mica were preparedby Robles et al..[55] The surface topography was measured of each sample employing a STM. From the topography measured with the STM, they determined the parameters that describe the surface roughness of each specimen in the scale of length set by λF (in gold, $\lambda F = 0.52$ nm). The typical grain texture of the films, measured with the STM, was displayed in the paper.

The STM not only provides the three-dimensional information about the topography of the sample, but it also gives the information about the spectroscopic properties and local variations of work functions. Further, STM can be used to operate in liquid solutions, the operation involves the immersion of the STM probe into the liquid media for in situ monitoring of redox processes on the sample.[56] Having these advantages, STM can be used to observe the in situ hydration process of the cementitious material in the future.

(h) Tomography

For the full understanding of nanomaterials' behavior, their control in manufacturing process and monitoring the service parameters need a characterization technique which provides a three-dimensional vision. A Characterization technique would be highly advantageous when it reveals the richness of microstructure in three dimensions rather than just providing two-dimensional projections. Various techniques, for example, X ray tomography and mechanical sectioning for three-dimensional reconstruction of microstructures have been carried out in the field of materials science for decades. Tomography instruments consist of Gantry bores, X-ray tubes, City scanners, Detection system, Independent Work Station, Control console, Advanced Reconstruction Display and Application Software.[58] Tomography in a wider meaning, includes all techniques that use sectional views (not only projections) as an intermediate step from which the three-dimensional object is re-assembled.[57]

11.4 TEST RESULTS AND DISCUSSION

1. The mortar compressive strength determined as per IS: 4031 shows a modest strength gain of 13% at 7 days and 32.55% increase in strength at 0.75% nS addition (optimized) at 28 days, with the rate of strength gain increasing up to 59.8% at 90 days, but then falling by 8.4% at 180 days at same optimization as shown in Table 11.2. The strength rate falling at a later age may be due to the fact that nanosilica particles gets agglomerated due to their low-weight percentages. It is seen that with increased concentration of nanosilica particles, the strength gain reverses and increases at later ages.
2. The Nanosilica was tested for XRD results as shown in Figure 11.3. No sharp peak was found for this nanomaterial thus confirming its amorphous nature.

TABLE 11.2 Strength (MPa) of OPC Mortar (% Increase w.r.t. Ordinary Cubes).

Sl. No.	% Nano additions in Cement (OPC)	Avg. 7 day cube strength (% increase)	Avg. 28 day cube strength (% increase)	Avg. 90 day cube strength (% increase)	Avg. 180 day cube strength (% increase)	Avg. 365 day cube strength (% increase)
1.	OPC(0%nS)	21.08	31.89	31.20	30.01	30.01
2.	OPC (0.5% nS)	23.85(13.14%)	35.51(11.35%)	41.3(32.7%)	27.47(−9.2%)	26.76(−4.29%)
3.	OPC (0.75% nS) (optimized)	23.85(13.14%)	42.27(32.55%)	49.85(59.8%)	32.52(8.4%)	31.5(4.96%)
4.	OPC (1.0 % nS)	25.07(18.93%)	37.36(17.15%)	42.98(37.7%)	33.68(12.2%)	32.41(8.0%)
5.	OPC (1.25% nS)	23.17(9.91%)	30.85(3.26%)	39.45(26.4%)	35.24(17.4%)	31.3(4.29%)
6.	OPC (1.5% nS)	23.81(12.95%)	37.79(18.5%)	33.42(7.12%)	31.23(4.07%)	29.12(−2.96%)

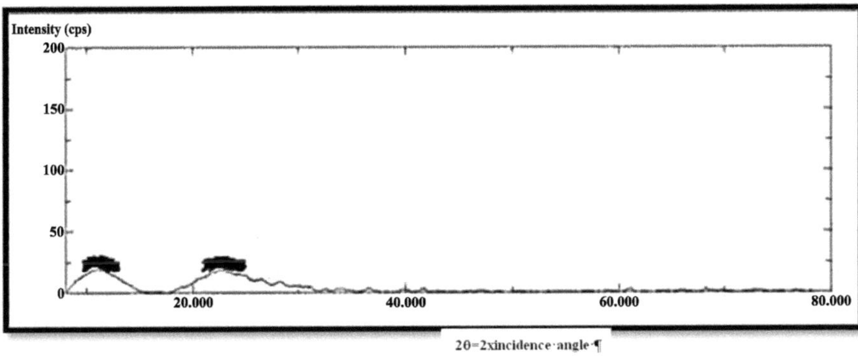

FIGURE 11.3 XRD image of nanosilica used.

3. From the SEM images as shown in Figure 11.4a and b, it is observed that nanoparticles improved the structure of the cement mortar matrix, resulting in a better bond between the sand particles and cement paste when compared with that of ordinary cement mortar. 0.75% nS system has shown a dense, compact, and uniform microstructure indicating the uniform distribution of the nanoparticles along with good bonding between cement and sand particles. Also with increased nS addition, there is an increased concentration of hydration products which are directly responsible for the strength gain of cement mortar with respect to ordinary mortar as shown in Table 11.2. This is attributable due to the fact that increased pozzolonic reaction has resulted in the creation of more C-S-H hydration products, thus directly contributing to the strength of cement mortar. Increased nucleation and more pore-filling effects have resulted in the nanosilica-added cement mortar matrix causing the gain in strength of nS-added cement mortar. Also, microcracks of the ordinary cement mortar as shown in Figure 11.4a get filled in the nS-added cement mortar as can be seen from Figure 11.4b. From Figure 11.5, it is observed that at longer term, the microstructure of cement mortar system becomes more refined with uniformity in hydration products though microcracks, and crevices are present, but with nanosilica addition, the microstructural system is filled up with more hydrated products which fill up the cracks and crevices, as can be seen in Figure 11.8 at 500 μm magnification. The energy dispersive X-ray spectroscopy (EDS or EDX) software plotting, as shown in Figure 11.6, shows a greater peak height for calcium

followed by oygen and then followed by silicon for ordinary cement mortar at 365 days, but with nanosilica-added cement mortar, there is a preponderence of silicon followed by oxygen and then calcium (see Figure 11.9) which means that more C-S-H products are formed here as already discussed in the previous sections. The elemental composition as shown in Figure 11.7 for ordinary cement mortar and Figure 11.10 for nS-added cement mortar reports the percentage weightage of Si = 1% for the former, while for the latter, it is 16% which confirms the presence of more hydrated C-S-H products in the nanosilica-added cement mortar when compared with ordinary cement mortars.

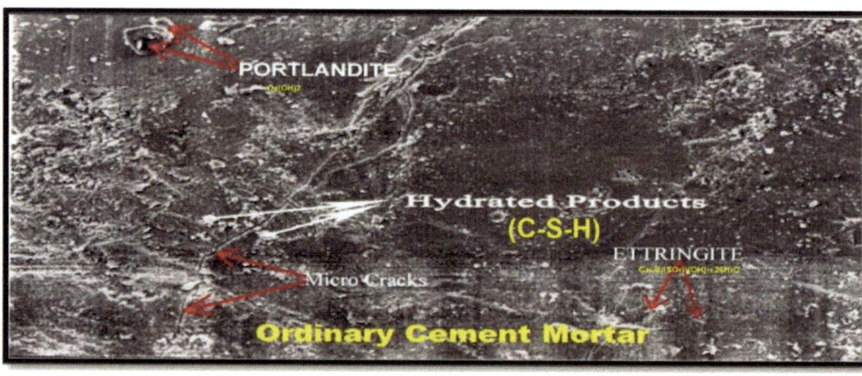

FIGURE 11.4(a) SEM image for ordinary cement mortar @ 28 days.

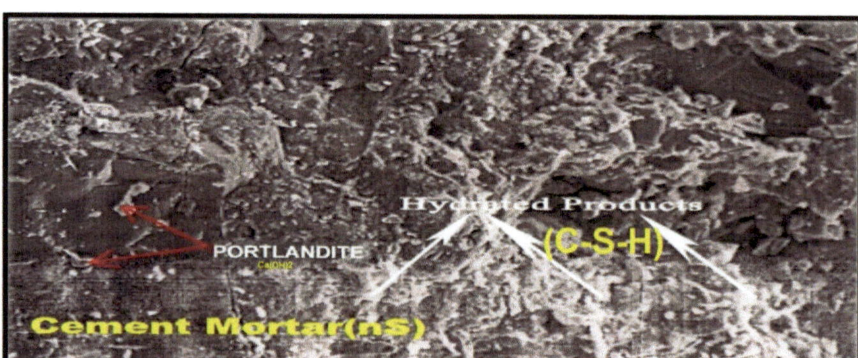

FIGURE 11.4(b) SEM image for nS-added cement mortar @ 28 days.

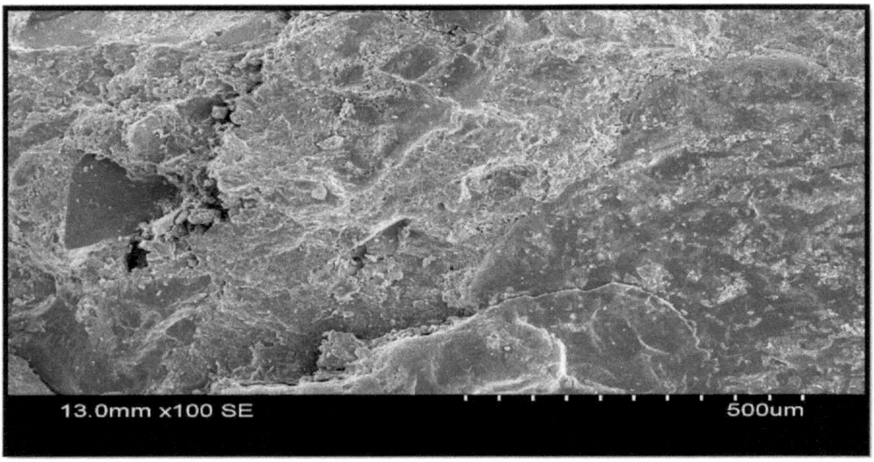

FIGURE 11.5 SEM image for ordinary cement mortar @ 365 days.

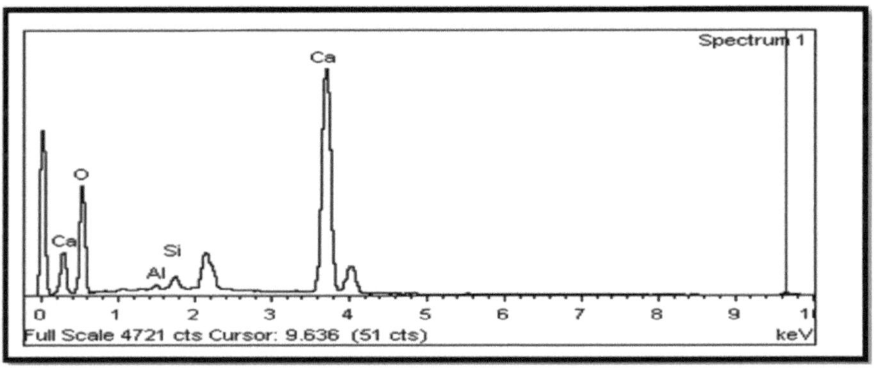

FIGURE 11.6 EDS spectrum of ordinary cement mortar at 365 days. Most of these peaks are X-rays given off as electrons return.

11.5 CONCLUSION

The characterization methods must provide complete structural information at all appropriate scales to describe where every atom starts, where each ends up and how and why they get there in a transformation processes. The further shrinkage of the dimensions and the introduction of new properties and material concepts demand for ever better resolution and stability. Nanoscopic analysis is essential for the characterization of nanomaterials

Spectrum processing: Ordinary Portland Cement(OPC)
Peak possibly omitted: 2.140 keV

Processing option: All elements analyzed (Normalized)
Number of iterations = 5

Standard:
O SiO2 1-Jun-1999 12:00 AM
Al Al2O3 1-Jun-1999 12:00 AM
Si SiO2 1-Jun-1999 12:00 AM
Ca Wollastonite 1-Jun-1999 12:00 AM

Element	Weight%	Atomic%
O K	53.28	73.79
Al K	0.45	0.37
Si K	1.08	0.85
Ca K	45.19	24.99
Totals	100.00	

FIGURE 11.7 Elemental composition of ordinary cement mortar at 365 days.

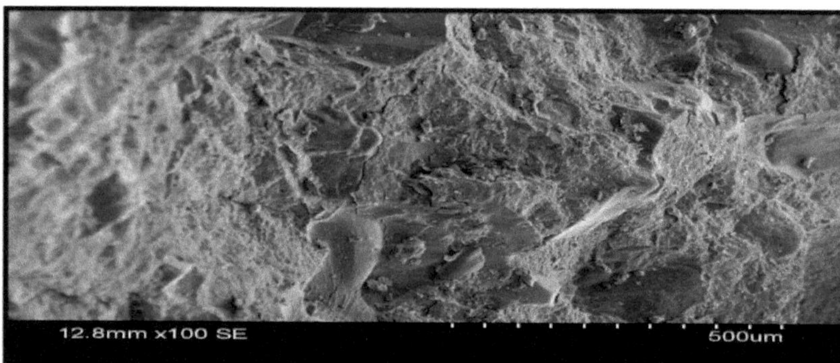

FIGURE 11.8 SEM image for nanosilica-added cement mortar @ 365 days.

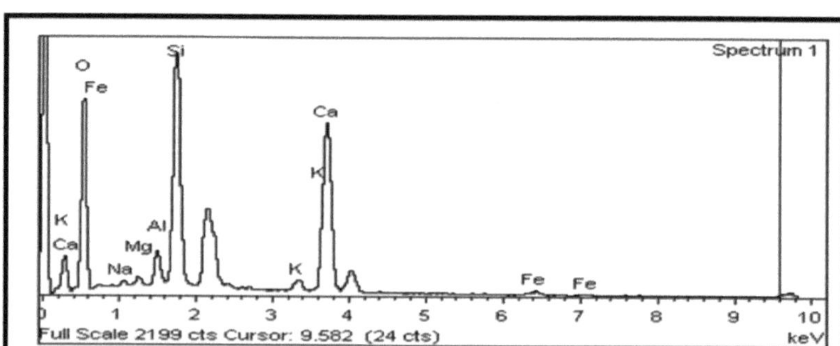

FIGURE 11.9 EDS spectrum of the nanosilica-added cement mortar @ 365 days. Most of these peaks are X-rays given off as electrons return.

Spectrum processing: n-SiO_2
Peak possibly omitted: 2.141 keV

Processing option : All elements analyzed (Normalized)
Number of iterations = 4

Standard:
O SiO2 1-Jun-1999 12:00 AM
Na Albite 1-Jun-1999 12:00 AM
Mg MgO 1-Jun-1999 12:00 AM
Al Al2O3 1-Jun-1999 12:00 AM
Si SiO2 1-Jun-1999 12:00 AM
K MAD-10 Feldspar 1-Jun-1999 12:00 AM
Ca Wollastonite 1-Jun-1999 12:00 AM
Fe Fe 1-Jun-1999 12:00 AM

Element	Weight%	Atomic%
O K	51.70	69.55
Na K	0.57	0.53
Mg K	0.55	0.49
Al K	2.29	1.83
Si K	16.34	12.52
K K	1.42	0.78
Ca K	25.32	13.60
Fe K	1.82	0.70
Totals	100.00	

FIGURE 11.10 Elemental composition of nanosilica-added cement mortar @ 365 days.

added structures. Previously, without the advent and study of nanomaterials, cement concrete had a bad reputation of strength gain at early ages. The higher strength gain of 13% for nanosilica-added cement mortar when compared with that of ordinary cement mortars at 7 days, has a potential benefit for application in fast-track construction projects where time constraint is in very high demand. However, as already pointed out, the long-term effects of optimized nanosilica addition (0.75% by weight of cement) to cement matrix composites have been found to be inconsistent with the strength development due to hydration at the standardized 28 days, not achieving the desired scales, with the rate of strength gain increasing up to 59.8% at 90 days, but then falling by 8.4% at 180 days at same optimization of 0.75% nS. At 365 days, the rate of strength gain had fallen to 4.96% but with the addition of more nanosilica, the strength gain revives to 8%. This when explained in terms of microstructural concepts, corroborates the fact that the agglomeration of nanosilica was the cause of this downfall in strength rate as shown in the elemental composition in Figure 11.10. The SEM image of nanosilica-added cement mortar in Figure 11.8 also shows uniform amount of hydrated products at 365 days and the cracks are filled up. The EDS spectrum of Figure 11.10 of nanosilica (nS)-added cement mortar shows a high peak of Silicon (Si) when compared with the EDS spectrum of Figure 11.6 of ordinary cement mortar composites. Thus it can be concluded that at all ages, the highly reactive nanosilica particles have invited more pozzolonic reaction along with nucleation thus resulting in more formation of hydration products like C-S-H. However, more research on microstructural studies, such as TEM and AFM is necessary for the characterization of nanomaterials in concrete.

ACKNOWLEDGMENTS

The authors would like to thank the Dr M. N. Dastur School of Materials Science and Engineering, Indian Institute of Engineering Science and Technology, Shibpur for providing the laboratory facilities in the present research.

KEYWORDS

- **cement**
- **concrete**
- **strength**
- **hydration**
- **nanomaterials**

REFERENCES

1. Khanzadi, M.; Tadayon, M.; Sepehri, H.; Sepehri, M. Influence of Nano-Silica Particles on Mechanical Properties and Permeability of Concrete. *2nd International Conference on Sustainable Construction Materials and Technologies*, June 28– 30, 2010, Italy.
2. Sobolev, K.; Flores, I.; Hermosillo, R.; Torres-Martinez, L. M. Nanomaterials and Nanotechnology for High-Performance Cement Composites. *Nanotechnology of Concrete: Recent Development and Future Perspectives*, November 7, 2006, Denver, USA.

3. Morsy, M. S.; Alsayed, S. H.; Aqel, M. Hybrid Effect of Carbon Nanotube and Nano Clay on Physico Mechanical of Cement Mortar. *Construct. Build. Mater.* **2011,** *25*, 145–149.
4. Givi, A. N. Surya Abdul Rashid, Influence of 15 and 80 Nano-SiO2 Particles Addition on Mechanical and Physical Properties of Ternary Blended Concrete Incorporating Rice Husk Ash. *Compos. (Part B)* **2011,** *42-s*, 562–569.
5. Sadrmomtazi, A.; Fasihi, A.; Haghi, A. K. Investigation of Mechanical and Physical Properties of Mortars Containing Silica Fume and Nano-Sio_2. *3rd International Conference on Concrete and Development*, At Tehran, Iran, Apr 2009.
6. Zaki, S. I.; Ragab, K. S. How Nano Technology Can Change Concrete Industry. SBEIDCO—1st *International Conference on Sustainable Built Environment Infrastructures in Developing Countries ENSET Oran* (Algeria), Oct 12–14, 2009.
7. Ashwani, V.; Rana, K.; Rana, S. B.; Kumari, A.; Kiran, V. Significance of Nano-technology in Construction Engineering. *Int. J. Recent Trends Eng.* Apr **2009,** 1 (4).
8. Pacheco-Torgal, F.; Jalali, S. Nanotechnology: Advantages and Drawbacks in the Field of Construction and Building Materials. *Construct. Build. Mater.* **2012,** *25* (2), 582–590.
9. Taylor, H. F. W. *Cement Chemistry*, 2nd ed.; Thomas Telford Ltd: London, 1997.
10. Kanniyappan, S. P.; Kumar, S.; Muthukumaran, T. Analyzing the Micro structural Properties of Nanomaterial in OPC by SEM, TEM, XRD and Corrosion Rate by TAFEL Techniques. *Int. Res. J. Eng. Technol.* Oct **2017,** *4*, *10*.
11. Sharif, A. Review on Advances in Nanoscale Microscopy in Cement Research. *Micron* **2016,** *80*, 45–58. DOI: 10.1016/j.micron.2015.09.010
12. Scrivener, K.; Snellings, R.; Lothenbach, B. *A Practical Guide to Microstructural Analysis of Cementitious Materials*; CRC Press: Boca Raton, FL, USA, 2015.
13. Li, H.; Xiao, H.-G.; Yuan, J.; Ou, J. Microstructure of Cement Mortar with Nano-Particles. *Compos. Part B.* **2004,** *35*, 185–189. DOI: 10.1016/S1359-8368 (03)00052-0
14. Liu, Z.; Fujita, N.; Miyasaka, K.; Han, L.; Stevens, S. M.; Suga, M.; Terasaki, O. A Review of Fine Structures of Nanoporous Materials as Evidenced by Microscopic Methods. *Microscopy* **2013,** *62* (1), 109–146.
15. Stevens, S. M.; Jansson, K.; Xiao, C.; Asahina, S.; Klingstedt, M.; Grüner, D.; Europe, J. E. O. L. An Appraisal of High Resolution Scanning Electron Microscopy Applied to Porous Materials. *JEOL News* **2009,** *44* (1), 17.
16. Newbury, D. E.; Joy, D. C.; Echlin, P.; Fiori, C. E.; Goldstein, J. I. *Scanning Electron Microscopy and X-ray Microanalysis*; Springer: New York, USA, 2003.
17. Goldstein, J. I.; Yakowitz, H. *Practical Scanning Electron Microscopy: Electron and Ion Microprobe Analysis*; Plenum Press: New York, USA, 1975.
18. Crewe, A. V.; Isaacson, M.; Johnson, D. A Simple Scanning Electron Microscope. *Rev. Sci. Instr.* **1969,** *40* (2), 241–246.
19. Crewe, A. V.; Wall, J.; Langmore, J. Visibility of Single Atoms. *Science* **1970,** *168* (3937), 1338–1340.
20. Meyer, J. C.; Girit, C. O.; Crommie, M. F.; Zettl, A. Imaging and Dynamics of Light Atoms and Molecules on Graphene. *Nature* **2008,** *454* (7202), 319–322.
21. Reimer, L.; Kohl, H. *Transmission Electron Microscopy: Physics of Image Formation*, Vol. 36; Springer Science & Business Media: New York, USA, 2008.

22. Williams, D. B.; Carter, C. B. The Transmission Electron Microscope. In *Transmission Electron Microscopy*; Springer: New York, US, 2009; pp 3–22.
23. Sun, Y.; Zhang, Z.; Wong, C. P. Study on Mono-Dispersed Nano-Size Silica by Surface Modification for Underfill Applications. *J. Colloid Interf. Sci.* **2005,** *292* (2), 436–444.
24. Zhang, F.; Jin, Q.; Chan, S. W. Ceria Nanoparticles: Size, Size Distribution, and Shape. *J. Appl. Phys.* **2004,** *95* (8), 4319–4326.
25. Lin, K. L.; Chang, W. C.; Lin, D. F.; Luo, H. L.; Tsai, M. C. Effects of Nano-SiO_2 and Different Ash Particle Sizes on Sludge Ash–Cement Mortar. *J. Environ. Manage.* **2008,** *88* (4), 708–714.
26. Borchert, H.; Shevchenko, E. V.; Robert, A.; Mekis, I.; Kornowski, A.; Grübel, G.; Weller, H. Determination of Nanocrystal Sizes: A Comparison of TEM, SAXS, and XRD Studies of Highly Monodisperse CoPt3 Particles. *Langmuir* **2005,** *21* (5), 1931–1936.
27. Ziel, R.; Haus, A.; Tulke, A. Quantification of the Pore Size Distribution (Porosity Profiles) in Microfiltration Membranes by SEM, TEM and Computer Image Analysis. *J. Membr. Sci.* **2008,** *323* (2), 241–246.
28. Hou, P.; Cheng, X.; Qian, J.; Zhang, R.; Cao, W.; Shah, S. P. Characteristics of Surface-Treatment of Nano-SiO2 on the Transport Properties of Hardened Cement Pastes with Different Water-to-Cement Ratios. *Cement Concr. Compos.* **2015,** *55*, 26–33.
29. Chaunsali, P.; Peethamparan, S. Influence of the Composition of Cement Kiln Dust on Its Interaction with Fly Ash and Slag. *Cement Concr. Res.* **2013,** *54*, 106–113.
30. Ramezanianpour, A. A.; Ghahari, S. A.; Esmaeili, M. Effect of Combined Carbonation and Chloride Ion Ingress by an Accelerated Test Method on Microscopic and Mechanical Properties of Concrete. *Construct. Build. Mater.* **2014,** *58*, 138–146.
31. Xin, H. L.; Niu, K.; Alsem, D. H.; Zheng, H. In Situ TEM Study of Catalytic Nanoparticle Reactions in Atmospheric Pressure Gas Environment. *Microsc. Microanal.* **2013,** *19* (06), 1558–1568.
32. Zhong, Q.; Inniss, D.; Kjoller, K.; & Elings, V. B. Fractured Polymer/Silica Fiber Surface Studied by Tapping Mode Atomic Force Microscopy. *Surf. Sci. Lett.* **1993,** *290* (1), L688–L692.
33. Magonov, S. N.; Elings, V.; Whangbo, M. H. Phase Imaging and Stiffness in Tapping-Mode Atomic Force Microscopy. *Surf. Sci.* **1997,** *375* (2), L385–L391.
34. Garcıa, R.; Pérez, R. Dynamic Atomic Force Microscopy Methods. *Surf. Sci. Rep.* **2002,** *47* (6), 197–301.
35. Liu, Q.; Tong, T.; Liu, S.; Yang, D.; Yu, Q. Investigation of Using Hybrid Recycled Powder from Demolished Concrete Solids and Clay Bricks as a Pozzolanic Supplement for Cement. *Construct. Build. Mater.* **2014,** *73*, 754–763.
36. Horszczaruk, E.; Mijowska, E.; Kalenczuk, R. J.; Aleksandrzak, M.; Mijowska, S. Nanocomposite of Cement/Graphene Oxide–Impact on Hydration Kinetics and Young's Modulus. *Construct. Build. Mater.* **2015,** *78*, 234–242.
37. Gibson, C. T.; Watson, G. S.; Myhra, S. Lateral Force Microscopy—a Quantitative Approach. *Wear* **1997,** *213* (1), 72–79.
38. Cain, R. G.; Reitsma, M. G.; Biggs, S.; Page, N. W. Quantitative Comparison of Three Calibration Techniques for the Lateral Force Microscope. *Rev. Sci. Instr.* **2001,** *72* (8), 3304–3312.

39. Zhou, Y.; Fan, H.; Fong, T.; Lopez, G. P. Compositional Mapping of Micropatterned, Mixed Self-Assembled Monolayers by Lateral Force Microscopy. *Langmuir* **1998,** *14* (3), 660–666.
40. Peled, A.; Castro, J.; Weiss, W. J. Atomic Force and Lateral Force Microscopy (AFM and LFM) Examinations of Cement and Cement Hydration Products. *Cement Concr. Compos.* **2013,** *36*, 48–55.
41. Binnig, G.; Rohrer, H. Scanning Tunneling Microscopy. *Surf. Sci.* **1983,** *126* (1), 236–244.
42. Griffith, J. E.; Kochanski, G. P. Scanning Tunneling Microscopy. *Annu. Rev. Mater. Sci.* **1990,** *20* (1), 219–244.
43. Hamers, R. J.; Tromp, R. M.; Demuth, J. E. Scanning Tunneling Microscopy of Si (001). *Phys. Rev. B* **1986,** *34* (8), 5343.
44. Bai, C. *Scanning Tunneling Microscopy and Its Application*, Vol. 32; Springer-Verlag Berlin Heidelberg: New York, 2000.
45. Chen, C. J. *Introduction to Scanning Tunneling Microscopy*, Vol. 227; Oxford University Press: New York, 1993.
46. Ishigami, M.; Chen, J. H.; Cullen, W. G.; Fuhrer, M. S.; Williams, E. D. Atomic Structure of Graphene on SiO_2. *Nano Lett.* **2007,** *7* (6), 1643–1648.
47. Stolyarova, E.; Rim, K. T.; Ryu, S.; Maultzsch, J.; Kim, P.; Brus, L. E.; Heinz, T. F.; Hybertsen, M. S.; Flynn, G. W. High-Resolution Scanning Tunneling Microscopy Imaging of Mesoscopic Graphene Sheets on an Insulating Surface. *Proc. Natl. Acad. Sci.* **2007,** *104* (22), 9209–9212.
48. Zhang, Y.; Brar, V. W.; Wang, F.; Girit, C.; Yayon, Y.; Panlasigui, M.; Zettl, A.; Crommie, M. F. Giant Phonon-Induced Conductance in Scanning Tunnelling Spectroscopy of Gate-Tunable Graphene. *Nat. Phys.* **2008,** *4* (8), 627–630.
49. Geringer, V.; Liebmann, M.; Echtermeyer, T.; Runte, S.; Schmidt, M.; Rückamp, R.; Lemme, M. C.; Morgenstern, M. Intrinsic and Extrinsic Corrugation of Monolayer Graphene Deposited on SiO_2. *Phys. Rev. Lett.* **2009,** *102* (7), 076102.
50. Deshpande, A.; Bao, W.; Miao, F.; Lau, C. N.; LeRoy, B. J. Spatially Resolved Spectroscopy of Monolayer Graphene on SiO_2. *Phys. Rev. B* **2009,** *79* (20), 205411.
51. Xu, K.; Cao, P.; Heath, J. R. Scanning Tunneling Microscopy Characterization of the Electrical Properties of Wrinkles in Exfoliated Graphene Monolayers. *Nano Lett.* **2009,** *9* (12), 4446–4451.
52. Ferrari, A. C.; Meyer, J. C.; Scardaci, V.; Casiraghi, C.; Lazzeri, M.; Mauri, F.; Piscanec, S.; Jiang, D.; Novoselov, K. S.; Roth, S.; Geim, A. K. Raman Spectrum of Graphene and Graphene Layers. *Phys. Rev. Lett.* **2006,** *97* (18), 187401.
53. Biedermann, L. B.; Bolen, M. L.; Capano, M. A.; Zemlyanov, D.; Reifenberger, R. G. Insights into Few-Layer Epitaxial Graphene Growth on 4 H-SiC (000 1¯) Substrates from STM Studies. *Phys. Rev. B* **2009,** *79* (12), 125411.
54. Li, X.; Cai, W.; An, J.; Kim, S.; Nah, J.; Yang, D.; Jung, I.; Tutuc, E.; Banerjee, S. K.; Colombo, L.; Ruoff, R. S. Large-Area Synthesis of High-Quality and Uniform Graphene Films on Copper Foils. *Science* **2009,** *324* (5932), 1312–1314.
55. Robles, M. E.; Gonzalez-Fuentes, C. A.; Henriquez, R.; Kremer, G.; Moraga, L.; Oyarzun, S.; Suarez, M. A.; Flores, M.; Munoz, R. C. Resistivity of Thin Gold Films on Mica Induced by Electron–Surface Scattering: Application of Quantitative Scanning Tunneling Microscopy. *Appl. Surf. Sci.* **2012,** *258* (8), 3393–3404.

56. Yagati, A. K.; Jung, M.; Kim, S. U.; Min, J.; Choi, J. W. Nanoscaled Redox Active Protein Adsorption on Au-Dot Arrays: An Electrochemical Scanning Probe Microscopic Investigation for Application in Nano-Biodevices. *Thin Solid Films* **2009,** *518* (2), 634–637.
57. Bushby, A. J.; P'ng, K. M.; Young, R. D.; Pinali, C.; Knupp, C.; Quantock, A. J. Imaging Three-Dimensional Tissue Architectures by Focused Ion Beam Scanning Electron Microscopy. *Nat. Protocols* **2011,** *6* (6), 845–858.
58. Mohamed Ali, S. T.; Hamad, M. M.; Ayad, C. E.; Abdalla, E. A.; Ahmed, A. S. Evaluation of the Technical Specifications of Computerized Tomography Scanners in Jazan. *Sudan Med. Monit* **2013,** *8*, 159–166.
59. Reddy, A. N.; Meena, T. An Experimental Study on Effect of Colloidal Nano-Silica on Tetranary Blended Concrete. *Adv. Concr. Construct.* **2019,** *7* (2), 107–115. DOI: https://doi.org/10.12989/acc.2019.7.2.107
60. Andrabi, S. B.; Singh, Er. J. Applications and Effects of Silica Fume and Nano Silica on the Compressive Strength of Concrete. *In. J. Trend Sci. Res. Dev.* June **2019,** *3* (4), 1273–1277.
61. Elboghdadi, A. S.; Elkady, H. M.; Salem, H. M.; Farahat, A. M. Coupled Effect of Nano Silica and Steel Fiber on Fresh and Hardened Concrete Properties. *Int. J. Modern Eng. Res.* July **2015,** *5* (7), 38–48.
62. Sidhardhan, S.; Karthikeyan, G. Effect of Incorporation of Nano-Sio2 in Mortar and Its Durability Studies. *Int. J. Innov. Res. Sci. Eng. Technol.* May **2019,** *8* (5), 5618–5623.
63. Boshehrian, A.; Hosseini, P. Effect of Nano-SiO2 Particles on Properties of Cement Mortar Applicable for Ferrocement Elements. *Concr. Res. Lett.* Mar **2011,** *2* (1).
64. Teja, G. R.; Reddy, A. N.; Meena, T. Experimental Study on the Strength and Durability of High Performance Nano Silica Concrete. *Int. J. Innov. Technol. Explor. Eng.* Mar **2019,** *8* (5), 1141–1144.

CHAPTER 12

Effect of Carbon Nanotube Content on Mechanical Properties of MWCNT/Epoxy-Based Nanocomposite

GAURAV SAPRA[1*], JASKARAN SINGH[1], PARVEEN KUMAR[2], MANU SHARMA[1], and RENU VIG[1]

[1]*UIET, Panjab University, Chandigarh 160014, India*

[2]*Nanotechnology Lab, H-1 Division, CSIR-Central Scientific Instruments Organisation, Chandigarh 160030, India*

[*]*Corresponding author. E-mail: gaurav.sapra@pu.ac.in*

ABSTRACT

In this study, different weight concentrations of Multiwalled Carbon Nanotube (MWCNT) are dispersed in epoxy resin to form nanocomposite samples. To study the effect of filler concentrations on mechanical properties of nanocomposite's various parameters such as tensile strength, stress–strain curve, and Young's modulus of elasticity and extension at break are studied. Characteristic results show that the tensile strength and Young's modulus of the MWCNT/epoxy nanocomposite samples increases whereas the extension at break decreases with the increase in weight percentage loading of MWCNT, Also, the physical and functional properties of the nanocomposite are studied using Raman spectroscopy. Results from Raman spectroscopy suggest that the structural defects in the MWCNT/epoxy nanocomposite matrix decreases with increase in the weight percentage loading of MWCNT in the nanocomposite. Study results give an insight

Advanced Microscopy: A Strong Analytical Tool in Materials Science. Merin Sara Thomas, Jozef T. Haponiuk, Sabu Thomas, & Anne George (Eds.)

to researchers regarding the effect of filler weight concentrations in the nanocomposite samples.

12.1 INTRODUCTION

Since the discovery of carbon nanotubes (CNTs) by Iijima[1] in the year 1991, the multiwalled CNTs (MWCNTs) have drawn the interest of researchers and scientists from all over the world. It is primarily because of their unusual and unique physical, mechanical,[2,3] electrical, and thermal properties.[4] CNTs are the strongest material discovered in terms of tensile strength and also the stiffest material in terms of Young's modulus of elasticity.[5–8] MWCNT occurs in two forms, that is, pristine and nanocomposite forms. Pristine form is also referred to as pure form of MWCNT. In nanocomposite form, MWCNT acts as filler reinforced into different substrates such as epoxy, polyaniline, polyvinylidene fluoride (PVDF), polymethyl methacrylate (PMMA), polyether ether ketone (PEEK), cement mortar, etc. The reinforcement of CNT into matrix materials enhances the mechanical properties of matrix.[9,10]

As CNTs are unidirectional materials and possess some characteristics and properties of fiber, they are increasingly being used as a filler content in nanocomposites to get high strength structural materials that have highly improved mechanical and physical properties[11–16] over conventional materials. The high-caliber properties of CNT polymer composites[17] are attributed to the fiber-like nature of CNT filler and strong binding nature of the polymer matrix. The high strength, high stiffness and lightweight properties of CNT polymer composites have made them an excellent choice for harsh loading condition applications such as aerospace components, military armors, etc.

To the best of the author's knowledge, there is no work available in literature where the effect of filler weight percentage in MWCNT/epoxy nanocomposite on its mechanical behavior has been studied. In this work, three different samples of MWCNT/epoxy nanocomposites with different weight fraction of fillers (4%, 6%, and 8%) are fabricated. Various mechanical parameters such as tensile strength, stress–strain curve, and Young's modulus of elasticity and extension at break are studied at different weight percentages of MWCNT in nanocomposite. Spectroscopic study of different nanocomposites samples is also reported in this study.

12.2 EXPERIMENTAL ANALYSIS

12.2.1 MATERIALS USED

12.2.1.1 FILLER

Pristine MWCNT synthesized with Chemical Vapor Deposition (CVD) technique[18] of purity higher than 99% and diameter of 10–20 nm, length 3–8 μm are used as filler for the composite, procured from Nanoshell, India. Different types of samples with filler loading 4%, 6%, and 8% by weight are prepared.

12.2.1.2 MATRIX

Epoxy resin system, Araldite cured with Aradur HY951 hardener is used as the matrix for the nanocomposites. Araldite used is a bisphenol-A based clear, pale yellow colored unmodified epoxy resin of medium viscosity. Aradur HY951 is an aliphatic amine hardener of low viscosity.

12.2.2 NANOCOMPOSITE FABRICATION

Thin films of MWCNT/epoxy nanocomposites are fabricated to effectively study the mechanical and morphological characteristics. Initially pristine MWCNT powder of different weight concentrations is dispersed in an epoxy resin system and stirred manually till the dispersion is uniform. The mixture is then bath sonicated for 30 min to achieve uniform dispersion. After sonication, Aradur HY 951 hardener is added to the mixture as a curing agent. The mixture is then spread on a 1-mm thick template made of teflon to get the nanocomposite films of the required shape and size. The nanocomposite films are then cured for one day at room temperature and postcuring is done at 100°C for 1 h.

12.2.3 EVALUATION OF MECHANICAL PROPERTIES OF MWCNT/ EPOXY NANOCOMPOSITES AT DIFFERENT LOADING

Tensile test of the MWCNT/epoxy nanocomposite samples is conducted by using Starrett FMS-1000 L2 tensile testing machine. For this, the samples are cut into dog-bone shaped films of dimensions (35 × 5 × 0.25 mm) as shown in Figure 12.1. Tensile test of different samples is conducted at displacement control mode with loading speed of 1 mm/min.

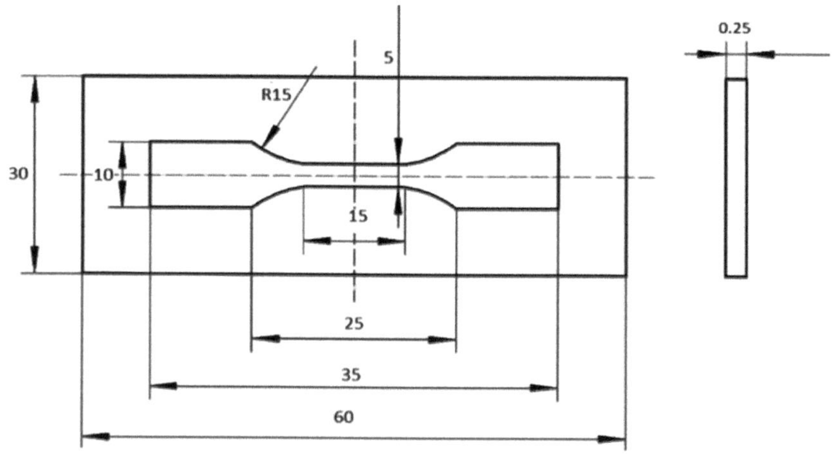

FIGURE 12.1 Dimensions of dog bone shaped MWCNT/epoxy nanocomposite sample.

12.2.4 SPECTROSCOPIC ANALYSIS

Spectroscopic analysis of the MWCNT/epoxy nanocomposite samples with different filler loading are done by analyzing the Raman spectra of all the samples using the Renishaw Invia confocal microscope.

12.3 RESULTS AND DISCUSSIONS

12.3.1 EFFECT OF MWCNT LOADING ON THE MECHANICAL PROPERTIES OF NANOCOMPOSITES

Tensile tests are conducted to study the effect of MWCNT loading on the mechanical properties of nanocomposite samples. Figure 12.2 shows the stress–strain curves derived from the tensile tests for neat epoxy and different concentrations (i.e., 4%, 6%, and 8% wt.) of MWCNT/epoxy nanocomposite samples. All samples curves show linearity before reaching their ultimate strength. All samples tested failed immediately after reaching ultimate strength indicating low toughness of the samples.

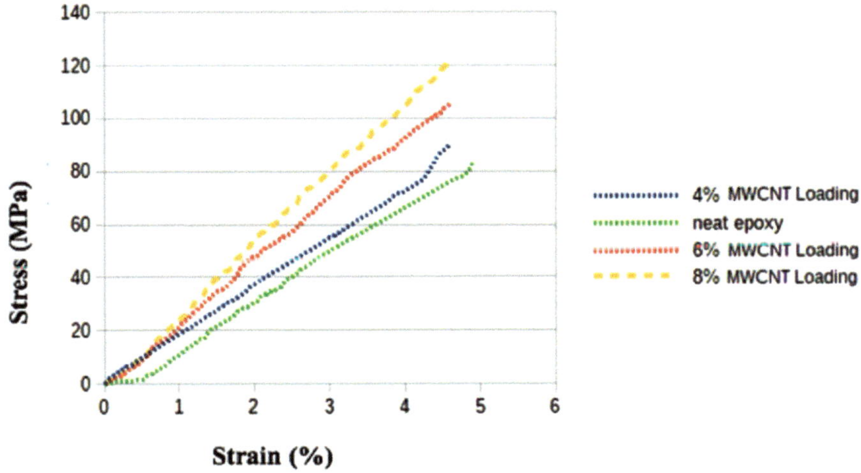

FIGURE 12.2 Stress versus strain curve for neat epoxy and different concentration (i.e., 4% wt., 6% wt., and 8% wt.) of MWCNT/epoxy nanocomposite samples.

The ultimate tensile strength of MWCNT/epoxy nanocomposites with increasing weight concentrations of filler is shown in Figure 12.3. It has been found experimentally that the tensile strength of the nanocomposites shows an increasing trend with increase in filler loading. The tensile strength of neat epoxy is found to be 83.4 MPa, which increases with increase in MWCNT content and reaches a value of 90, 105, and 122.2 MPa at filler concentration of 4%, 6%, and 8%, respectively. This relative improvement in tensile strength of nanocomposite in addition to MWCNT is due to strong chemical bonding of MWCNT reinforced epoxy composite.[19] The addition of MWCNT plays an important role in enhancing the polymer matrix framework, which improves its overall mechanical strength.

Figure 12.4 shows the relation of Young's modulus with increasing percentage of filler content in the nanocomposites. It has been found that the Young's modulus of the MWCNT epoxy nanocomposite samples increases with the increase in filler content.

The Young's Modulus for neat epoxy resin is found to be 1.43 GPa, which increases with the increase in weight concentration of the filler. It attains the value of 1.75, 2.32, and 2.64 GPa at 4%, 6%, and 8% wt. concentration of MWCNT, respectively. This improvement in Young's modulus can be attributed to tougher reinforcement due to stronger chemical bonding interactions between MWCNT. Young's modulus is calculated by the following mathematical expression:

$$\text{Young's Modulus} = \frac{\text{Stress}}{\text{Strain}} \tag{12.1}$$

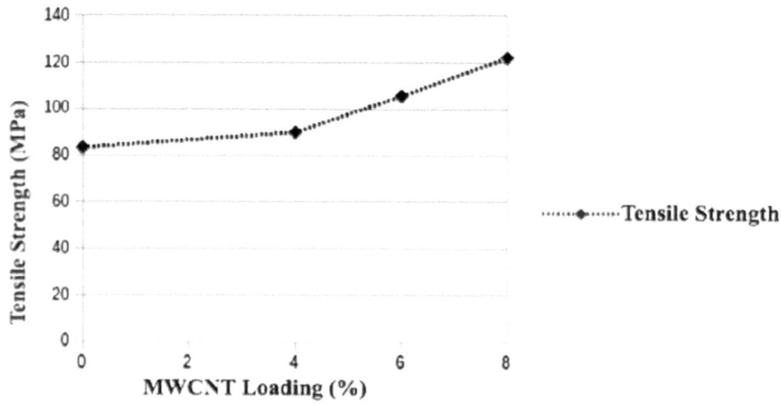

FIGURE 12.3 Plot of tensile strength of MWCNT/epoxy nanocomposite sample at different percentage loading of MWCNTs.

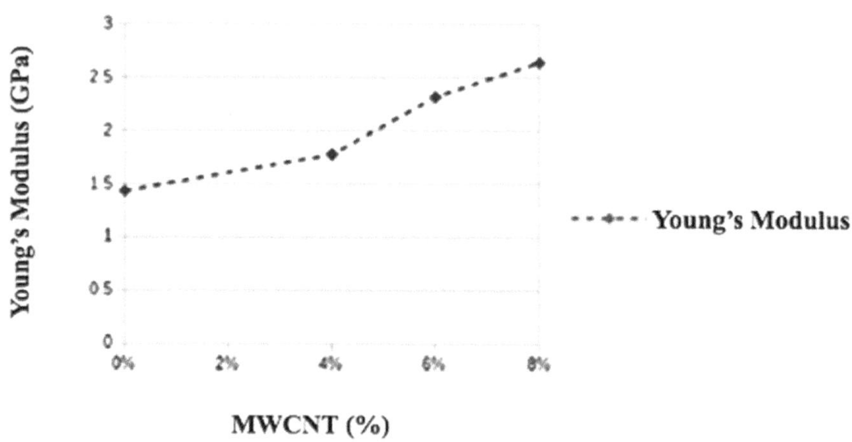

FIGURE 12.4 Plot of Young's modulus of MWCNT/epoxy nanocomposite sample at different percentage loading of MWCNTs.

The *extension at break* of the nanocomposite samples is also calculated. It is also referred to as an extension required for breaking the material. Mathematically. it is calculated by the following expression:

$$\text{Extension at break} = \frac{\text{Elongation at break}}{\text{Initial length}} \tag{12.2}$$

Figure 12.5 shows *extension at break* of MWCNT/epoxy nanocomposite samples decreases with increase in MWCNT loading. The extension for 4% nanocomposite sample is found to be 2.89 mm, which decreases to 1.82 mm for 8% wt. concentration. This is due to the fact that the extension at break depends on the flexibility of the molecular chain, and the addition of MWCNT in nanocomposite restricts the conformation change of the polymer matrix, which limits the extension.[20]

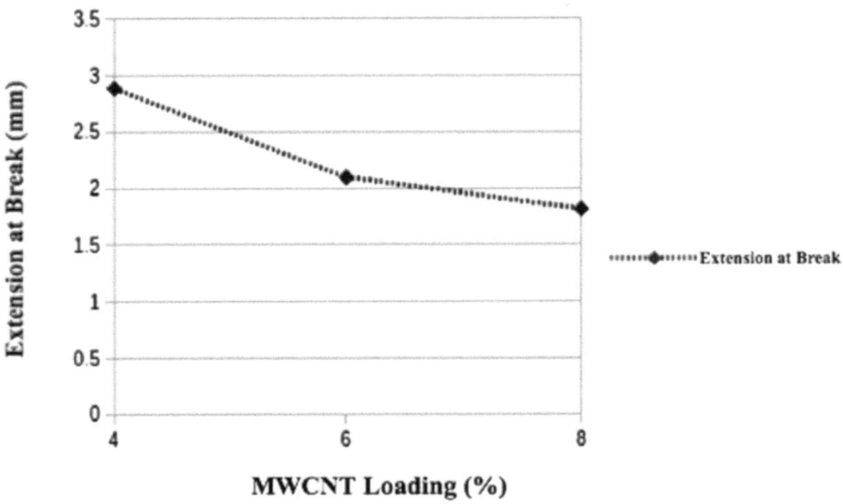

FIGURE 12.5 Extension at break of MWCNT/epoxy nanocomposite sample at different percentage loading of MWCNTs.

12.3.2 SPECTROSCOPIC CHARACTERIZATION

Spectroscopic characterization of MWCNT/epoxy nanocomposite samples of different concentrations is done using Raman spectroscopy. The results of the Raman spectral analysis for MWCNT/epoxy nanocomposites with different filler content are shown in Figure 12.6. Intensity ratio 'ρintensity ratio of all three different filler content samples is calculated by using following mathematical expression:

$$\rho = \frac{I_D}{I_G} \tag{12.3}$$

Where ρ is intensity ratio, I_D is intensity of D peaks and I_G is intensity of G peaks.

Intensity ratio is considered as an important factor in determining the defects and disorders in carbon nanostructures. In Raman spectra, D peak is due to the presence of disorder in graphitic structure and G peak is due to tangential stretching of carbon–carbon bonds in graphite structure. On increasing the MWCNT weight concentration in nanocomposite, the intensity of D and G band increases. For pristine MWCNT D band peak is at 1360 cm^{-1} and G band peak is at 1600 cm^{-1}.[21] For 8% wt. MWCNT/nanocomposite sample D and G peaks are more close to pristine MWCNT in comparison to other two samples. The intensity ratio shows a significant decrease from 0.94 for 4% samples to 0.917 and 0.90 for 6% and 8% wt. samples, respectively. This shows that the structural defects in nanocomposite samples decrease with the increase in filler concentration of nanocomposites.

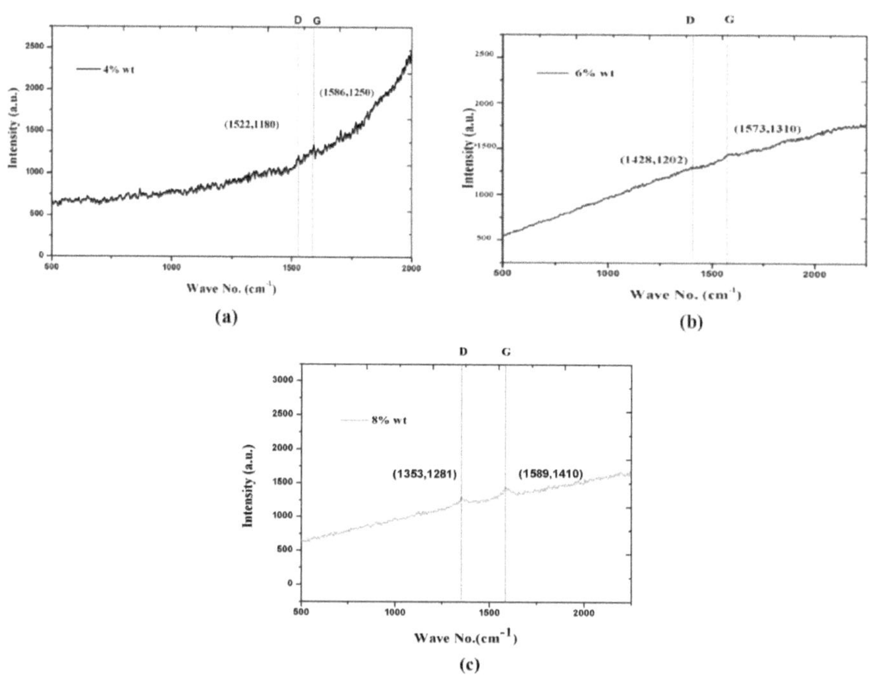

FIGURE 12.6 Spectroscopic characterization of MWCNT/epoxy nanocomposite using Raman Spectroscopy at different filler concentrations of (a) 4% wt. (b) 6% wt., and (c) 8% wt.

12.4 CONCLUSION

In this study, different filler concentrations (4%, 6%, and 8% by wt.) of MWCNT/epoxy nanocomposites samples are fabricated by using bath sonication techniques. Experiment results show that the tensile strength of nanocomposites increases with the increase in filler loading. The tensile strength of neat epoxy is found to be 83.4 MPa, which increases to 90, 105, and 122.2 MPa at filler weight concentrations of 4%, 6%, and 8%, respectively. Similarly, the Young's modulus for neat epoxy resin is 1.43 GPa also increases to 1.75, 2.32, and 2.64 GPa at filler weight concentrations of 4%, 6%, and 8%, respectively. However, the *extension at break* of MWCNT/epoxy nanocomposite samples shows a decreasing trend with increase in MWCNT loading. The extension at break for 4% wt. nanocomposite sample is found to be 2.89 mm, which decreases to 1.82 mm at 8% wt. of filler concentration. It is been observed through Raman spectroscopy that the structural defects decrease with the increase in filler concentration. It is concluded from the result that to enhance the mechanical properties of nanocomposite the weight concentration of filler must be kept low.

KEYWORDS

- **multiwall carbon nanotube (MWCNT)**
- **composite**
- **epoxy**
- **mechanical properties**

REFERENCES

1. Iijima, S. Helical Microtubules of Graphitic Carbon. *Nature* **1991,** *354*, 56–58.
2. Mintmire, J. W.; White, C. T. Electronic and Structural Properties of Carbon Nanotubes. *Carbon* **1995,** *33*, 893–902.
3. Arash, B.; Wang, Q.; Varadan, V. Mechanical Properties of Carbon Nanotube/Polymer Composites. *Sci. Rep.* **2014,** *4*, 1–8.

4. Biercuk, M. J.; Llaguno, M. C.; Radosavljevic, M.; Hyun, J. K.; Johnson, A. T.; Fischer, J. E. Carbon Nanotube Composites for Thermal Management. *Appl. Phys. Lett.* **2002,** *80*, 2767–2769.
5. Li, Y.; Wang, S.; Wang, Q.; Xing, M. A Comparison Study on Mechanical Properties of Polymer Composites Reinforced by Carbon Nanotubes and Graphene Sheet. *Compos. Part B-Eng.* **2018,** *133*, 35–41.
6. Cui, X.; Han, B.; Zheng, Q.; Yu, X.; Dong, S.; Zhang, L.; and Ou, J. Mechanical Properties and Reinforcing Mechanisms of Cementitious Composites with Different Types of Multiwalled Carbon Nanotubes. *Compos. Part A Appl. Sci. Manuf.* **2017,** *103*, 131–147.
7. Essawy, H.; Fathy, N.; Tawfik, M.; El-Sabbagh, S.; Ismail, N.; Youssef, H. Fabrication of Single-Walled Carbon Nanotubes from Vulcanized Scrap Rubber via Thermal Chemical Vapor Deposition. *RSC Adv.* **2017,** *7*, 12938–12944.
8. Hahm, M. G.; Leela Mohana Reddy, A.; Cole, D. P.; Rivera, M.; Vento, J. A.; Nam, J.; Jung, H. Y.; Kim. Y. L.; Narayanan, N. T.; Hashim, D. P.; Galande, C. Carbon Nanotube–Nanocup Hybrid Structures for High Power Supercapacitor Applications. *Nano Lett.* **2012,** *12,* 5616–5621.
9. Bakshi, S. R.;Lahiri, D and Agarwal,A. Carbon Nanotubes: Reinforced Metal Matrix Composites-A Review. *Int. Matr. Rev.* **2010,** *55*, 41–64.
10. Quan, D.; Urdaniz, J. L.; Ivankovic, A. Enhancing Mode-I and Mode-II Fracture Toughness of Epoxy and Carbon Fibre Reinforced Epoxy Composites Using Multi-Walled Carbon Nanotubes. *Mater. Des.* **2018,** *143*, 81–92.
11. Zhou, Y. X.; Wu, P. X.; Cheng, Z. Y.; Ingram, J.; Jeelani, S. Improvement in Electrical, Thermal and Mechanical Properties of Epoxy by Filling Carbon Nanotube. *Express Polym. Lett.* **2008,** *2*, 40–48.
12. Treacy, M. J.; Ebbesen, T. W.; Gibson, J. M. Exceptionally High Young's Modulus Observed for Individual Carbon Nanotubes. *Nature* **1996,** *381*, 678–680.
13. Zhao, J.; Wu, L.; Zhan, C.; Shao, Q.; Guo, Z.; Zhang, L. Overview of Polymer Nanocomposites: Computer Simulation Understanding of Physical Properties. *Polymer* **2017,** *133*, 272–287.
14. Zheng, Y.; Li, Y.; Li, Z.; Wang, Y.; Dai, K.; Zheng, G.; Shen, C. The Effect of Filler Dimensionality on the Electromechanical Performance of Polydimethylsiloxane Based Conductive Nanocomposites for Flexible Strain Sensors. *Compos. Sci. Technol.* **2017,** *139*, 64–73.
15. Li, H. X.; Zare, Y.; Rhee, K. Y. The Percolation Threshold for Tensile Strength of Polymer/CNT Nanocomposites Assuming Filler Network and Interphase Regions. *Mater. Chem. Phys.* **2018,** *207*, 76–83.
16. Radzuan, N. A. M.; Zakaria, M. Y.; Sulong, A. B. and Sahari, J. The Effect of Milled Carbon Fibre Filler on Electrical Conductivity in Highly Conductive Polymer Composites. *Compos. Part B- Eng.* **2017,** *110*, 153–160.
17. Byrne, M. T.; Gunko, Y. K. Recent Advances in Research on Carbon Nanotube–Polymer Composites. *Adv. Mater.* **2010,** *22*, 1672–1688.
18. Thostenson, E. T.; Ren, Z.; Chou, T. W. Advances in the Science and Technology of Carbon Nanotubes and Their Composites: A Review. *Compos. Sci. Technol.* **2001,** *61*, 1899–1912.

19. Haque, A.; Ramasetty, A. Theoretical Study of Stress Transfer in Carbon Nanotube Reinforced Polymer Matrix Composites. *Compos. Struct.* **2005,** *71*, 68–77.
20. Nikpur, K.; Chen, Y. F.; Kardos, J. L. Fracture Toughness of Unidirectional Short-Fiber Reinforced Epoxy Composites. *Compos. Sci. Technol.* **1990,** *38*, 175–191.
21. Bokobza, L.; Zhang, J. Raman Spectroscopic Characterization of Multiwall Carbon Nanotubes and of Composites. *EXPRESS Polym. Lett.* **2012,** *6*, 601–608.

CHAPTER 13

Dyes Adsorption by Nanomaterials: Fixed-Bed Column Studies Using Nano Magnesium Oxide-Loaded Activated Carbon

VENKATA RATNAM MYNENI[1*,4],
NAGAMALLESWARA RAO KANIDARAPU[2],
SENTHIL KUMAR KANDASAMY[3], and MEENA VANGALAPATI[4]

[1]*Department of Chemical Engineering, Mettu University, Mettu, Ethiopia*

[2]*Centre for Disaster Mitigation and Management,
Vellore Institute of Technology, Vellore, Tamil Nadu, India*

[3]*Department of Chemical Engineering, Kongu College of Engineering, Tamil Nadu, India*

[4]*Department of Chemical Engineering, A.U. College of Engineering, Vizag, Andhra Pradesh, India*

[*]*Corresponding author. E-mail: mvratnam81@gmail.com*

ABSTRACT

Rapid removal rates with high pollutant binding strength allow nanostructured materials ideal for dyes adsorption. In this analysis, methylene blue (MB) column adsorption experiments were performed using nano-MgO-loaded activated carbon (n-MgO/AC). The synthesized n-MgO/AC adsorbent was adequately characterized by several analytical methods. The potential effect of variables on the adsorption of columns, that is, MB flow rate, n-MgO/AC

Advanced Microscopy: A Strong Analytical Tool in Materials Science. Merin Sara Thomas, Jozef T. Haponiuk, Sabu Thomas, & Anne George (Eds.)

bed height, and MB inlet concentration was examined in-depth. Usually, with an improvement in MB flow rate (4–12 mL/min), n-MgO/AC bed height (2.5–7.5 cm) and MB concentration (20–60 mg/L) decrease in break time from 210 to 30 min, improvement from 210 to 720 min and decrease from 390 to 60 min, respectively. For forecasting breakthrough curves, the experimental results under different operating conditions were fitted with models, such as Thomas, modified dose–response/Yan, and Bed-depth service time (BDST). Breakthrough time, MB percent removal, the zone of mass transfer (MTZ), the empty contact time of the bed (EBCT), and dye uptake (q_{exp}) were reported. The compatibility of the experimental results with the dynamic mathematical models indicates the suitability of these models in a breakthrough curve prediction. After three cycles of adsorption–desorption, the n-MgO/AC adsorbent regeneration studies showed almost 80% regeneration efficiency. Hence, the potentiality of the n-MgO/AC composite as adsorbent was confirmed.

HIGHLIGHTS

- The composite nanoadsorbent n-MgO/AC was synthesized in the laboratory.
- The synthesized n-MgO/AC adsorbent was tested for its adsorptive ability by conducting column studies with methylene blue dye (MB) as a target pollutant.
- The regeneration studies show that the prepared adsorbent is effective for practical applications.

GRAPHICAL ABSTRACT

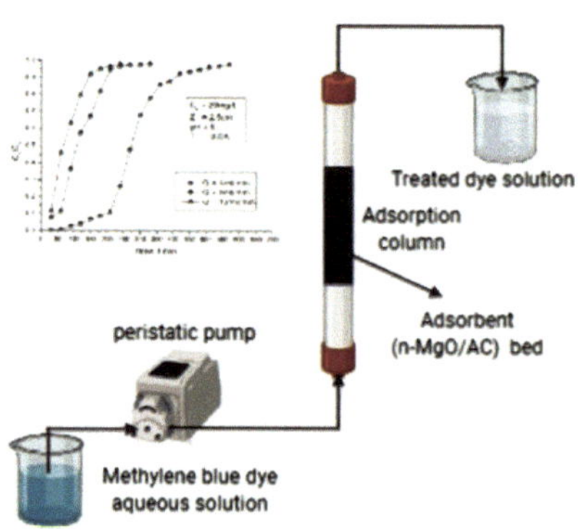

NOMENCLATURE

a, a'	Old and new slopes for the BDST model
A	Area above the breakthrough curve, m^2
b, b'	Old and new intercepts for the BDST model
C_o	Inlet concentration of dye (MB) in aqueous solution, mg/L
C_1	outlet concentration of dye (MB) in aqueous solution, mg/L
C_b	Breakthrough concentration mg/L
C_t	Dye (MB) concentration in aqueous solution at the given time 't', mg/L
k_{BA}	Bohart-Adams rate constant (L/mg/mi)
k_{Th}	Thomas rate constant (L/mg/mi)
k_{YN}	Yoon-Nelson rate constant (min^{-1})
m	Mass of the adsorbent n-MgO/AC, g
M	Total MB dye sent into the column, mg
N_o	Saturation concentration, mg/L
Q	Volumetric flow rate of MB solution, mL/min
q_t	Amount of adsorbate (MB dye) adsorbed per amount of adsorbent (n-MgO/AC) at the given time 't', mg/g
q_{cal}	Amount of adsorbate (MB dye) adsorbed per amount of adsorbent (n-MgO/AC) at equilibrium (calculated), mg/g
q_{exp}	Amount of adsorbate (MB dye) adsorbed per amount of adsorbent (n-MgO/AC) at equilibrium (Experimental), mg/g
q_{total}	Total amount of dye adsorbed, mg
q_o	Maximum capacity of the n-MgO/AC adsorbent, mg/g
q_d	Desorption capacity, mg/g
R^2	Linear regression coefficient
S	Adsorption column surface area, m^2
t	Time, min
t_b	Time of breakthrough, min
t_e	Exhaust time, min.
t_{Total}	Total time, min
U	Superficial velocity, m/min.
T	Temperature, K
τ	Time required for 50% adsorbate (MB dye) breakthrough, min
Z	Bed height, cm
Z_0	Critical bed depth, cm
χ^2	Nonlinear regression coefficient

13.1 INTRODUCTION

The fundamental challenge for the scientific society is to cut back the pollution ensuing from industrialization to safeguard the environment.[1] Industrial water necessities have grown staggeringly (up to 20% of the available freshwater) and the wastewater containing numerous toxic substances has been discharged heavily into the water streams. Leather, textile, food, paper, plastics, and cosmetics industries usually make extensive use of dyes and pigments extensively for a broad range of applications.[2] The increased use of dyes remains a serious environmental issue, with almost 10% of the seven lakh tons of dyes produced worldwide entering the water bodies.[3] The acute problem with dyes is not only that they are carcinogenic, but even at fairly low concentrations, they exhibit inhibitory effects on the photosynthesis process of plants disrupting the aquatic ecosystem.[4,5] Dyes being organic are resistant to biodegradation and release toxic by-products due to incomplete degradation. The complex structure of dyes appears to be another prime reason that makes dyes the most difficult, troublesome pollutant to treat through traditional methods.

Adsorption has long been considered as a recommended technique for decolorizing different coloring substances thanks to its cost-effectiveness, efficient separation, and the broad choice in adsorbent selection with no secondary pollutant generation.[5–8] The larger surface area and high porosity of adsorbents are deemed the principal features of the adsorbents. The main industrial adsorbent is activated carbon, which will rapidly adsorb contaminants from water, because it is extremely inert, thermally stable, and is effective over wider pH.[4] The carboxyl groups, carbonyl groups, and phenolic groups in activated carbon are proved to be responsible for the adsorptive capability of activated carbon.[9] However, the difficulty in separation and regeneration limits the usage for many applications.[10] While developing new sorbents, the uptake capacity improvement is a topic of enormous interest together with factors, such as price, availability, surface activity, and toxic nature.

Nanotechnology has penetrated through all allied branches of science because of the diverse and unique properties exhibited by nanomaterials.[11] The superior properties exhibited by metal oxide nanoparticles (MONPs) make them appropriate materials for several environmental and energy applications.[11–13] MONPs exhibit higher surface uptake capacity with shorter diffusion route as the binding sites reside solely on the surface.[14] The laser ablation synthesized nano-nickel oxide demonstrated the MB uptake of 10,584.6 mg/g, the highest reported to date.[15] Among the several MONPs, nano-magnesium oxide (n-MgO) is proven to be a destructive adsorbent for treating gaseous pollutants and is often received a key consideration for wastewater decontamination.[16–20] Mainly, the reasons for this rise in attention are the moderate cost of magnesium oxide, chemical stability, higher surface area, and nontoxicity.[21] Contaminant removal studies with nano-MgO had shown substantially higher adsorption capacity.[22–25]

The scientific community's enthusiasm for hybrid nanomaterials is due to their excellent energy and environmental applications.[11,26] The hybrid nanocomposites synthesis typically involves impregnating the nanoparticles onto several porous carriers, such as activated carbon,[4,9,27–30] silica,[31–33] zeolites,[34,35] and polymers.[36,37] These composites have convincingly shown the combined characterization of the outstanding flow characteristics of porous carriers acting as supports and the specific affinities between nanoscale materials with potential toxins. During this study, nano-magnesium oxide and activated carbon composite (n-MgO/AC) was synthesized by simple precipitation, and its ability to get rid of MB from aqueous solutions was assessed. Analyzing the adsorption of the dyes in column mode is necessary, as column studies were more widely employed in practice and less studied compared with the batch studies.[38] The adsorption factors have been analyzed and the result has been modeled for the estimation of column performance using the Thomas, Yan, and Bed-depth service time (BDST) models.

13.2 MATERIALS AND METHODS

13.2.1 CHEMICALS

All the chemicals which include magnesium nitrate hexahydrate GR ($MgNO_3 \cdot 6H_2O$), ammonium hydroxide (NH_4OH), polyvinyl pyrrolidone (PVP), activated carbon, and ethanol procured were of superior grade and used without any further purification. The lab grade methylene blue was used for stock solution preparation. NaOH (0.1M) and HCl (0.1M) were prepared for pH adjustments. The solutions required were prepared using double-distilled water.

13.2.2 ADSORBENT (N-MGO/AC) SYNTHESIS AND CHARACTERIZATION

The chemical precipitation process remains an easy and successful approach in today's research for nano-metal oxides synthesis. The precursor used in this study is 0.5M of magnesium nitrate hexahydrate solution, complimented 2 g of PVP acting as a size control agent. For proper mixing of PVP with the precursor, the solution is magnetically stirred for 30 min. The pH of the solution is monitored until it reaches 10.5 upon the dropwise addition of 0.5M ammonium hydroxide. The mixture is continually stirred at room temperature until the formation of a white precipitate. The obtained precipitate was stored for 24 h under room temperature conditions, and then filtered, washed with water and ethanol. The washed precipitate is air-dried at 100°C for 4 h. The resulting $Mg(OH)_2$ sample is subjected to calcination at 500°C for 4 h to get MgO nanoparticles.

The hybrid adsorbent n-MgO/AC employed for MB adsorption was prepared in ethanol media by dispersing 1 g nano-MgO and 5 g activated carbon. The mixture was continuously agitated for 12 h at room temperature at 300 rpm to ensure the homogenous loading of nano-MgO on the activated carbon sheet. The leftover ethanol was removed from the prepared composite by heating the composite at 80°C. The residual composite was water-washed and filtered repeatedly to eliminate the signs of any ethanol present and held for overnight drying at ambient conditions. The morphological studies for the synthesized n-MgO and n-MgO/AC were done with scanning electron microscopy (FE-SEM Carl Zeiss NTS GmbH, Germany). Regaku X-ray diffractometer with Cu-Kα radiation (λ = 1.5418 A°, scan rate: 0.05° s^{-1}, 2Θ range: 10°–80°) was used for the structural analysis. The FTIR spectra of n-MgO, untreated n-MgO/AC, and MB treated n-MgO/AC samples were carried out with Bruker FTIR spectrometer (range: 4000–400 cm^{-1}).

13.2.3 FIXED-BED COLUMN STUDIES

From a process modeling perspective, MB effluent concentration–time profiles (breakthrough curves) are used for describing the dynamic behavior of an adsorption column.[39] Typically, the inlet flow rate of MB effluent, bed height (adsorbent mass), and effluent concentration in the feed liquid are the significant process parameters to be evaluated for assessing the sorption column performance.[5,10,14,40–48] The breakthrough curve shape as well as the time of breakthrough is essential for assessing the adsorption column dynamic characteristics.[14,38] The time at which MB removal of 10% and 95% levels concerning inlet MB concentration is attained is called the Breakthrough time (t_b) and adsorbent bed exhaust time (t_e), respectively.

For the present studies, a packed-bed glass column (15 cm length with 25 mm diameter) filled with the adsorbent n-MgO/AC is employed for sorption studies. The optimal *pH* for the uptake of MB ions as determined by batch studies[17] is maintained for the experimentation and a precision peristaltic pump is employed for pumping the MB solution through the column from bottom to the top. The MB solution temperature was maintained at 30°C using a thermostatic bath. The absorbance of MB samples collected was measured with UV-Spectrophotometer at 664 nm. All experiments were conducted three times and the mean values were considered for analysis.

13.2.4 COLUMN DATA MODELING AND ANALYSIS

The MB solution treated (V_{eff}) is measured as:

$$V_{eff} = Qt_e \tag{13.1}$$

The following eq 13.2 can be employed for finding the q_{total}, adsorbed quantity of MB in *mg* for the specified flow rate, MB dye concentration.

$$q_{total} = \frac{Q{*}A}{1000} = \frac{Q}{1000} * \int_{t=o}^{t=t_{total}} (C_o - C_t).dt \tag{13.2}$$

The complete quantity of MB present (mg) in the solution that is being sent to the column:

$$M = \frac{C_o Q t_{total}}{1000} \tag{13.3}$$

Hence, the total MB removal percentage can be found as given below:

$$\% \text{ Removal} = \frac{q_{total}}{M} * 100 \tag{13.4}$$

Equilibrium of MB uptake per unit n-MgO/AC mass, q_{exp} (mg/g) will be calculated as:

$$q_{exp} = \frac{q_{total}}{m} \tag{13.5}$$

The empty contact time of the bed, EBCT (min) was computed by the division of the bed volume with the flow rate

$$\text{EBCT} = \frac{\text{bed volume}}{\text{flowrate}} \tag{13.6}$$

MTZ (cm), the zone of mass transfer is calculated as follows:

$$\text{MTZ} = Z(1 - \frac{t_b}{t_e}) \tag{13.7}$$

13.2.5 COLUMN BREAKTHROUGH MODELS

The fixed-bed adsorption column's successful design requires the use of quantitative mathematical models to predict the concentration–time profiles (breakthrough curves). Table 13.1 summarizes the significant mathematical models used in the literature for modeling fixed-bed adsorption units. Numerous mathematical models are found in the literature, such as Bohart-Adams[49] (the oldest among the available models), Thomas model[50], Yoon-Nelson model[51], Yan model/modified dose–response model.[52] Lee et al.[39] compared different fixed-bed adsorptio n models by experimenting with adsorption of phosphate with slag filter media as an adsorbent, proposing that all three models be, that is, Thomas, Bohart-Adams, and Yoon-Nelson are equivalent to each other. In another study, Chu et al.[53] established the mutual relation between Thomas parameters (k_{Th} and q_0) and Bohart–Adams parameters (k_{BA} and N_0).

$$k_{BA} = k_{Th} \tag{13.12}$$

$$q_0 = \frac{N_0 ZS}{1000\text{m}} \tag{13.13}$$

Similarly, the Yoon–Nelson parameters (k_{YN} and τ) and the Bohart–Adams parameters (k_{BA} and N_0) can be related as:

$$k_{YN} = k_{BA} C_0 \tag{13.14}$$

$$\tau = \frac{N_0 Z}{C_0 U} = \frac{N_0 ZS}{1000 C_0 Q} \tag{13.15}$$

As noted in the previous literature, any model among Thomas, Bohart-Adams, and Yoon-Nelson is adequate for the approximation of the remaining parameters of the models, which is why the present study takes a similar approach. Nonlinear regression was used to predict the breakthrough curves and to estimate model constants.

TABLE 13.1 Different Models Used for Fixed-Bed Adsorption Studies.

Model	**Description**	**Mathematical representation**	**Eq. no.**
Bohart-Adams	"The adsorption rate exhibits a direct proportionality relation with the adsorbent residual capacity and the adsorbate concentration" Assumption: There is no instant equilibrium	$\frac{C_1}{C_0}=\frac{1}{e^{k_{BA}N_0\left(\frac{Z}{U}\right)-k_{BA}C_0t}+1}$	(13.8)
Thomas	Assumption: Adsorption kinetics is of second-order and the equilibrium is explained by Langmuir isotherm	$\frac{C_1}{C_0}=\frac{1}{e^{k_{Th}q_0(\frac{m}{Q})-k_{Th}C_0t}+1}$	(13.9)
Yoon-Nelson	The model is developed assuming that "The reduction in adsorption probability for each adsorbate molecule is in direct relation to adsorbate breakthrough probability onto the adsorbent"	$\frac{C_1}{C_0}=\frac{1}{e^{k_{YN}(\tau-t)}+1}$	(13.10)
Yan/Modified dose–response	Over a range of operational periods, the model is having the potential of minimizing the subsequent error while using the Thomas model	$\frac{C_1}{C_0}=1-\frac{1}{\left(\frac{C_0Qt}{q_0m}\right)^a+1}$	(13.11)

13.2.6 BED-DEPTH SERVICE TIME MODEL

The BDST model described by Hutchins[54] is the linearized version of the Bohart–Adams model **as** represented by the following equation:

$$t_b=\frac{N_oZ}{C_oU}-\frac{\ln\left(\frac{C_0}{C_b}-1\right)}{k_{BA}C_0} \quad (13.16)$$

The BDST model correlates the bed height (Z) with the service time (t_b).

The minimum bed depth required for achieving the expected effluent MB concentration, that is, critical bed depth (Z_0), is calculated using the rearranged form of eq. 13.16 at time $t_b = 0$

$$Z_o = \frac{U}{k_{BA}N_0}\ln\left(\frac{C_0}{C_b}-1\right) \tag{13.17}$$

The evaluation of BDST constants will help in predicting the adsorbent efficiency at any condition without further experimentation.[24]

For simplification, the BDST model can be represented as

$$t = ax + b \tag{13.18}$$

where $a = slope\,of\ BDST\ model = \frac{N_o}{C_oU}$ and $b = intercept\ of\ BDST\ model = \frac{\ln\left(\frac{C_0}{C_b}-1\right)}{k_{BA}C_0}$.

Equation 13.18 indicates that the BDST plot slope "*a*" changes with changes in the flow rate/velocity, concentration, and the BDST plot intercept "*b*" changes with changes in concentration.

For different flow rates in the column: $\frac{a}{a'} = \frac{U}{U'} = \frac{Q}{Q'}$ (13.19)

where *a* and *a'* are the old and new slopes, *U* and *U'* are the old and new influent velocities, *Q* and *Q'* are the old and new flow rates, respectively.

For other concentrations in the column:

$$\frac{a}{a'} = \frac{C_0}{C_0'} \tag{13.20}$$

$$\frac{b}{b'} = \frac{C_0'\ln(C_0-1)}{C_0\ln(C_0'-1)} \tag{13.21}$$

where b, C_o are the old intercept and influent concentration, respectively, and b', C_o' is the new intercept and new influent concentration.

13.2.7 REGENERATION OF N-MGO/AC

The successful application of any newly developed adsorbent depends on its reusability. To assess the reusability of the n-MgO/AC, desorption experiments are conducted.

The n-MgO/AC desorbed is obtained from the equation

$$q_d = \frac{Q}{1000}\int_0^t C_t\,dt \tag{13.22}$$

The regeneration efficiency is estimated as:

$$RE = \frac{q_d}{q} \times 100 \tag{13.23}$$

13.3 RESULTS AND DISCUSSION

13.3.1 THE ADSORBENT (N-MGO/AC) CHARACTERIZATION

The synthesized n-MgO (Fig.13.1) shows peaks between 660 and 500 cm^{-1}, which is a confirmation of nano-MgO formation[55,56] and the stretching vibrations within the 400–800 cm^{-1} range relate to M–O and M–OH.[57] The peaks between 1000 and 1600 cm^{-1} reflect the carbonate species, that is, C–O–C at 1025 cm^{-1}, C–OH at 1425 cm^{-1}, and C=C at 1569 cm^{-1}.[58–61]

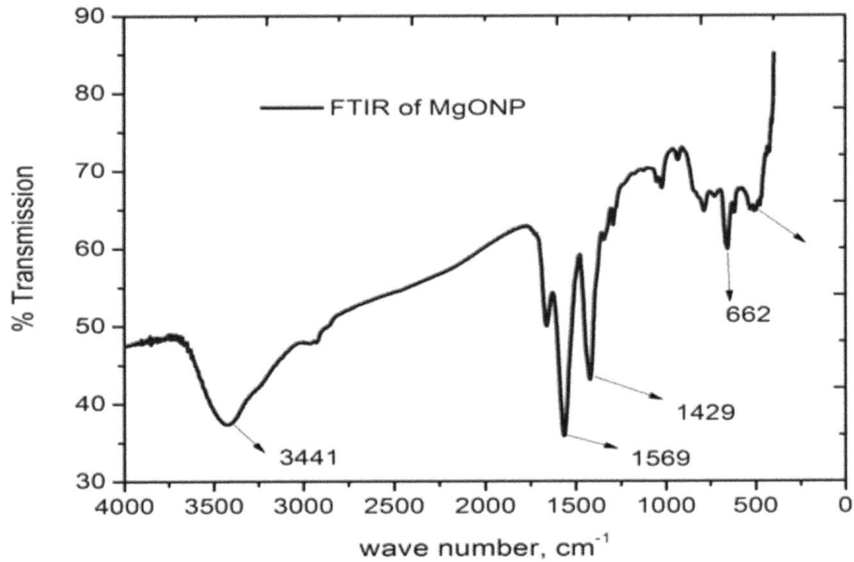

FIGURE 13.1 FTIR spectra of n-MgO.

Since the MgO has adsorptive nature toward atmospheric carbon dioxide and water molecules, there exists a low intense absorption band at 2374 cm^{-1}.[60] The presence of the –OH group was confirmed with an intense peak at 3441 cm^{-1}. The FTIR study of the untreated n-MgO/AC and MB-treated is shown in Figure 13.2a. For untreated adsorbent n-MgO/

AC, the –OH stretching vibrations of alcohols, phenols, and carboxylic acids present on the surface of active carbon is evident from the band at 3697 cm^{-1}.[62,63] The peaks below 1000 cm^{-1} confirm the presence of Mg-O functional groups in the prepared composite.[56] The peaks at 1423 and 1497 cm^{-1} suggest the presence of functional groups of oxygen, that is, C=O symmetric and asymmetric modes and C-C vibrations, respectively.[22,57,61,62] The frequency at 2954–2835 cm^{-1} relates to the vibrations of the C–H bond. It seems that π-π stacking interactions between n-MgO/AC and MB may be an alternative responsible factor of adsorption, apart from the electrostatic forces.

The diffraction studies were used to identify the crystalline structure, and the n-MgO XR pattern is presented in Figure 13.3a agrees well with the MgO Periclase phase cubic crystal system (JCPDS No: # 96-900-0501). The intense peaks at 2θ values of 38.1°, 42.85°, and 62.4° reflect the (111), (200), and (202) lattice planes, respectively. The particle size for the intense peak at 42.85° was calculated from the FWHM calculations based on the Debby-Scherer equation and was obtained as 20.11 nm. Hence, the XRD studies confirm good crystalline n-MgO synthesis. The n-MgO/AC adsorbent XRD analysis (Fig.13.3b) which exhibits peaks at 38°, 62°, and 74° reveals the planes of (111), (202), and (311). It is evident from Figure 13.3b that after composting with activated carbon, the crystalline nature of MgO has been intact. However, little hydration of MgO to $Mg(OH)_2$ was also observed from Figure 13.3b, which is in agreement with the previous literature.[59,64,65] The electron microscopy image shown in Figure 13.3c represents the cubic structured morphology of the synthesized n-MgO and Figure 13.3d represents the distribution of n-MgO onto the AC surface.

13.3.2 MECHANISM OF ADSORPTION

The sorption mechanism is regulated by the physical/chemical properties of adsorbents. Since MB is a planar molecule, the π–π stacking interaction between the aromatic spine of MB dye, activated carbon surface leads to easier adsorption.[66] In addition to stacking interaction, the electrostatic force of attraction between positively charged MB molecules can lead to adsorption of the negatively charged activated carbon surface. The bandgap energy value of the synthesized n-MgO was calculated using the cut-off wavelength obtained from the UV analysis. The UV-VIS spectra of the calcined MgO sample were observed in the wavelength range of

200–800 nm.. The MgO samples were dispersed into absolute ethanol for absorbance measurement by the UV spectral analysis. For the n-MgO test sample, the graph between absorbance and wavelength was represented in Figure 13.2b. The absorbance was first found to decline sharply with an increased wavelength near the edge of the band edge (265 nm) indicating the nanostructure, followed by a constant absorption coefficient suggesting the uniformity of the particle size. There might be surface defects caused by oxygen vacancy, which can be evident from the low bandgap of 4.8 eV for synthesized n-MgO (7.8 eV bandgap for bulk MgO) explains its enhanced surface activity. Also, the adsorption appears to be very pH-dependent caused by the loss of nano-MgO at too acidic conditions, apart from the protonation of functional groups.

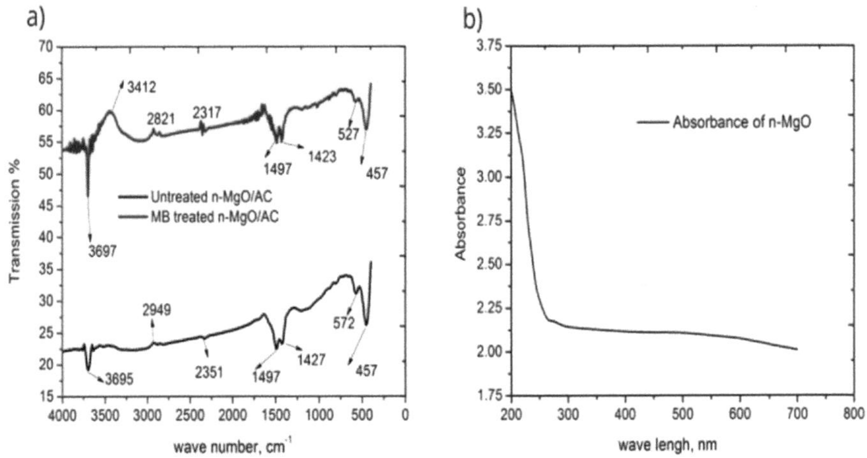

FIGURE 13.2(a) FTIR of n-MgO/AC before and after MB adsorption. (b) Absorbance plot of n-MgO.

13.3.3 COLUMN STUDIES

13.3.3.1 STUDIES ON THE EFFECT OF FEED SOLUTION FLOW RATE

Studies showed that the MB adsorption is dependent on the flow rate of the solution.[14,38] Figure 13.4a explains the same for the experimentation done at 4, 8, and 12 mL/min solution flow rates with the initial concentration of MB and n-MgO/AC bed height kept constant at 20 mg/L and 2.5

cm respectively. The results regarding breakthrough time, exhaust time, uptake capacity, MTZ, and MB removal % were tabulated in Table 13.2. The observation from Figure 13.4a is that breakthrough was rapid (210–30 min) with a change in flow rate from 4 to 12 mL/min. Faster breakthrough was observed as more MB molecules come into contact with n-MgO/AC at higher flow rates. The turbulence generated at higher flow rates results in a decrease in the external bulk mass transfer resistance to the adsorbent surface, which increases the movement of MB molecules to the adsorbent surface.[67] There will be higher adsorption efficiency due to the enhanced dye molecules diffusion through the internal boundary layer of the adsorbent resulting from the longer interactions between adsorbate molecules and the adsorbent at smaller flow rates.[48] When compared with smaller MB flow rates, a saturation of adsorbent bed had happened quickly at larger MB flow rates. Several studies reported a similar trend at different flow rates.[14,48,68–70]

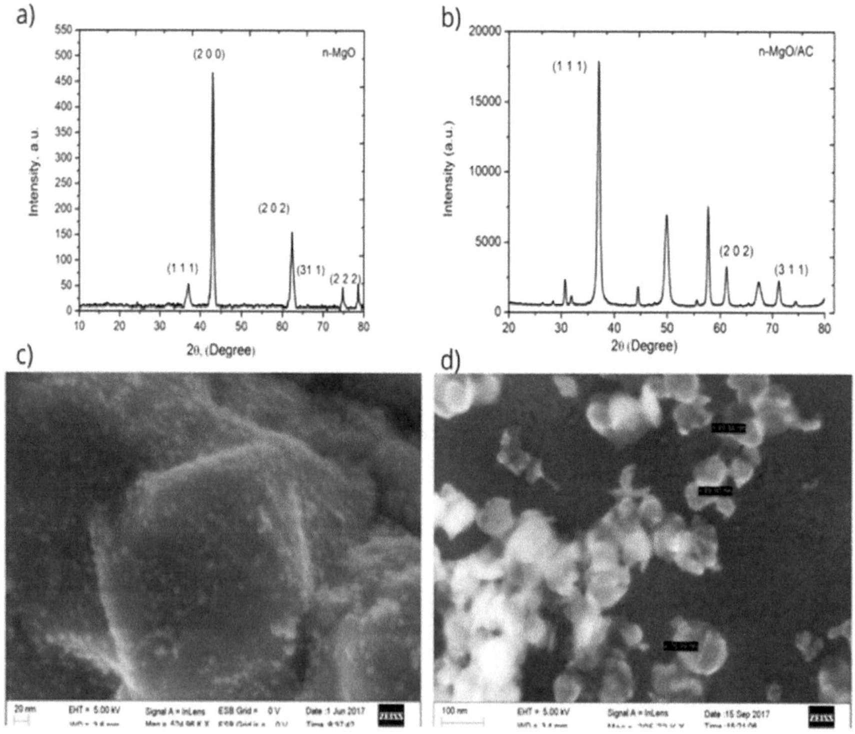

FIGURE 13.3 XRD pattern of (a) n-MgO, (b) n-MgO/AC, SEM image of (c) n-MgO, and (d) n-MgO/AC, respectively.

TABLE 13.2 Adsorption of MB onto the n-MgO/AC Adsorbent.

Process parameters	**V_{eff} (mL)**	**t_b (min)**	**t_e (min)**	**q_{exp} (mg·g^{-1})**	**% MB removal**	**EBCT (min)**	**MTZ (cm)**
Flow rate Q, mL/min (C_0 = 20 mg/L, Z = 2.5 cm, pH = 6, T = 303 K)							
4	2311	210	480	21.93	48	3.07	1.406
8	2642	60	210	16.95	32.1	1.534	1.786
12	2888	30	180	15.68	27.24	1.023	2.083
n-MgO/AC Bed height, Z cm (C_0 = 20 mg/L, Q = 4 mL/min, pH = 6, T = 303 K)							
5	4200	390	990	25.74	61.28	6.135	3.03
7.5	5770	720	1380	26.48	68.96	9.2	3.587
MB Inlet concentration, C_0 mg/L (Q = 4 mL/min, Z = 5 cm, pH = 6, T = 303 K)							
40	2286	150	450	19.98	43.82	6.135	3.33
60	1680	60	330	17.08	33.1	6.135	4.09

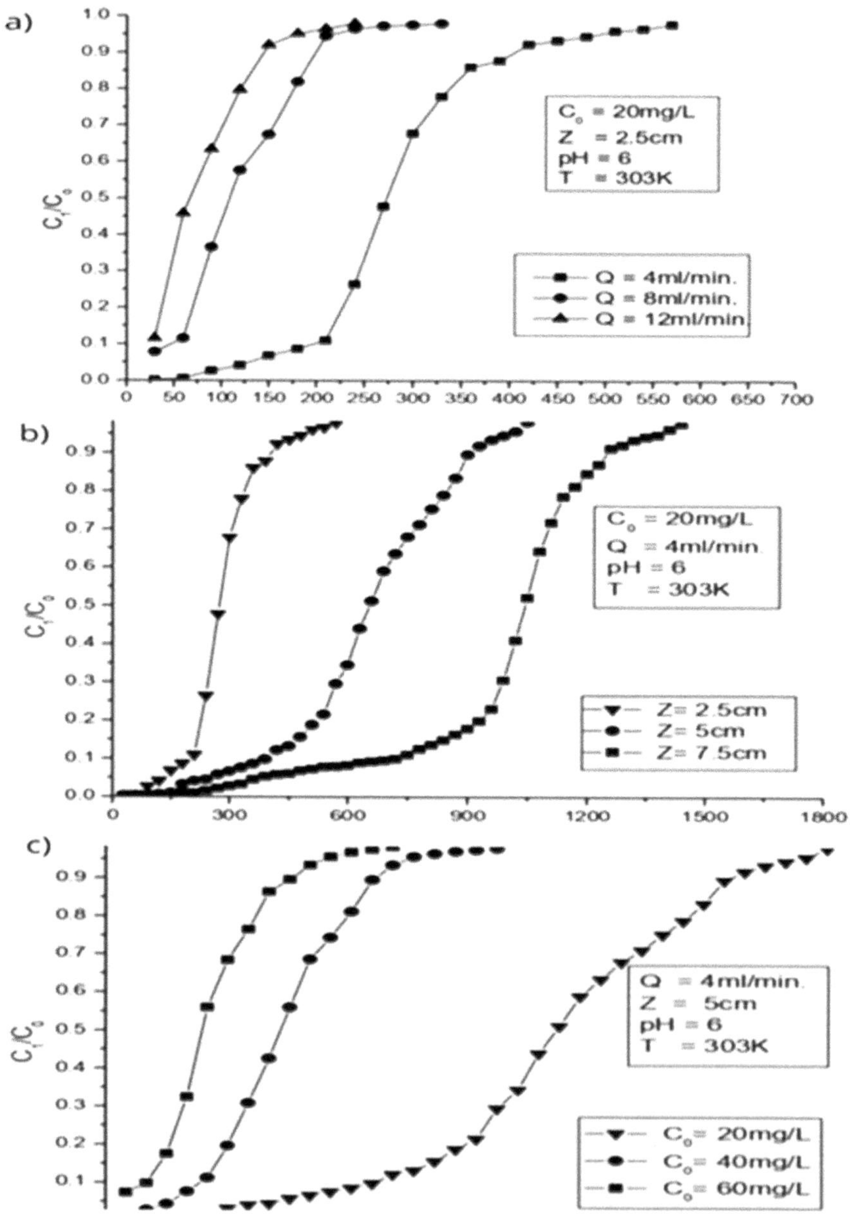

FIGURE 13.4 Effect of (a) feed solution flow rate, (b) n-MgO/AC dosage, (c) MB inlet concentration on breakthrough curve.

13.3.3.2 STUDIES ON EFFECT OF ADSORBENT (N-MGO/AC) BED HEIGHT

The column's dye accumulation was based mainly on the amount of adsorbent within the column.[71] Three different n-MgO/AC bed heights (2.5, 5, and 7.5 cm) were used for plotting the MB breakthrough curves. Since, there is higher active sites availability for adsorption, the bed exhaust time and EBCT increased with enhanced bed height.[72] The experimental results (Fig.13.4b) show that an enhancement in n-MgO/AC bed height resulted in a substantial prolongation of both the breakthrough times (210–720 min) and the exhaust times (480–1380 min). With the rise in n-MgO/AC bed height, the surface available for adsorption increased, resulting in greater contact time, and hence the higher uptake. The resulting breakthrough curves at different bed depths exhibit analogous shape. Similar phenomena are represented in earlier studies.[73] The MB uptake has been enhanced from 21.93 to 26.48 mg/g with n-MgO/AC height altered from 2.5 to 7.5 cm. However, a marginal increase in the MB uptake of 2.87% was observed (25.74–26.48 mg/g), with a change in bed height from 7.5 to 5 cm, compared with an increase of 17.3% with a change in bed height from 2.5 to 5 cm. A probable n-MgO/AC particle aggregation might have resulted in the reduced available total surface for adsorption.[74] Hence, the bed height of 5 cm was considered to be the optimal bed height for further studies.

13.3.3.3 STUDIES ON THE EFFECT OF METHYLENE BLUE INITIAL CONCENTRATION

The concentration changes of the adsorbate would have a significant impact on the breakthrough curves.[75] In this analysis, three different concentrations of MB (20, 40, and 60 mg/L) were employed in this study, and Figure 13.4c explains the breakthrough curves at these concentrations. The slope of the breakthrough curve increased and breakthrough time (390–60 min) and exhaust time (990–330 min) were reduced, and the volume handled was reduced (420–1680 mL) as a result of the increased initial MB concentration. The observed decrease in adsorption capacity is attributed to the saturation of n-MgO/AC adsorption sites n the internal and external surfaces.[74]

13.3.4 ANALYSIS OF BREAKTHROUGH CURVE MODELING USING THOMAS AND YAN MODEL

Thomas and Yan's models were used for simulating the breakthrough curves. The recorded data of C_1/C_0 with time is fitted with simulated data using nonlinear regression for variable flow rates, bed heights, and concentrations (Figs. 13.5a, 13.5b, 13.5c). The model parameters together with the regression coefficients R-square (R^2), Chi-square (χ^2) values are given in Table 13.3a. A decent correlation was obtained between the experimental data and the Thomas model generated. With a rise in flow rate (4–12 mL/min), there is a negative change in uptake capacity (q_0) from 22.25 to 17.66 mg/g, and an increase in the Thomas constant (k_{Th}) values was observed, which is in agreement with the previous studies.[42,48,75–77] Experiments performed at different n-MgO/AC bed heights (2.5–7.5 cm) with flow rate fixed at 4 mL/min have resulted in decreased rate constant (k_{Th}) values and enhanced uptake capacity (q_0). Later, the studies on the increase of MB concentration (20–60 mg/L) resulted in the uptake capacity reduction from 26.60 to 18.02 mg/g. Based on the Thomas model parameters obtained, Bohart-Adams parameters (N_o, K_{BA}) and Yoon-Nelson model parameters (τ, K_{YN}) were obtained using eqs 13.12–13.15 and are presented in Table 13.3b. It was observed from Figure 13.5a–c and Table 13.3a, that the Yan model was good enough in breakthrough curve simulation.

13.3.5 BED-DEPTH SERVICE TIME MODEL

The constant adsorption rate "k_{BA}" and the adsorption capacity per unit bed volume "N_0" were determined from the slope and intercept of the BDST curve for a flow rate of $Q = 4$ mL/min (shown in Table 13.4a). The same for a different flow rate of $Q = 8$ mL/min were computed using eqs. 13.20 and 13.21. Table 13.4b compares the experimental and predicted service times found from the new slope and intercept. It is evident from Table 13.4b that both experimental and predicted values of service time are matching with each other, hence, the BDST model can be employed for the successful adsorption column design at an unchecked flow rate and concentration conditions by altering the BDST constants *a*, *b*.

TABLE 13.3a Thomas Model and Yan Model Parameters.

Model/process parameters	Thomas model				Yan/modified dose–response model				q_{exp} (mg/g)
	$k_{Th} \times 10^{-3}$ ($L \cdot min^{-1} \cdot mg^{-1}$)	q_o (mg/g)	R^2	χ^2	a	q_o (mg/g)	R^2	χ^2	
Flow rate Q, mL/min (Inlet concentration $C_0 = 20$ mg/L, Bed height, $Z = 2.5$ cm, pH = 6, $T = 303$ K)									
4	1.161	22.25	0.993	0.0779	6.42	21.98	0.996	5.817	21.93
8	1.35	18.62	0.991	0.0376	3.28	17.53	0.9891	0.301	16.95
12	1.652	17.66	0.982	0.0446	2.57	16.07	0.9917	0.008	15.68
n-MgO/AC Bed height, Z cm (Inlet concentration $C_0 = 20$ mg/L, Flow rate $Q = 4$ mL/min, pH = 6, $T = 303$ K)									
5	0.426	26.6	0.998	0.1192	5.75	26.16	0.9927	13.92	25.74
7.5	0.5	27.72	0.984	1.926	10.8	27.63	0.9784	10.02	26.48
MB Inlet concentration, C_0 mg/L (Flow rate $Q = 4$ mL/min, Bed height, $Z = 5$ cm, pH = 6, $T = 303$ K)									
40	0.419	20.83	0.998	0.0176	4.42	20.18	0.997	2.274	19.98
60	0.362	18.02	0.993	0.0265	3.38	17.15	0.9941	0.9628	17.08

TABLE 13.3b Bohart-Adams and Yoon-Nelson Model Parameters Obtained from the Thomas Model.

Model/process parameters	Bohart-Adams model		Yoon-Nelson model	
	$K_{BA} \times 10^{-3}$ ($L \cdot min^{-1} \cdot mg^{-1}$)	N_o (mg/L)	K_{YN} (min^{-1})	τ (min)
Flow rate Q, mL/min (Inlet concentration $C_0 = 20$ mg/L, Bed height, $Z = 2.5$ cm pH = 6, $T = 303$ K)				
4	1.161	1813.4	0.0232	278.15
8	1.35	1517.5	0.027	116.39
12	1.652	1439.3	0.034	73.59
n-MgO/AC Bed height, Z cm (Inlet concentration $C_0 = 20$ mg/L, Flow rate $Q = 4$ mL/min, pH = 6, $T = 303$ K)				
5	0.426	2167.9	0.0085	665.07
7.5	0.5	2259.2	0.01	1039.6
MB Inlet concentration, C_0 mg/L (Flow rate $Q = 4$ mL/min, Bed height, $Z = 5$ cm, pH = 6, $T = 303$ K)				
40	0.419	1697.6	0.0168	260.4
60	0.362	1468.6	0.0217	150.18

TABLE 13.4a Bed-Depth Service Time (BDST) Model Parameters for MB Adsorption Using n-MgO/AC.

C_1/C_0	Slope (a, min·cm^{-1})	Intercept (b, min)	k_{BA} (L/mg/mi)	N_0 (mg/L)	R^2
0.1	102	−70	−0.0016	1662.396	0.9435
0.9	172	−6.666	0.0165	2803.256	0.9824

Flow rate Q = 4 mL/min, Concentration C_o = 20 mg/L.

TABLE 13.4b Comparison of Experimental and Predicted Service Times.

C_1/C_0	Slope (a', min·cm^{-1})	Intercept (b', min)	Service time (min) Experimental	Service time (min) Predicted
0.1	51	−70	60	57.5
0.9	86	−6.666	210	208.334

Flow rate Q = 8 mL/min, Concentration C_o = 20 mg/L.

13.3.6 DESORPTION STUDIES

The adsorbent regeneration is important from an economic point of view and for the practical use of the sorbent for several cycles.[78] The experiments were performed with 0.1*M* HCl used as an eluting substance. After adsorption, the n-MgO/AC loaded with MB ($Q = 4$ mL/min, $C_0 = 60$ mg/L, $Z = 5$ cm, pH = 6, and $T = 303$ K) was taken out from the column and oven-dried at 50°*C* for 12 h and placed in the bed again with 0.1*M* HCl solution pumped through the column at 4 mL/min. The adsorption–desorption process was carried out in three cycles. The adsorbed and desorbed dye quantity and the regeneration efficiencies were estimated using eqs 13.22 and 13.23, respectively. The regeneration efficiencies obtained were 84%, 81.6%, and 78.9%, respectively for the three cycles. The reduction in regeneration efficiency might be due to the MgO hydration and stronger attachment between n-MgO/AC and MB.

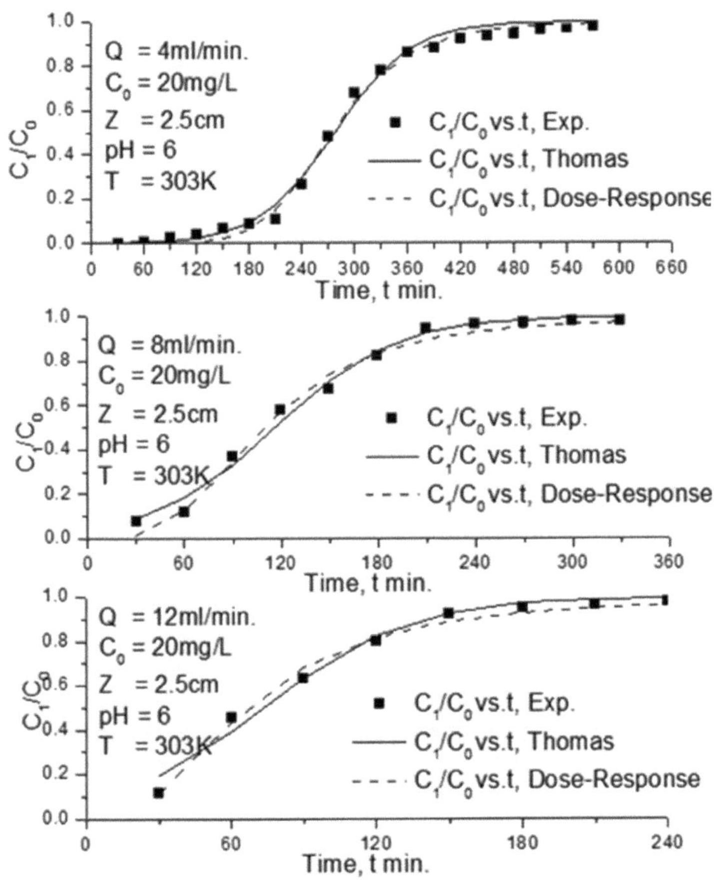

FIGURE 13.5(a) Experimental and simulated breakthrough curves at different flow rates.

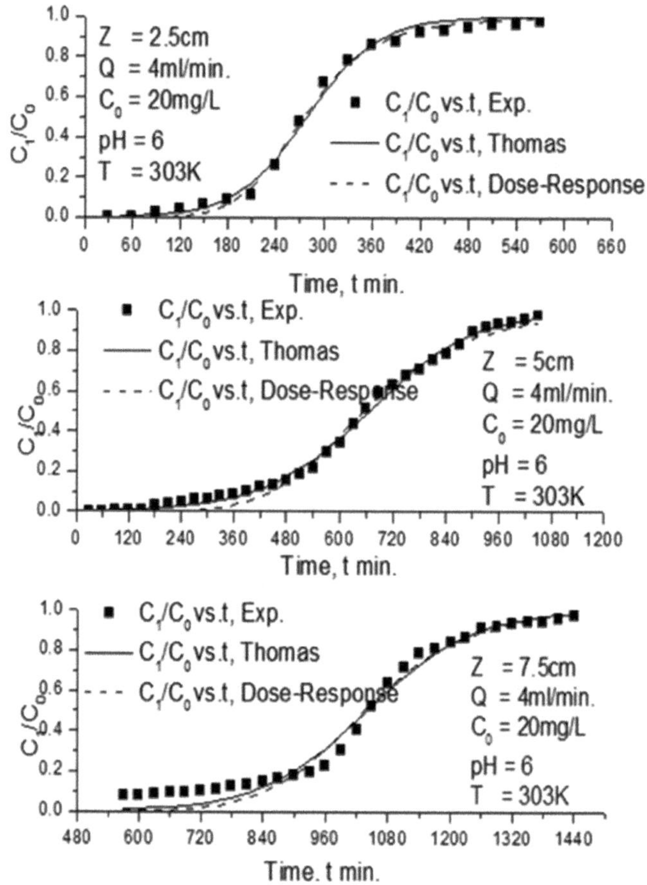

FIGURE 13.5(b) Experimental and simulated breakthrough curves at different bed heights.

13.4 CONCLUSIONS

Through this study, fixed-bed sorption studies were performed for the removal of MB employing a hybrid nanocomposite adsorbent n-MgO/AC. The MB removal was strongly affected by the variation in flow rates, bed heights, and inlet MB concentrations. Thomas and Yan's models were employed for explaining the column adsorption behavior through breakthrough curve analysis. The successful application of the BDST model shows that it can be used for the adsorption column design. The spent nanocomposite n-MgO/AC was successfully regenerated.

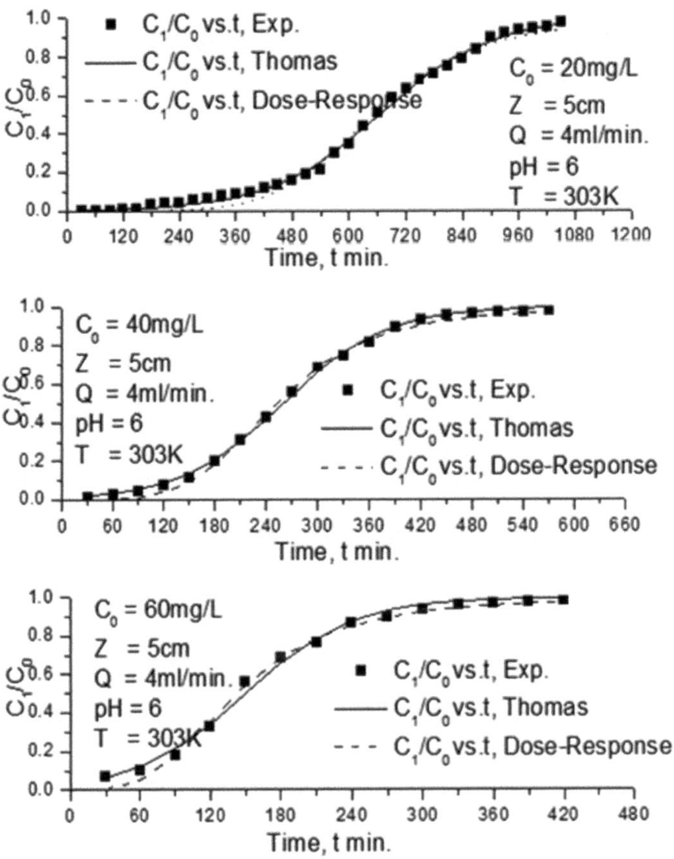

FIGURE 13.5(c) Experimental and simulated breakthrough curves at different inlet MB concentrations.

KEYWORDS

- **nano-MgO**
- **activated carbon**
- **column adsorption**
- **methylene blue**
- **breakthrough curves**

REFERENCES

1. Vafakish, B.; Wilson, L. D. *Surface-Modified Chitosan : An Adsorption Study of a " Tweezer-Like " Biopolymer with Fluorescein*, 2019; pp 468–484.
2. Abbasi, S.; Hasanpour, M.; Ahmadpoor, F.; Sillanpää, M.; Dastan, D.; Achour, A. Application of the Statistical Analysis Methodology for Photodegradation of Methyl Orange Using a New Nanocomposite Containing Modified TiO_2 Semiconductor with SnO_2. *Int. J. Environ. Anal. Chem.* **2019,** *00* (00), 1–17.
3. Ratnam, M. V.; Karthikeyan, C.; Rao, K. N.; Meena, V. Magnesium Oxide Nanoparticles for Effective Photocatalytic Degradation of Methyl Red Dye in Aqueous Solutions: Optimization Studies Using Response Surface Methodology. *Mater. Today Proc.* **2020,** 26 (2), 2308–2313.
4. Deb, A.; Kanmani, M.; Debnath, A.; Bhowmik, K. L.; Saha, B. Ultrasonic Assisted Enhanced Adsorption of Methyl Orange Dye onto Polyaniline Impregnated Zinc Oxide Nanoparticles: Kinetic, Isotherm and Optimization of Process Parameters. *Ultrason. Sonochem.* **2019,** *54* (August 2018), 290–301.
5. Afroze, S.; Sen, T. K.; Ang, H. M. Adsorption Performance of Continuous Fixed Bed Column for the Removal of Methylene Blue (MB) Dye Using Eucalyptus Sheathiana Bark Biomass. *Res. Chem. Intermed.* **2016,** *42* (3), 2343–2364.
6. Akbari, A.; Arsalani, N.; Eftekhari-Sis, B.; Amini, M.; Gohari, G.; Jabbari, E. Cube-Octameric Silsesquioxane (POSS)-Capped Magnetic Iron Oxide Nanoparticles for the Efficient Removal of Methylene Blue. *Front. Chem. Sci. Eng.* **2019,** *13* (3), 563–573.
7. Singh, N. B.; Nagpal, G.; Agrawal, S.; Rachna. Water Purification by Using Adsorbents: A Review. *Environ. Technol. Innov.* **2018.**
8. Zhou, K.; Li, L.; Ma, X.; Mo, Y.; Chen, R.; Li, H.; Li, H. Activated Carbons Modified by Magnesium Oxide as Highly Efficient Sorbents for Acetone. *RSC Adv.* **2018,** *8* (6), 2922–2932.
9. Pooralhossini, J.; Ghaedi, M.; Zanjanchi, M. A.; Asfaram, A. Ultrasonically Assisted Removal of Congo Red, Phloxine B and Fast Green FCF in Ternary Mixture Using Novel Nanocomposite Following Their Simultaneous Analysis by Derivative Spectrophotometry. *Ultrason. Sonochem.* **2017,** *37*, 452–463.
10. Jain, S. N.; Tamboli, S. R.; Sutar, D. S.; Jadhav, S. R.; Marathe, J. V.; Shaikh, A. A.; Prajapati, A. A. Batch and Continuous Studies for Adsorption of Anionic Dye onto Waste Tea Residue: Kinetic, Equilibrium, Breakthrough and Reusability Studies. *J. Clean. Prod.* **2020,** *252*, 119778.
11. Boruah, P. K.; Borthakur, P.; Das, M. R. *Magnetic Metal/Metal Oxide Nanoparticles and Nanocomposite Materials for Water Purification*; Elsevier Inc., 2018.
12. Liu, J.; Wang, N.; Zhang, H.; Baeyens, J. Adsorption of Congo Red Dye on Fe x Co 3-x O 4 Nanoparticles. *J. Environ. Manage.* **2019,** *238* (01), 473–483.
13. Sahoo, J. K.; Paikra, S. K.; Mishra, M.; Sahoo, H. Amine Functionalized Magnetic Iron Oxide Nanoparticles: Synthesis, Antibacterial Activity and Rapid Removal of Congo Red Dye. *J. Mol. Liq.* **2019,** *282*, 428–440.
14. Lee, S. Y.; Shim, H. E.; Yang, J. E.; Choi, Y. J.; Jeon, J. Continuous Flow Removal of Anionic Dyes in Water by Chitosan-Functionalized Iron Oxide Nanoparticles Incorporated in a Dextran Gel Column. *Nanomaterials* **2019,** *9* (8).

15. Li, L. H.; Xiao, J.; Liu, P.; Yang, G. W. Super Adsorption Capability from Amorphousization of Metal Oxide Nanoparticles for Dye Removal. *Sci. Rep.* **2015,** *5*, 1–6.
16. Nassar, M. Y.; Mohamed, T. Y.; Ahmed, I. S.; Samir, I. *MgO Nanostructure via a Sol-Gel Combustion Synthesis Method Using Different Fuels: An Efficient Nano-Adsorbent for the Removal of Some Anionic Textile Dyes*; Elsevier B.V., 2017; Vol. 225.
17. Myneni, V. R.; Punugoti, T.; Kala, N. S.; Kanidarapu, N. R.; Vangalapati, M. Modelling and Optimization of Methylene Blue Adsorption onto Magnesium Oxide Nanoparticles Loaded onto Activated Carbon (MgONP-AC): Response Surface Methodology and Artificial Neural Networks. *Mater. Today Proc.* **2019,** *18*, 4932–4941.
18. Tan, K. B.; Vakili, M.; Horri, B. A.; Poh, P. E.; Abdullah, A. Z.; Salamatinia, B. Adsorption of Dyes by Nanomaterials: Recent Developments and Adsorption Mechanisms. *Sep. Purif. Technol.* **2015,** *150*, 229–242.
19. Moussavi, G.; Mahmoudi, M. Removal of Azo and Anthraquinone Reactive Dyes from Industrial Wastewaters Using MgO Nanoparticles. *J. Hazard. Mater.* **2009,** *168* (2–3), 806–812.
20. Cao, C. Y.; Qu, J.; Wei, F.; Liu, H.; Song, W. G. Superb Adsorption Capacity and Mechanism of Flowerlike Magnesium Oxide Nanostructures for Lead and Cadmium Ions. *ACS Appl. Mater. Interfaces* **2012,** *4* (8), 4283–4287.
21. Kumar, A.; Kumar, J. On the Synthesis and Optical Absorption Studies of Nano-Size Magnesium Oxide Powder. *J. Phys. Chem. Solids* **2008,** *69* (11), 2764–2772.
22. Mahmoud, H. R.; Ibrahim, S. M.; El-Molla, S. A. Textile Dye Removal from Aqueous Solutions Using Cheap MgO Nanomaterials: Adsorption Kinetics, Isotherm Studies and Thermodynamics. *Adv. Powder Technol.* **2016,** *27* (1), 223–231.
23. Mohan, S.; Kumar, V.; Singh, D. K.; Hasan, S. H. *Effective Removal of Lead Ions Using Graphene Oxide-MgO Nanohybrid from Aqueous Solution: Isotherm, Kinetic and Thermodynamic Modeling of Adsorption*; Elsevier B.V., 2017; Vol. 5.
24. Mohan, S.; Singh, D. K.; Kumar, V.; Hasan, S. H. Modelling of Fixed Bed Column Containing Graphene Oxide Decorated by MgO Nanocubes as Adsorbent for Lead(II) Removal from Water. *J. Water Process Eng.* **2017,** *17*, 216–228.
25. Myneni, V. R.; Kanidarapu, N. R.; Vangalapati, M. Methylene Blue Adsorption by Magnesium Oxide Nanoparticles Immobilized with Chitosan (CS-MgONP): Response Surface Methodology, Isotherm, Kinetics and Thermodynamic Studies. *Iran. J. Chem. Chem. Eng.* **2019.**
26. Wang, C.; Zhao, J.; Wang, S.; Zhang, L.; Liu, N.; Zhang, B. Selective Capture Models and Mechanisms of Pb(II) from Wastewater Using Tannic-Functionalized Nickel-Iron Oxide Nanoparticles. *Colloids Surfaces A Physicochem. Eng. Asp.* **2019,** *570* (January), 265–273.
27. Asfaram, A.; Ghaedi, M.; Hajati, S.; Goudarzi, A.; Dil, E. A. Screening and Optimization of Highly Effective Ultrasound-Assisted Simultaneous Adsorption of Cationic Dyes onto Mn-Doped Fe_3O_4-Nanoparticle-Loaded Activated Carbon. *Ultrason. Sonochem.* **2017,** *34*, 1–12.
28. Asfaram, A.; Ghaedi, M.; Yousefi, F.; Dastkhoon, M. Experimental Design and Modeling of Ultrasound Assisted Simultaneous Adsorption of Cationic Dyes onto ZnS: Mn-NPs-AC from Binary Mixture. *Ultrason. Sonochem.* **2016,** *33*, 77–89.
29. Bagheri, A. R.; Ghaedi, M.; Asfaram, A.; Jannesar, R.; Goudarzi, A. Design and Construction of Nanoscale Material for Ultrasonic Assisted Adsorption of Dyes:

Application of Derivative Spectrophotometry and Experimental Design Methodology. *Ultrason. Sonochem.* **2017,** *35*, 112–123.
30. Darvishi Cheshmeh Soltani, R.; Safari, M.; Mashayekhi, M. Sonocatalyzed Decolorization of Synthetic Textile Wastewater Using Sonochemically Synthesized MgO Nanostructures. *Ultrason. Sonochem.* **2016,** *30*, 123–131.
31. Hassan, H.; Salama, A.; El-ziaty, A. K.; El-Sakhawy, M. New Chitosan/Silica/Zinc Oxide Nanocomposite as Adsorbent for Dye Removal. *Int. J. Biol. Macromol.* **2019,** *131*, 520–526.
32. Zhang, Y.; Xia, K.; Liu, X.; Chen, Z.; Du, H.; Zhang, X. Synthesis of Cationic-Modified Silica Gel and Its Adsorption Properties for Anionic Dyes. *J. Taiwan Inst. Chem. Eng.* **2019,** *102*, 1–8.
33. Abukhadra, M. R.; Shaban, M. Recycling of Different Solid Wastes in Synthesis of High-Order Mesoporous Silica as Adsorbent for Safranin Dye. *Int. J. Environ. Sci. Technol.* **2019,** *16* (11), 7573–7582.
34. Mahmoodi, N. M.; Saffar-Dastgerdi, M. H. Zeolite Nanoparticle as a Superior Adsorbent with High Capacity: Synthesis, Surface Modification and Pollutant Adsorption Ability from Wastewater. *Microchem. J.* **2019,** *145*, 74–83.
35. Piri, F.; Mollahosseini, A.; Khadir, A.; Milani Hosseini, M. Enhanced Adsorption of Dyes on Microwave-Assisted Synthesized Magnetic Zeolite-Hydroxyapatite Nanocomposite. *J. Environ. Chem. Eng.* **2019,** *7* (5), 103338.
36. Hemine, K.; Skwierawska, A.; Kernstein, A.; Kozłowska-Tylingo, K. Cyclodextrin Polymers as Efficient Adsorbents for Removing Toxic Non-Biodegradable Pimavanserin from Pharmaceutical Wastewaters. *Chemosphere* **2020,** *250*, 126250.
37. Chen, H.; Zhou, Y.; Wang, J.; Lu, J.; Zhou, Y. Polydopamine Modified Cyclodextrin Polymer as Efficient Adsorbent for Removing Cationic Dyes and Cu2+. *J. Hazard. Mater.* **2020,** *389*, 121897.
38. Patel, H. Fixed-Bed Column Adsorption Study: A Comprehensive Review. *Appl. Water Sci.* **2019,** *9* (3).
39. Lee, C. G.; Kim, J. H.; Kang, J. K.; Kim, S. B.; Park, S. J.; Lee, S. H.; Choi, J. W. Comparative Analysis of Fixed-Bed Sorption Models Using Phosphate Breakthrough Curves in Slag Filter Media. *Desalin. Water Treat.* **2015,** *55* (7), 1795–1805.
40. Shahid, M. K.; Kim, Y.; Choi, Y. G. Magnetite Synthesis Using Iron Oxide Waste and Its Application for Phosphate Adsorption with Column and Batch Reactors. *Chem. Eng. Res. Des.* **2019,** *148*, 169–179.
41. Charumathi, D.; Das, N. Packed Bed Column Studies for the Removal of Synthetic Dyes from Textile Wastewater Using Immobilised Dead *C. tropicalis*. *Desalination* **2012,** *285*, 22–30.
42. de Franco, M. A. E.; de Carvalho, C. B.; Bonetto, M. M.; Soares, R. de P.; Féris, L. A. Removal of Amoxicillin from Water by Adsorption onto Activated Carbon in Batch Process and Fixed Bed Column: Kinetics, Isotherms, Experimental Design and Breakthrough Curves Modelling. *J. Clean. Prod.* **2017,** *161*, 947–956.
43. Dhoble, R. M.; Maddigapu, P. R.; Rayalu, S. S.; Bhole, A. G.; Dhoble, A. S.; Dhoble, S. R. *Removal of Arsenic(III) from Water by Magnetic Binary Oxide Particles (MBOP): Experimental Studies on Fixed Bed Column*; Elsevier B.V., 2017; Vol. 322.

44. Dichiara, A. B.; Harlander, S. F.; Rogers, R. E. Fixed Bed Adsorption of Diquat Dibromide from Aqueous Solution Using Carbon Nanotubes. *RSC Adv.* **2015,** *5* (76), 61508–61512.
45. Han, R.; Wang, Y.; Zhao, X.; Wang, Y.; Xie, F.; Cheng, J.; Tang, M. Adsorption of Methylene Blue by Phoenix Tree Leaf Powder in a Fixed-Bed Column: Experiments and Prediction of Breakthrough Curves. *Desalination* **2009,** *245* (1–3), 284–297.
46. Lin, X.; Huang, Q.; Qi, G.; Shi, S.; Xiong, L.; Huang, C.; Chen, X.; Li, H.; Chen, X. Estimation of Fixed-Bed Column Parameters and Mathematical Modeling of Breakthrough Behaviors for Adsorption of Levulinic Acid from Aqueous Solution Using SY-01 Resin. *Sep. Purif. Technol. 174*, 222–231.
47. Low, L. W.; Teng, T. T.; Morad, N.; Azahari, B. Optimization of the Column Studies into the Adsorption of Basic Dye Using Tartaric Acid-Treated Bagasse. *Desalin. Water Treat.* **2014,** *52* (31–33), 6194–6205.
48. Maheshwari, U.; Gupta, S. Removal of Cr(VI) from Wastewater Using Activated Neem Bark in a Fixed-Bed Column: Interference of Other Ions and Kinetic Modelling Studies. *Desalin. Water Treat.* **2016,** *57* (18), 8514–8525.
49. Bohart, G. S.; Adams, E. Q. Some Aspects of the Behavior of Charcoal with Respect to Chlorine. *J. Am. Chem. Soc.* **1920,** *42* (3), 523–544.
50. Thomas, H. C. Heterogeneous Ion Exchange in a Flowing System. *J. Am. Chem. Soc.* **1944,** *66* (10), 1664–1666.
51. Yoon, Y. H. E. E.; Nelson, J. H. Application of Gas Adsorption Kinetics I. A Theoretical Model for Respirator Cartridge Service Life. *Am. Ind. Hyg. Assoc. J.* **1984,** *45* (8), 509–516.
52. Yan, G.; Viraraghavan, T.; Chen, M. A New Model for Heavy Metal Removal in a Biosorption Column. *Adsorpt. Sci. Technol.* **2001,** *19* (1), 25–43.
53. Chu, K. H. Fixed Bed Sorption: Setting the Record Straight on the Bohart-Adams and Thomas Models. *J. Hazard. Mater.* **2010,** *177* (1–3), 1006–1012.
54. Hutchins, A. R. New Method Simplifies Design of Activated Carbon Systems. *Chem. Eng.* **1973,** *80*, 133–138.
55. Niu, H.; Yang, Q.; Tang, K.; Xie, Y. Large-Scale Synthesis of Single-Crystalline MgO with Bone-like Nanostructures. *J. Nanoparticle Res.* **2006,** *8* (6), 881–888.
56. Jeevanandam, J.; Chan, Y. S.; Danquah, M. K. Biosynthesis and Characterization of MgO Nanoparticles from Plant Extracts via Induced Molecular Nucleation. *New J. Chem.* **2017,** *41* (7), 2800–2814.
57. Imani, M. M.; Safaei, M. Optimized Synthesis of Magnesium Oxide Nanoparticles as Bactericidal Agents. *J. Nanotechnol.* **2019,** *2019*.
58. Haldorai, Y.; Shim, J. J. An Efficient Removal of Methyl Orange Dye from Aqueous Solution by Adsorption onto Chitosan/MgO Composite: A Novel Reusable Adsorbent. *Appl. Surf. Sci.* **2014,** *292*, 447–453.
59. Heidarizad, M.; Şengör, S. S. Synthesis of Graphene Oxide/Magnesium Oxide Nanocomposites with High-Rate Adsorption of Methylene Blue. *J. Mol. Liq.* **2016,** *224*, 607–617.
60. Jadhav, A. H.; Lim, A. C.; Thorat, G. M.; Jadhav, H. S.; Seo, J. G. Green Solvent Ionic Liquids: Structural Directing Pioneers for Microwave-Assisted Synthesis of Controlled MgO Nanostructures. *RSC Adv.* **2016,** *6* (38), 31675–31686.

61. Balakrishnan, G.; Velavan, R.; Mujasam Batoo, K.; Raslan, E. H. Microstructure, Optical and Photocatalytic Properties of MgO Nanoparticles. *Results Phys.* **2020,** *16* (February), 103013.
62. Raveesha, H. R.; Nayana, S.; Vasudha, D. R.; Begum, J. P. S.; Pratibha, S.; Ravikumara, C. R.; Dhananjaya, N. The Electrochemical Behavior, Antifungal and Cytotoxic Activities of Phytofabricated MgO Nanoparticles Using *Withania somnifera* Leaf Extract. *J. Sci. Adv. Mater. Devices* **2019,** *4* (1), 57–65.
63. Karimi, M. A.; Hatefi-Mehrjardi, A.; Askarpour Kabir, A.; Zaydabadi, M. Synthesis, Characterization, and Application of MgO/ZnO Nanocomposite Supported on Activated Carbon for Photocatalytic Degradation of Methylene Blue. *Res. Chem. Intermed.* **2015,** *41* (9), 6157–6168.
64. Srivastava, V.; Sharma, Y. C.; Sillanpää, M. Green Synthesis of Magnesium Oxide Nanoflower and Its Application for the Removal of Divalent Metallic Species from Synthetic Wastewater. *Ceram. Int.* **2015,** *41* (5), 6702–6709.
65. Yousefi, S.; Ghasemi, B.; Tajally, M.; Asghari, A. Optical Properties of MgO and $Mg(OH)_2$ Nanostructures Synthesized by a Chemical Precipitation Method Using Impure Brine. *J. Alloys Compd.* **2017.**
66. Ma, F.; Yu, L.; Zhou, L.; Jin, M. X.; Yang, J. S.; Luan, Y. H.; Tang, H. B.; Fan, Z. W.; Yuan, J. H. C. Enhanced Adsorption Removal of Methyl Orange and Methylene Blue from Aqueous Solution by Alkali-Activated Multiwalled Carbon Nanotubes. *Appl. Mater. Interfaces* **2012,** 5749–5760.
67. Dawood, S.; Sen, T. K.; Phan, C. Performance and Dynamic Modelling of Biochar and Kaolin Packed Bed Adsorption Column for Aqueous Phase Methylene Blue (MB) Dye Removal. *Environ. Technol. (United Kingdom)* **2019,** *40* (28), 3762–3772.
68. Djelloul, C.; Hamdaoui, O. Dynamic Adsorption of Methylene Blue by Melon Peel in Fixed-Bed Columns. *Desalin. Water Treat.* **2015,** *56* (11), 2966–2975.
69. Ren, X.; Zhang, X.; Zhang, L.; Han, R. Biosorption of Methylene Blue by Natural and Chemical Modified Wheat Straw in Fixed-Bed Column. *Desalin. Water Treat.* **2013,** *51* (22–24), 4514–4523.
70. Yagub, M. T.; Sen, T. K.; Afroze, S.; Ang, H. M. Fixed-Bed Dynamic Column Adsorption Study of Methylene Blue (MB) onto Pine Cone. *Desalin. Water Treat.* **2015,** *55* (4), 1026–1039.
71. Raj, R. A.; Manimozhi, V.; Saravanathamizhan, R. Activation of Petroleum Coke Using KOH and Fixed Bed Adsorption Studies for Acid Red Dye Removal from Aqueous Solution. *Pet. Sci. Technol.* **2020,** *0* (0), 1–9.
72. Yusuf, M.; Song, K.; Li, L. Fixed Bed Column and Artificial Neural Network Model to Predict Heavy Metals Adsorption Dynamic on Surfactant Decorated Graphene. *Colloids Surfaces A Physicochem. Eng. Asp.* **2020,** *585.*
73. Ealias, A. M.; Saravanakumar, M. P. Application of Protein-Functionalised Aluminium Nanosheets Synthesised from Sewage Sludge for Dye Removal in a Fixed-Bed Column: Investigation on Design Parameters and Kinetic Models. *Environ. Sci. Pollut. Res.* **2020,** *27* (3), 2955–2976.
74. Mouldar, J.; Hatimi, B.; Hafdi, H.; Joudi, M.; Belghiti, M. E. A.; Nasrellah, H.; El Mhammedi, M. A.; El Gaini, L.; Bakasse, M. Pyrrhotite Ash Waste for Capacitive Adsorption and Fixed-Bed Column Studies: Application for Reactive Red 141 Dye. *Water. Air. Soil Pollut.* **2020,** *231* (5).

75. Ataei-Germi, T.; Nematollahzadeh, A. Bimodal Porous Silica Microspheres Decorated with Polydopamine Nano-Particles for the Adsorption of Methylene Blue in Fixed-Bed Columns. *J. Colloid Interface Sci.* **2016,** *470*, 172–182.
76. Dabbagh, R.; Sharifipoor, S.; Keshtkar, A.; Vafajoo, L. Removal of Zinc (II) from Synthetic Effluent Using Seaweeds: A Review of Modeling of Fixed-Bed Columns. *Desalin. Water Treat.* **2016,** *57* (51), 24509–24518.
77. Khan, T. A.; Nazir, M.; Khan, E. A.; Riaz, U. Multiwalled Carbon Nanotube-Polyurethane (MWCNT/PU) Composite Adsorbent for Safranin T and Pb(II) Removal from Aqueous Solution: Batch and Fixed-Bed Studies. *J. Mol. Liq.* **2015,** *212*, 467–479.
78. Maatar, W.; Boufi, S. Poly(Methacylic Acid-Co-Maleic Acid) Grafted Nanofibrillated Cellulose as a Reusable Novel Heavy Metal Ions Adsorbent. *Carbohydr. Polym.* **2015,** *126* (June), 199–207.

CHAPTER 14

Pattern Formation Studies During Different Milk Processes

ASWINI HARINDRAN and V. MADHURIMA

Department of Physics, School of Basic and Applied Sciences, Central University of Tamil Nadu, Tamil Nadu, India

Corresponding author. E-mail: aswiniharindran@gmail.com

ABSTRACT

Dairy milk and its products are one of the most consumable product groups of the food industry. Hence, identification of various stages in milk processing is of importance from both a basic science standpoint and from an industrial point of view. This chapter discusses the patterns formed by dried milk droplets during different processes, such as dilution, coagulation, and staling. The earlier studies on dried milk droplets mainly focused on spray drying for milk powder production, while here we utilize the droplet patterns imaged using confocal laser scanning microscopy as a fingerprint to the various stages of dilution, coagulation, and spoiling of milk colloids. As the mechanism involved in each process is different, the pattern formation also differs. The patterns obtained here are analyzed and explained using two mechanisms—the coffee-ring effect and the Marangoni effect and these indicate the physicochemical changes occurring within the system. Thus, the dried milk patterns can be used to detect adulteration, spoilage, and can be further modified and used for industrial purposes. This method proves to be an easier, cost-effective and quicker method to detect the various stages during different milk processes with an advantage of smaller sample size (few microliters).

Advanced Microscopy: A Strong Analytical Tool in Materials Science. Merin Sara Thomas, Jozef T. Haponiuk, Sabu Thomas, & Anne George (Eds.)

14.1 INTRODUCTION

Evaporation of sessile droplets containing insoluble particles is of great importance as they have potential applications in medical diagnosis, ink-jet printers, nanotechnology, colloidal self-assembly, microfluidics and more.[1–6] This method is nowadays widely used in the field of medical diagnosis as it is a low cost method and a convenient tool compared with other techniques.[7–10] During drying of droplets, various kinds of patterns are formed, and the differences in the morphology of each pattern define the different physical or physicochemical processes experienced by the particular system.[11,12] The droplet drying patterns are mainly dependent upon the nature of the substrate and its temperature, the environmental temperature, nature of the fluid, concentration, shape and size of the particles.[13,14] When a sessile droplet or liquid droplet is placed on a glass substrate, the volatile liquid within the droplet begins to evaporate, unless the atmosphere around the droplet is saturated by the corresponding liquid vapor. The pattern evolution basically consists of two regimes known as constant radius regime (CRR) and constant contact angle regime (CCAR).[14,15] In CRR, the contact line is pinned to the substrate, hence, a constant radius is maintained and the contact angle decreases,[16] whereas in CCAR, there is a decrease in the radius of the droplet with constant contact angle and depinning occurs.[17] This is because, the contact line recedes as the liquid evaporates and is observed at a critical contact angle. After attaining this critical contact angle, the radius of the drop starts reducing and the contact angle remains constant. In most cases, these two modes occur successively, and in some cases, mixed modes can also be seen. When evaporation takes place, heat and mass transfer occurs in the droplet and the phenomena related to this are the coffee-ring effect and the Marangoni effect.[18] The coffee ring is formed as a result of capillary flow and is observed when the dispersed particles in the droplet are deposited on the contact line. When the particles are pinned to the contact line, the fluid begins to flow from the central portion of the droplet to the edges as the evaporation at the edges are comparably higher than that at the center. This outward flow is referred to as capillary flow. Hence, due to the capillary flow within the droplet, the suspended particles in the drop are carried to the edges, finally leaving a ring pattern after evaporation.[19,20] On the other hand, Marangoni flow is driven by the surface tension gradient at the drop–air interface, that is, during the fluid evaporation, the concentration of the suspended particles varies locally, hence, generating a surface tension gradient. Hence, the liquid from the lower surface tension region will flow to a higher surface tension region creating Marangoni convection.[21,22] Researchers are nowadays controlling the flow regimes within the droplet by controlling various factors responsible for the drying patterns, and hence manipulating the patterns accordingly to their needs. Hence, understanding these processes is necessary to study and analyze the patterns obtained and also it gives an idea of how to use the deposition patterns usefully.

This work discusses the different patterns obtained during milk dilution, coagulation, and staling processes. The drying patterns were imaged using confocal laser scanning microscopy (CLSM). As all the three are different mechanisms, the patterns obtained were also different from each other. The difference in the patterns attributes to the physical changes occurred in them. The phenomena contributing to variations in each pattern are identified and discussed. In addition to that an easier way to distinguish the different-sized particles within the same fluid using droplet drying technique will be discussed.

14.2 EXPERIMENTAL

Pasteurized homogenized standardized milk with 4.5% fat content was procured from a local vendor. For dilution, milk was diluted with deionized water and a 10 mL solution of each concentration varying from 100% pure to 10% pure milk solution was prepared. For the purpose of imaging, each 10 mL solution of the sample was stained using 0.5 mL of Rhodamine dye solution (2.5 mg/mL). A sessile droplet of each sample was placed on a precleaned glass slide with a needle and syringe. The droplet was dried at room temperature (25°C) for 15 min and then imaged.

For coagulation process, 40 mL of the milk was heated up to 40°C to which 5 mL of the commercial curd at ambient temperature was added. 1 mL of Rhodamine dye solution (2.5 mg/mL) was added to the 45 mL milk with curd solution for imaging. At each 30 min interval, the droplets were made and the droplets were dried on a precleaned glass substrate at room temperature. The experiment was repeated and the images are replicable.

For staling process, 40 mL of the milk was stained with 1 mL of Rhodamine dye solution (2.5 mg/mL) at room temperature. The droplets were made on a precleaned glass substrate at each 1 h interval for 23 h. After drying for 24 h, the imaging was done. The experiment was repeated and the images are replicable.

The drying patterns of all the samples were imaged using confocal laser scanning microscope from Leica, TCS SP8 model with 0.1 μm resolution and 10 X objective.

14.3 RESULTS AND DISCUSSIONS

The pattern formation during dilution, coagulation, and staling of milk is studied by imaging the dried droplets during each process. In each process, due to various physicochemical changes occurring in them, the patterns vary with respect to time in the case of both coagulation and staling, whereas in the case of dilution, the difference in patterns occurs due to the difference in the concentration of milk present. Thus, the pattern formation enables us to determine the respective significance of the system and the corresponding structural change occurred in them. These pattern formations are mainly due to the coffee-ring effect and Marangoni effect. These are discussed in this section.

14.3.1 DILUTION

The drying patterns of varying concentration of milk droplets with water are discussed here. All the patterns obtained after drying basically comprised of three regions, (1) an outer ring, (2) a uniformly packed region, and (3) an empty space/bump in the central portion. The formation and the significance of these patterns are discussed.

The drying patterns of all the 10 concentrations of milk diluted with water are imaged using CLSM[23] and are shown in Figure 14.1. It can be seen from Figure 14.1a, the 100% dried milk droplet has a thick, wide, and denser outer ring. As the concentration decreases, the thickness and width of the outer ring decreases and also the outer ring becomes less dense. The formation of the outer ring in each case is the result of the coffee-ring effect.[24] Milk consists of several colloidal particles with varying sizes. Hence, during evaporation, the smaller particles, as a result of capillary flow, move to the edges with the liquid comparably faster than the larger particles. Then the particles get pinned at the edges and consequently, the liquid at the edges evaporates faster when compared with that at the center. The local vapor pressure of the liquid in the droplet will not be uniform and hence the local evaporation rate at the central portion will be slower compared with that at the edge. As a result of this, a non-uniform pattern is observed with most of the particles deposited at the edges.

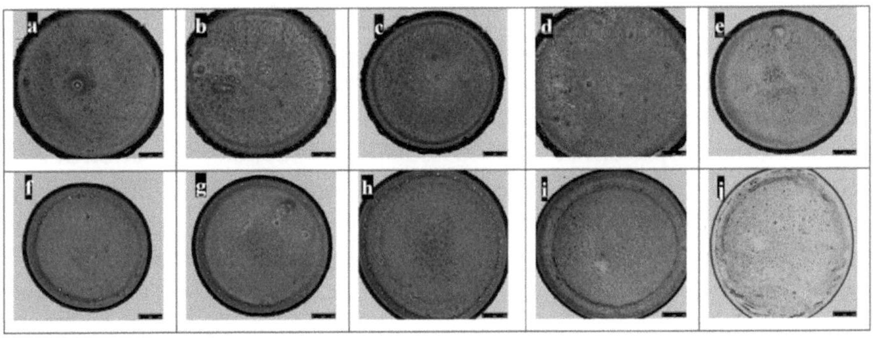

FIGURE 14.1 Dried droplet patterns of diluted milk at various concentrations. (a) 100%, (b) 90%, (c) 80%, (d) 70%, (e) 60%, (f) 50%, (g) 40%, (h) 30%, (i) 20%, (j) 10%.

The second region observed in all the patterns was a uniform distribution of different-sized particles together. As mentioned above, due to the coffee-ring effect, concentration variation and a difference in local evaporation occur. This in turn creates a surface tension gradient within the droplet during drying. Consequently, a circular flow is generated from the base toward the edges and then again it flows back to the base. This is called Marangoni convection. Here, the circular flows counteract the outward capillary flow, thus resulting in the formation of a uniform deposit within the droplet. This region consists of both the larger particles and smaller particles evenly distributed, resulting in a uniform pattern.

The third region is either an empty space or a bump depending upon the concentration of milk. For all concentrations of the milk except 10% and 20% (Fig. 14.1a–h), a bump is observed within the droplet pattern, whereas an empty space is majorly observed in the patterns of 10% and 20% (Fig. 14.1i and j) samples. At higher concentrations, the particles are large in number with various sizes. Hence, during evaporation, all the particles do not get transferred to the edges and the heavier or larger particles move toward the central portion due to the surface tension force of the liquid. Also, during the evaporation, all the smaller particles will not get transferred to the edges. The heavier particles restrict the movement of smaller particles to the edges and they get adsorbed[25] to the substrate. This also contributes to a uniform pattern to an extent. But as the heavier particles are accumulated at the center, a bump is observed.[26] At lower concentrations, as the milk is diluted, the number of particles present will be very less compared with that at higher ones. Subsequently, the particles during evaporation get rearranged by capillary flow and surface tension gradient, resulting in a non-uniform pattern. After the evaporation, the remaining particles get adsorbed to the substrate non-uniformly resulting in the formation of empty spaces in the patterns.

In the drying pattern of 50%, 40%, 30%, and 20% concentration of milk, concentric rings were observed. At these concentrations, different-sized particles become distinguishable.[24] Thus, due to capillary flow, similar-sized particles move to the edges with the same velocity and get anchored at the edges. Next, adjacent to that ring, concentric rings of much larger sized particles with lesser velocities get anchored. The negatively charged particles pinned to the contact line will also get adsorbed to the positive glass substrate as a result of attractive forces. Other than this, the particles at the edges get attracted to each other resulting in the aggregation of particles within each ring. All these together contribute to the formation of concentric rings and hence each ring attributes to the presence of similarly sized particles.[19] Milk consists of proteins, whey proteins, fats, lactose, etc., and when diluted, the particle concentration becomes less dense. Thus, during drying, the smallest ones, such as lactose and salts with whey proteins (average particle size ~ 1 nm) move to the edge, the second ring is formed by casein–calcium phosphates (average particle size ~ 100 nm) and a third layer is formed by fat globules with an average particle size of nearly 1000 nm.[27,28]

14.3.2 COAGULATION

The dried droplet patterns of milk during coagulation process are imaged at every half an hour interval and the significance of these patterns at each interval is discussed here. The coagulation referred here is driven by bacteria, hence, it is lactic acid-based coagulation. The three stages of lactic acid/bacteria-based milk coagulation include (1) hydrolysis of κ casein present in the milk, (2) flocculation of casein micelles, and (3) coagulation followed by syneresis.[29–31] The patterns obtained here are basically due to coffee-ring effect and Marangoni effect. Also, when the coagulation process begins, the structural changes occurring within the system also contribute to the pattern formation. The dried droplet patterns of milk during coagulation at each 30 min interval[23] are shown in Figure 14.2. It can be noticed that the pattern in Figure 14.2a is very much different from Figure 14.2n. This is due to the conversion of the initial emulsion to a final gel system. So the changes occurring during this conversion is studied using the dried droplet patterns obtained using CLSM.

Initially, the dried droplets consist of networks interconnected with each other with a thick outer ring. As the process continues, the patterns differ and networks start rearranging from the earlier, that is, in Figure 14.2e, suddenly the pattern forms into more of a thread-like structures connected with each other. This is just after the first stage of coagulation process. The first stage of coagulation process corresponds to the cleavage of κ caseins due to hydrolysis, and thus the collapse of the structure occurs.[29] This then further paves the path to the second stage known as flocculation process. Figure 14.2e depicts the collapse of the casein micelles leaving strands in the observed pattern. Subsequently, the particles come together as they lose their stability, and formation of flocs happens further. As flocculation takes place, the micellar portion of proteins gets converted into aqueous phase. From Figure 14.2f onwards, we can observe the thick spherical particles interconnected with each other, and as coagulation process takes place, the spherical flocs are observed. Now, from Figure 14.2j onwards, the spherical flocs can be seen to be attaching with the neighboring flocs, and thus, this process continues as is shown till Figure 14.2m. In Figure 14.2m, all the flocs gets interconnected with each other resulting into a thick compact gel. Later, due to syneresis, the gel structure gets rearranged and that can be seen in Figure 14.2n. Once this stage occurs, then the gel becomes stable.

It can be seen from the dried droplet images that the thickness of the outer ring decreases as the coagulation process takes place. The formation of the outer ring is due to the coffee-ring effect.[13] During evaporation, the smaller particles present in the sample as a result of capillary flow move to the edges with the liquid comparably faster than the larger particles and get pinned at the edges and consequently, the liquid at the edges evaporates faster compared with that at the center. The local vapor pressure of the liquid in the droplet will not be uniform and hence the local evaporation rate at the central portion will be slower compared with that at the edge. As a result of this, a non-uniform pattern is observed with most of the particles deposited at the edges resulting into a thicker outer ring. But as coagulation occurs, after the final stage, due to syneresis[32] the amount of water thrown out of the gel increases compared with the initial stage. Also almost all the particles participate in the gel formation process. Hence, the concentration of particles will be much less compared with the initial stage resulting in the formation of a thin outer ring which is observed from Figure 14.2l onwards.

A circular flow is generated from the base toward the edges, and then again it flows back to the base. This is called Marangoni convection and it is due to the surface tension gradient within the droplet during drying.[14] This surface tension gradient is a resultant of the concentration variation and a difference in the local evaporation occurred during coffee-ring formation. As a result of this effect, the spherical particles can be seen evenly distributed resulting in a uniform pattern as it is seen from Figure 14.2e–h. This is because, the circular flows counteract the outward capillary flow, thus leading to the formation of a uniform deposit within the droplet. After Figure 14.2h, as coagulation process continues, the formation of heavier particles (coagulant) occurs. This can be explained using the Marangoni effect. During evaporation, all the particles do not get transferred to the edges and the heavier or larger particles move toward the central portion due to the surface tension force, and these heavier particles restrict the movement of smaller particles to the edges and they get adsorbed to the substrate resulting into the image of the thick gel/coagulant.

14.3.3 STALING

The staling of milk occurs due to the growth of certain pathogenic bacteria.[33] These bacteria ferment the lactose present in the milk to lactic acid. Also some bacteria produce gas accompanied with acid production and these degrade the protein, fats, and carbohydrates present in the milk. Hence, it is very important to know whether the milk that we consume is fresh or degraded as stale milk which can create serious health issues. In order to find the freshness of the milk, we imaged the dried droplet of milk at each 1 h interval till 23 h.[23] The changes in the patterns formation is discussed with the help of coffee-ring effect and Marangoni effect. These structural changes can be used to detect the onset of staling.

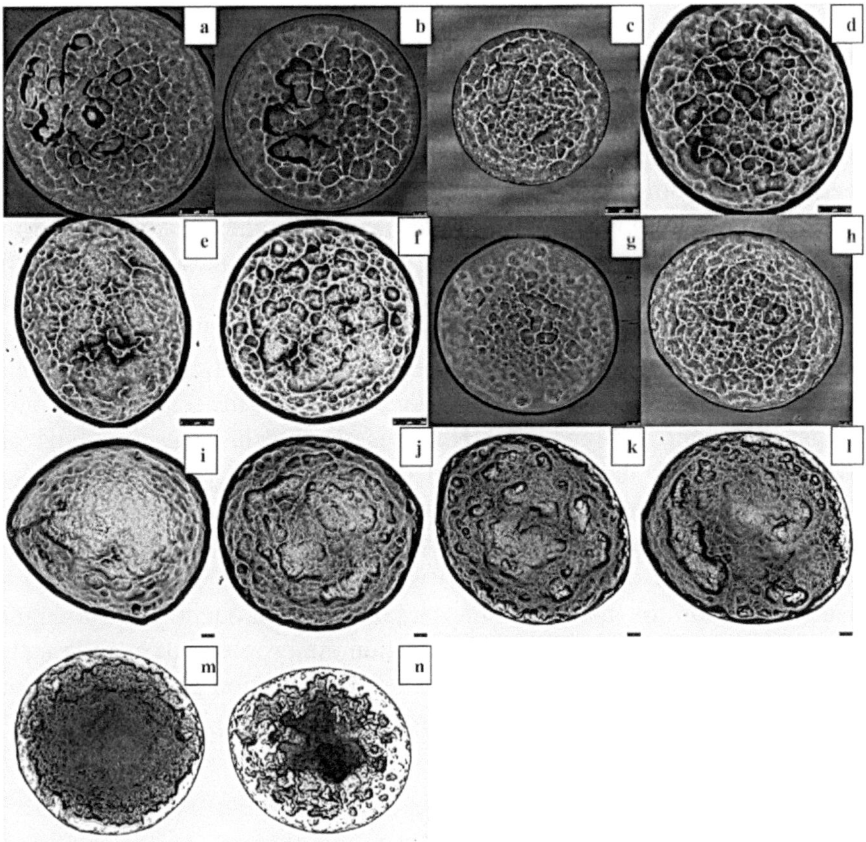

FIGURE 14.2 Dried droplet patterns of milk during coagulation. (a) 0 min, (b) 30 min, (c) 60 min, (d) 90 min, (e) 120 min, (f) 150 min, (g) 180 min, (h) 210 min, (i) 240 min, (j) 270 min, (k) 300 min, (l) 330 min, (m) 330 min, (n) 360 min.

In the beginning, as milk is fresh, it is a stable emulsion. Hence, a stable structure can be seen (Fig. 14.3(0)). The milk is left open at room temperature and the readings are taken after each 1 h. The patterns remain to be similar till Figure 14.3(5), while in Figure 14.3(6), thick flocs are observed which are separated from each other. At Figure 14.3(7), the separated flocs can be observed clearly. Thus, this confirms that some physicochemical changes have taken place within the system. So Figure 14.3(6) is the time at which the milk solids undergo a period of instability.[34] From this time onwards, the microbial growth and activity begins leading to the degradation of proteins, fats, carbohydrates, etc., and as a result of this, the structure oscillates between smaller floc and larger coagulants. From Figure 14.3(8) onwards, it could be said that the network formation is observed and this network begins to grow until the milk completely gets coagulated. When the network formation reaches at its maximum implying the formation of a precipitant leaving out free water, the milk gets spoiled completely. The coagulant/precipitant formed is observed in Figure 14.3(15). The inner thicker portion is the precipitant and the absence of enough particles at other portions of the pattern is due to the presence of free water which evaporated during drying.

The patterns formed during staling process are similar to those formed during coagulation, hence, the same principle of coffee-ring effect and Marangoni effect come into the picture resulting in the formation of outer rings and inner portions.[13,14,24] Here, as the mechanisms involved are slightly different, the patterns formed are also different, but the reason for the pattern formation is the same. Initially, thicker and wider outer ring can be observed, and as milk stales, the outer ring becomes very thin with less number of particles. This is similar to that of coagulation, and the explanation for this is also the same. Similarly, due to the Marangoni effect a uniform pattern is observed inside the droplet, but as milk stales due to precipitation and aggregation, the inner portion gets much thicker. This is also similar to that of the coagulation process.

14.4 CONCLUSION

The drying patterns of milk droplets during different processes, such as dilution, coagulation, and staling were studied and the differences in the morphology of each pattern were explained by the coffee-ring effect and Marangoni effect. The coffee-ring effect provided an understanding about the formation of outer rings to all the patterns, whereas the variation in the inner structure in each process was explained with the Marangoni effect. Patterns observed during the dilution of milk with various concentrations of water showed a clear difference between the pure and the diluted milk. Thus, this can be used as a technique for quality check applications. Also, for diluted samples (50%, 40%, 30%, and 20%) concentric rings were observed which implied the presence of different-sized particles, thus showing a method to identify/separate different-sized particles. Similarly, in the case of coagulation, the three stages of coagulation process were detected from the dried patterns, and in the case of staling process, the

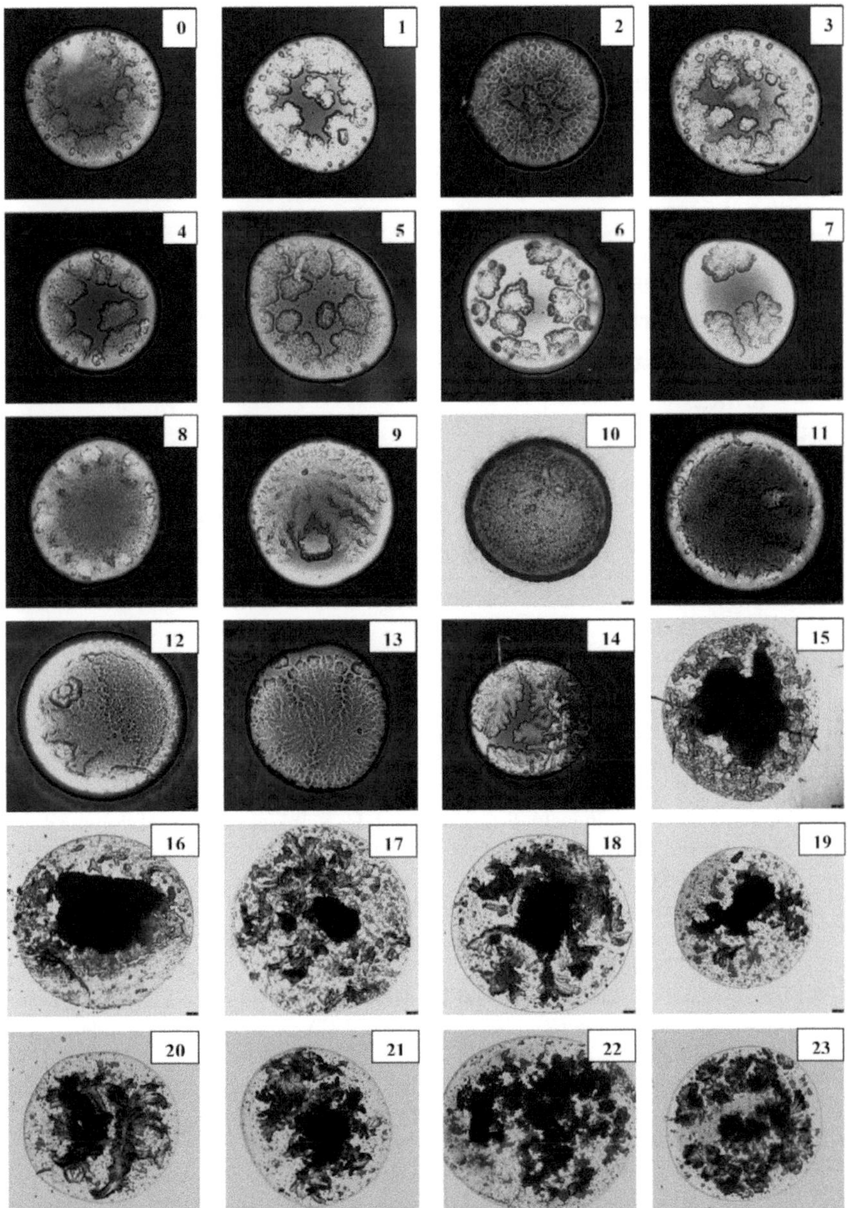

FIGURE 14.3 Dried droplet patterns of milk during staling at (0) Initial stage (1) 1 h, (2) 2 h, (3) 3 h, (4) 4 h, (5) 5 hr, (6) 6 hr, (7) 7 h, (8) 8 h, (9) 9 h, (10) 10 h, (11) 11 h, (12) 12 h, (13) 13 h, (14) 14 h, (15) 15 h, (16) 16 h, (17) 17 h, (18) 18 h, (19) 19 h, (20) 20 h, (21) 21 h, (22) 22 h, and (23) 23 h.

onset of staling was detected from the patterns. Thus, droplet drying patterns prove to be a versatile tool which can be used according to the applications required.

KEYWORDS

- **pattern formation**
- **droplet drying**
- **confocal laser scanning microscopy**
- **milk**
- **dilution**
- **coagulation**
- **staling**

REFERENCES

1. Chhasatia, V. H.; Sun, Y. Interaction of Bi-Dispersed Particles with Contact Line in an Evaporating Colloidal Drop. *Soft Matter* **2011,** *7* (21), 10135. doi: 10.1039/c1sm06393f.
2. De Gans, B. B.; Duineveld, P. C.; Schubert, U. S. Inkjet Printing of Polymers : State of the Art and Future Developments. *Adv. Mater.* **2004,** *16* (3), 203–213. doi: 10.1002/adma.200300385.
3. Park, J.; Moon, J. Control of Colloidal Particle Deposit Patterns within Picoliter Droplets Ejected by Ink-Jet Printing. *Langmuir* Apr. **2006,** *22* (8), 3506–3513. doi: 10.1021/la053450j.
4. Tarasevich, Y. Y. Mechanisms and Models of the Dehydration Self-Organization in Biological Fluids. *Physics-Uspekhi* **2004,** *47* (7), 717. [Online]. Available: http://stacks.iop.org/1063-7869/47/i=7/a=R05.
5. Esmonde-White, K. A.; Esmonde-White, F. W. L.; Morris, M. D.; Roessler, B. J. Characterization of Biofluids Prepared by Sessile Drop Formation. *Analyst* **2014,** *139* (11), 2734–2741. doi: 10.1039/C3AN02175K.
6. Bhardwaj, R.; Fang, X.; Somasundaran, P.; Attinger, D. Self-Assembly of Colloidal Particles from Evaporating Droplets: Role of DLVO Interactions and Proposition of a Phase Diagram. *Langmuir* **2010,** *26* (11), 7833–7842. doi: 10.1021/la9047227.
7. Sobac, B.; Brutin, D. Desiccation of a Sessile Drop of Blood: Cracks, Folds Formation and Delamination. *Colloids Surfaces A Physicochem. Eng. Asp.* Apr. **2014,** *448*, 34–44. doi: 10.1016/j.colsurfa.2014.01.076.

8. Yakhno, T. A.; Sedova, O. A.; Sanin, A. G.; Pelyushenko, A. S. On the Existence of Regular Structures in Liquid Human Blood Serum (plasma) and Phase Transitions in the Course of Its Drying. *Tech. Phys.* Apr. **2003,** *48* (4), 399–403. doi: 10.1134/1.1568479.
9. Ruoyang, C.; Zhang, L.; Zang, D.; Shen, W. Blood Drop Patterns: Formation and Applications. *Adv. Colloid Interface Sci.* **2016,** *231.* doi: 10.1016/j.cis.2016.01.008.
10. Yakhno, T. A. et al. The Informative-Capacity Phenomenon of Drying Drops. *IEEE Eng. Med. Biol. Mag.* **2005,** *24* (2), 96–104. doi: 10.1109/MEMB.2005.1411354.
11. Yakhno, T. Salt-induced Protein Phase Transitions in Drying Drops. *J. Colloid Interface Sci.* Feb. **2008,** *318* (2), 225–230. doi: 10.1016/j.jcis.2007.10.020.
12. Parsa, M.; Harmand, S.; Sefiane, K. Mechanisms of Pattern Formation from Dried Sessile Drops. *Adv. Colloid Interface Sci.* Apr. **2018,** *254,* 22–47. doi: 10.1016/j.cis.2018.03.007.
13. Mampallil, D. Some Physics inside Drying Droplets. *Resonance* Feb. **2014,** *19* (2), 123–134. doi: 10.1007/s12045-014-0016-z.
14. Zang, D.; Tarafdar, S.; Tarasevich, Y. Y.; Dutta Choudhury, M.; Dutta, T. Evaporation of a Droplet: From Physics to Applications. *Phys. Rep.* Apr. **2019,** 804, 1–56. doi: 10.1016/j.physrep.2019.01.008.
15. Sefiane, K. Patterns from Drying Drops. *Adv. Colloid Interface Sci.* Apr. **2014,** *206,* 372–381. doi: 10.1016/j.cis.2013.05.002.
16. Wang, F.-C.; Wu, H.-A. Pinning and Depinning Mechanism of the Contact Line during Evaporation of Nano-Droplets Sessile on Textured Surfaces. *Soft Matter* **2013,** *9* (24), 5703. doi: 10.1039/c3sm50530h.
17. Erbil, H. Y.; McHale, G.; Newton, M. I. Drop Evaporation on Solid Surfaces: Constant Contact Angle Mode. *Langmuir* Apr. **2002,** *18* (7), 2636–2641. doi: 10.1021/la011470p.
18. Tarafdar, S.; Tarasevich, Y. Y.; Dutta Choudhury, M.; Dutta, T.; Zang, D. Droplet Drying Patterns on Solid Substrates: From Hydrophilic to Superhydrophobic Contact to Levitating Drops. *Adv. Condensed Matter Phys.* **2018,** *2018.* doi: 10.1155/2018/5214924.
19. Zhang, W.; Yu, T.; Liao, L.; Cao, Z. Ring Formation from a Drying Sessile Colloidal Droplet. *AIP Adv.* Oct. **2013,** *3* (10), 102109. doi: 10.1063/1.4824741.
20. Shen, X.; Ho, C.-M.; Wong, T.-S. Minimal Size of Coffee Ring Structure. *J. Phys. Chem. B* Apr. **2010,** *114* (16), 5269–5274. doi: 10.1021/jp912190v.
21. Hu, H.; Larson, R. G. Marangoni Effect Reverses Coffee-Ring Depositions. *J. Phys. Chem. B* Apr. **2006,** *110* (14), 7090–7094. doi: 10.1021/jp0609232.
22. Majumder, M. et al. Overcoming the "Coffee-Stain" Effect by Compositional Marangoni-Flow-Assisted Drop-Drying. *J. Phys. Chem. B* Jun. **2012,** *116* (22), 6536–6542. doi: 10.1021/jp3009628.
23. Harindran, A.; Hashmi, S.; Madhurima, V. Pattern Formation of Dried Droplets of Milk during Different Processes and Classifying them Using Artificial Neural Networks. *J. Dispers. Sci. Technol.* (in press). doi: 10.1080/01932691.2021.1880927.
24. Malla, L. K.; Bhardwaj, R.; Neild, A. Analysis of Profile and Morphology of Colloidal Deposits Obtained from Evaporating Sessile Droplets. *Colloids Surfaces A Physicochem. Eng. Asp.* Apr. **2019,** *567* (November 2018), 150–160. doi: 10.1016/j.colsurfa.2019.01.028.
25. Crivoi, A.; Duan, F. Elimination of the Coffee-Ring Effect by Promoting Particle Adsorption and Long-Range Interaction. *Langmuir* **2013,** *29* (39), 12067–12074. doi: 10.1021/la402544x.

26. Ryu, S.; Kim, J. Y.; Kim, S. Y.; Weon, B. M. Drying-mediated Patterns in Colloid-Polymer Suspensions. *Sci. Rep.* **2017,** *7* (1), 1079. doi: 10.1038/s41598-017-00932-z.
27. Arzola-Álvarez, C. et al. Particle Size Distribution and Chemical Composition of Total Mixed Rations for Dairy Cattle: Water Addition and Feed Sampling Effects. *J. Dairy Sci.* **2010,** *93* (9), 4180–4188. doi: https://doi.org/10.3168/jds.2009-2952.
28. Jhanwar, A.; Ward, R. E. Particle Size Distribution and Lipid Composition of Skim Milk Lipid Material. *Int. Dairy J.* **2014,** *36* (2), 110–117. doi: https://doi.org/10.1016/j.idairyj.2014.01.010.
29. Carlson, A.; Hill, C. G.; Olson, N. F. Kinetics of Milk Coagulation: I. The Kinetics of Kappa Casein Hydrolysis in the Presence of Enzyme Deactivation. *Biotechnol. Bioeng.* Apr. **1987,** *29* (5), 582–589. doi: 10.1002/bit.260290507.
30. Carlson, A.; Hill, C. G.; Olson, N. F. Kinetics of Milk Coagulation: II. Kinetics of the Secondary Phase: Micelle Flocculation. *Biotechnol. Bioeng.* Apr. **1987,** *29* (5), 590–600. doi: 10.1002/bit.260290508.
31. Dalgleish, D.; Alexander, M.; Corredig, M. Studies of the Acid Gelation of Milk Using Ultrasonic Spectroscopy and Diffusing Wave Spectroscopy. *Food Hydrocoll.* Sep. **2004,** *18* (5), 747–755. doi: 10.1016/j.foodhyd.2003.12.001.
32. Maćej, O.; Denin-Djurdjević Snežana Jovanović, J. Influence of Selected Factors on Induced Syneresis. *J. Agric. Sci.* **2004,** *49* (2), 205–217.
33. Samaržija, D.; Zamberlin, Š.; Pogačić, T. Psychrotrophic Bacteria and Their Negative Effects on Milk and Dairy Products Quality. *Mljekarstvo* **2012,** *62* (2), 77–95.
34. Madhurima, V.; Harindran, A. Probing the Onset of Structural Instabilities as a Tool for Detection of Staling of Dairy Milk : A Permittivity and Conductivity Study. **2018,** 050005. doi: 10.1063/1.5032660.

Index